The Challenges of Water Management and Governance in Cities

The Challenges of Water Management and Governance in Cities

Special Issue Editors

Kees van Leeuwen
Jan Hofman
Peter Driessen
Jos Frijns

MDPI • Basel • Beijing • Wuhan • Barcelona • Belgrade

MDPI

Special Issue Editors

Kees van Leeuwen
Utrecht University
The Netherlands

Jan Hofman
University of Bath
UK

Peter Driessen
Utrecht University
The Netherlands

Jos Frijns
KWR Water Research Institute
The Netherlands

Editorial Office
MDPI
St. Alban-Anlage 66
4052 Basel, Switzerland

This is a reprint of articles from the Special Issue published online in the open access journal *Water* (ISSN 2073-4441) from 2018 to 2019 (available at: https://www.mdpi.com/journal/water/special_issues/Challenges_Water_Management_Governance_Cities)

For citation purposes, cite each article independently as indicated on the article page online and as indicated below:

LastName, A.A.; LastName, B.B.; LastName, C.C. Article Title. *Journal Name* **Year**, *Article Number*, Page Range.

ISBN 978-3-03921-150-0 (Pbk)
ISBN 978-3-03921-151-7 (PDF)

Cover image courtesy of Kees van Leeuwen.

Contents

About the Special Issue Editors

Kees van Leeuwen is Chief Science Officer and Principal Scientist at KWR Water Research Institute (KWR) and Professor of Water Management and Urban Development at the Copernicus Institute of Sustainable Development of the Faculty of Geosciences at KWR Water Research Institute and Utrecht University (UU). His research focus is on sustainable urban water management & risk assessment of (drinking) water contaminants. He worked at the European Commission as Director at the Joint Research Centre (JRC) in Italy. He is experienced in managing complex multi-stakeholder processes in the science-policy interface on areas of chemicals, health and the environment and has a track record of putting science into regulatory practice. Currently, he coordinates the City Blueprint Action Group of the European Innovation Partnership on Water.

Jan Hofman is Professor of Water Science and Engineering in the Department of Chemical Engineering at the University of Bath (UoB) where he is Director of the campus-wide Water Innovation and Research Centre (WIRC). He is also Co-Director of the GW4 Water Security Alliance, an alliance of water centres at the Universities of Bristol, Bath, Cardiff and Exeter. Prof Hofman is Co-Director of the EPSRC Centre for Doctoral Training in Water Informatics: Science and Engineering and the NERC Centre for Doctoral Training in Freshwater Biosciences and Sustainability. He is and has been active in international leadership roles in the International Water Association (IWA) and the Water Europe (The European Technology Platform for Water). In the latter, he is currently leading the Working groups on Urban Water Pollutions, and Water Security.

Peter Driessen is Vice-Dean for Research and Professor of Environmental Governance at the Copernicus Institute of Sustainable Development of the Faculty of Geosciences of UU. His research focus is on governance assessment in several empirical fields connected to major sustainability challenges, such as climate change mitigation and adaptation, sustainable urban development and water governance. Professor Driessen recently coordinated the EU project STAR-FLOOD, i.e., "Strengthening And Redesigning European FLOOD risk practices: Towards appropriate and resilient flood risk governance arrangements".

Jos Frijns is the Resilience Management & Governance team leader at KWR. He works on sustainability themes such as water reuse and the water-energy nexus, with a main focus on the organisational process, citizen participation and strategy and knowledge development. He also gained extensive experience as a process and project manager in (international) consultancy and research projects. Jos is co-coordinator of the EU H2020 project NextGen on water in the circular economy (2018-2022) and has recently been appointed visiting fellow at Cranfield Water Science Institute (UK).

Preface to "The Challenges of Water Management and Governance in Cities"

Global population growth is urban growth and, therefore, most of the water-related challenges and solutions reside in cities. Unless water management and water governance processes are significantly improved within a decade or so, cities are likely to face serious and prolonged water insecurity, urban floods and/or heat stress, that may result in social instability and, ultimately, in massive migration. Aging water infrastructure, one of the most expensive infrastructures in cities, are a relevant challenge in order to address amongst other things Sustainable Development Goal (SDG) 6: clean water and sanitation, SDG 11: sustainable cities and communities and SDG 13: climate action.

Cities and their hinterlands face many challenges. In many places, good water governance is the main bottleneck. Cities require a long-term strategy and a multilevel water governance approach. Research has shown how important it is to involve the civil society and private parties early on in this process to create success. Collaboration among cities and regions by sharing best practices for rapid implementation is crucial not only to cope with SDG6 but also with many of the other SDGs. The choice of good governance arrangements has important consequences for economic performance, for the well-being of citizens and for the quality of life in urban areas. The better governance arrangements work in coordinating policies across jurisdictions and policy fields, the better the outcomes. Rapidly-changing global conditions will make future water governance more complex than ever before in human history, and expectations are that water governance and water management will change more during the next 20 years compared to the past 100 years.

To address these challenges, approaches need to be developed for a directed transition to more sustainable, resilient urban water services, including all stakeholders. In this Special Issue of Water, the focus is on practical concepts and tools for water management and water governance in cities. The contributors to this Special Issue provide a series of papers to create further awareness and solutions by presenting examples of integrated approaches, advanced water management practices and water governance strategies. This Special Issue contains 17 different contributions and includes a detailed introduction followed by 16 peer-reviewed papers. We have grouped these papers into four categories: (1) introduction to urban water challenges, (2) integrated assessment methods, (3) water management practices, and (4) water governance strategies.

Kees van Leeuwen, Jan Hofman, Peter Driessen, Jos Frijns
Special Issue Editors

water

MDPI

Editorial

The Challenges of Water Management and Governance in Cities

Kees van Leeuwen [1,2,*], Jan Hofman [2], Peter P. J. Driessen [3] and Jos Frijns [1]

1   KWR Watercycle Research Institute, Groningenhaven 7, 3430 BB Nieuwegein, The Netherlands;
     jos.frijns@kwrwater.nl
2   Department of Chemical Engineering, Water Innovation and Research Centre, University of Bath, Claverton
     Down, Bath BA2 7AY, UK; J.A.H.Hofman@bath.ac.uk
3   Copernicus Institute of Sustainable Development, Utrecht University, Princetonlaan 8a, 3508TC Utrecht,
     The Netherlands; p.driessen@uu.nl
*   Correspondence: kees.van.leeuwen@kwrwater.nl; Tel.: +31-30-606-9617

Received: 30 May 2019; Accepted: 3 June 2019; Published: 5 June 2019

Abstract: Combined impacts of sea-level rise, river flooding, increased frequency and magnitude of extreme rainfall, heatwaves, water scarcity, water pollution, ageing or lacking infrastructures for water, wastewater and solid waste in rapidly urbanising regions in the world call for improved water management and governance capacity in cities to accelerate the transition to water-wise cities. The sixteen contributions to this Special Issue create further awareness and present solutions on integrated approaches, advanced water management practices and water governance strategies. It is concluded that cities require a long-term strategy and a multilevel water governance approach. Research has shown how important it is to involve the civil society and private parties early on in this process to create success. Collaboration among cities and regions by sharing best practices for rapid implementation are crucial to cope with nearly all Sustainable Development Goals.

Keywords: water governance; urban water management; resilience; sustainable development goals

1. Introduction

Global population growth is urban growth and, therefore, most of the water-related challenges and solutions can be found in cities. Unless water management and water governance processes are significantly improved within a decade or so, cities are likely to face serious and prolonged water insecurity, urban floods, and/or heat stress, which may result in social instability and, ultimately, massive migration. Aging water infrastructures are among the most expensive infrastructures in cities and a relevant challenge in order to address Sustainable Development Goal (SDG) 6: clean water and sanitation, SDG 11: sustainable cities and communities, and SDG 13: climate action. In fact, many of the SDGs are water-related, directly or indirectly, as shown in Figure 1.

The choice of good governance arrangements has important consequences for economic performance, for the well-being of citizens, and for the quality of life in urban areas. The better governance arrangements work in coordinating policies across jurisdictions and policy fields, the better the outcomes. Rapidly-changing global conditions will make future water governance more complex than ever before in human history, and expectations are that water governance and water management will change more during the next 20 years compared to the past 100 years.

To address these challenges, approaches need to be developed for a directed transition to more sustainable, resilient urban water services, including all stakeholders. In this Special Issue of *Water*, the focus is on practical concepts and tools for water management and water governance in cities. Sixteen peer-reviewed papers were selected for this Special Issue. We have grouped these papers into four categories:

- Introduction to urban water challenges;
- Integrated assessment methods;
- Water management practices; and
- Water governance strategies.

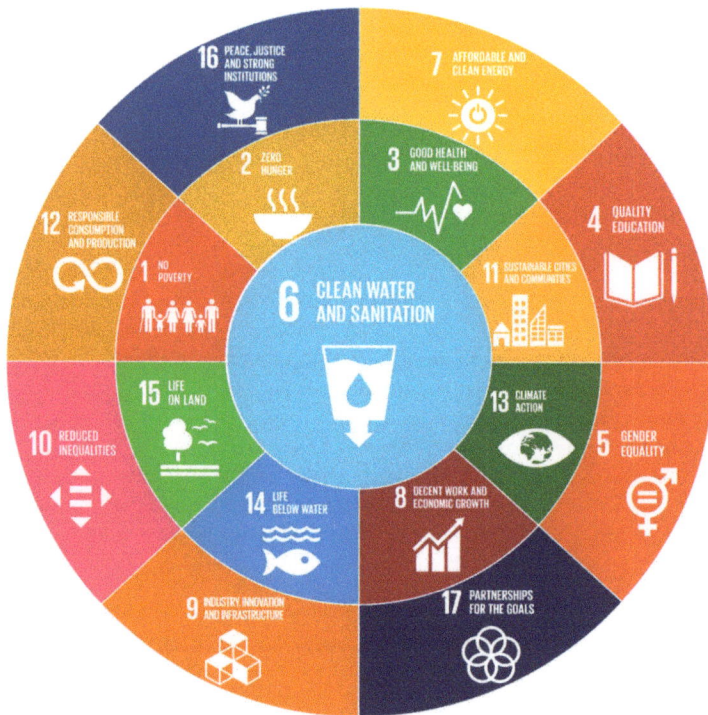

Figure 1. The water-centric 17 Sustainable Development Goals [1].

This Special Issue starts with two policy papers of the international organisations UNESCO and OECD, presenting a summary of their most recent work on policy solutions for sustainable water resources management in urban areas. Both organisations stress the importance of integrated methodologies to assess the urban water challenges across a range of temporal and spatial scales.

The following set of papers present such integrated assessment methods and their application for sustainable water resources management, water-sensitive urban design, urban water reuse, and sustainable wastewater management systems. These papers address the importance of enhancing governance capacity to implement systems for water management in cities.

The third group includes papers that present water management practices to increase water security under climate change conditions. Experiences with stormwater management, urban drainage systems, rainwater harvesting, and flood risk control are analysed and lessons learned are shared.

The urgency of the challenges related to urbanisation and climate change calls for adaptive water governance. In the final group of papers, multi-actor governance strategies are presented to take care of flood resilience, regional water supply and urban watershed management.

The following section summarises the contributions according to this categorisation.

2. Contributed Papers

2.1. Introduction to Urban Water Challenges

Makarigakis and Jiminez-Cisneros [1] provide an overview of the global urban water challenges. To achieve water security, UNESCO is developing tools for science-based decision making, promotes international cooperation through networking, enhances the science policy interface and facilitates education and capacity development.

The OECD developed a water governance indicator framework that cities can use to identify whether water governance conditions are in place and function or need improvement. The framework is composed of 36 indicators, measuring the what (policy framework), the who (institutions in charge) and the how (co-ordination tools for water policies). Romano and Akhmouch [2] report that the OECD framework can provide a global picture on the water governance system, rather than focusing on specific dimensions (e.g., transparency) or specific functions (e.g., water supply and sanitation). They advocate an institutional framework that encompasses accessible information and adequate capacity, sufficient funding and transparency and integrity, meaningful stakeholder engagement and coherence across sectoral policies.

2.2. Integrated Assessment Methods

The second group of papers present integrated assessment methods and their application for a variety of urban water management practices.

Kim et al. [3] examined the status of integrated water resources management of Seoul using the city blueprint approach. which consists of three different frameworks: (1) the trends and pressures framework, (2) the city blueprint framework and (3) the water governance capacity framework. The results indicate that nutrient recovery from wastewater, stormwater separation, and operation cost recovery of water and sanitation services are priority areas for Seoul. Furthermore, the local sense of urgency, behavioural internalisation, consumer willingness to pay and financial continuation are identified as barriers limiting Seoul's governance capacity.

Following the recent drought period, the City of Cape Town is restructuring its policy to include climate change adaptation strategies. Madonsela et al. [4] describe an evaluation of the water governance processes required to implement water-sensitive urban design in Cape Town. The analysis revealed that smart monitoring, community knowledge and experimentation with alternative water management technologies are important when considering uncertainties and complexities in the governance of urban water challenges.

The transformation to widespread application of water-reuse systems requires major changes in the way water is governed. Through the systematic assessment of the city of Sabadell (Spain), Šteflová et al. [5] identified the main barriers, opportunities and transferable lessons that can enhance governance capacity to implement systems for non-potable reuse of treated wastewater in cities. It was found that continuous learning, the availability and quality of information, the level of knowledge, and strong agents of change are the main capacity-building priorities. On the other hand, awareness, multilevel network potential and implementing capacity are already well-established.

Benavides et al. [6] developed a sustainability assessment method for wastewater management in Latin America that is multi-scalar (considering several territorial scales or spatial boundaries in one same study) and multidimensional (considering the different dimension of sustainability). This approach allowed making visible issues that are not shown by single scale analysis, namely, the interconnections of the technical system (waste water treatment) with ecological systems (watershed) and social systems (public administration, community dynamics, social perception).

Lahmouri et al. [7] analysed greenhouse gas (GHG) emissions and compared possible water reclamation with resource recovery scenarios in the town of Leh in India: a centralised scheme, a partly centralised combined with a decentralised scheme, and a household-level approach. Potential sources of reduction of GHG emissions through sludge and biogas utilisation have been identified and quantified to seize their ability to mitigate the carbon footprint of the water and wastewater sector. The study showed that decentralising wastewater management has the least carbon footprint during both construction and operation phases. These results have implications for cities worldwide.

2.3. Water Management Practices

This group of papers looks at urban water management practices that deal with the consequences of climate change such as increased precipitation and flood risks.

Zhang et al. [8] present the concept of a sponge city in Beijing, which allows storm water to be managed with natural infiltration, natural retention and detention, and natural cleaning facilities. It is based on natural and ecological laws and provides "elasticity" in adaptation to environmental changes and response to natural disasters.

One of the crucial elements in the sizing of a stormwater reservoir is determination of duration time and intensity of rainfall. The outcome is, however, affected by significant uncertainty of runoff modelling. Szelag et al. [9] analysed the effect of the uncertainty of a rainfall–runoff model, showing that the desired capacities of the stormwater reservoir were overestimated when uncertainty was neglected.

Haghighatafshar et al. [10] have aligned the engineering of drainage systems with urban planning and design. They introduce a conceptual model of mesoscale sustainable drainage systems (SuDS) that complies with hydraulic, hydrologic and social–ecological functions.

Implementing rainwater harvesting could contribute to the protection against damage caused by increasing precipitation frequency and intensity. Hofman-Claris et al. [11] calculated the total costs of ownership for decentralised drinking water supply from harvested rainwater. In the Netherlands, the amount of rainwater that can be harvested in the city district only covers about 50% of the demand, and the application of rainwater harvesting for drinking water production is currently not economically feasible.

Nicklin et al. [12] assessed the cost of inaction in relation to pluvial flood damages in Rotterdam and Leicester, concluding that investment in flood protection is an economically beneficial approach for cities.

2.4. Water Governance Strategies

The fourth group of papers present governance strategies dealing with urban water challenges through an interdisciplinary, collaborative and network approach.

Based on international comparative research on flood risk governance, Driessen et al. [13] derived key governance strategies that secure the necessary capacities to resist, to absorb and recover, and to transform and adapt. Taking diversification and alignment of flood risk management approaches as an important starting point, adaptive flood risk governance also requires a delicate balancing act between legal certainty and flexibility.

Strategic placement of green infrastructure has the potential to maximise water quality benefits and ecosystem services. Shifflett et al. [14] examined the factors that influence a multi-stakeholder watershed approach to planning, implementing and evaluating green infrastructure techniques in Cincinnati. Green infrastructure planning benefitted from governance strategies that include stakeholder engagement and collaboration.

For effective water governance, the coordination of multiple actors across different institutional levels is important. In a Swiss region, Lieberherr et al. [15] observed the importance of reputational power, i.e., a higher degree of coordination took place when the actors responsible for water supply regarded potential coordination partners as important. Likewise, democratic legitimacy is important, i.e., the stronger the region's capacity to steer, the stronger the coordination.

Tosun et al. [16] looked into transnational city networks on climate change adaptation and showed how these networks embraced goals related to urban water management. The main impact of city networks is to provide a forum for validating and optimising the design of policies and measures and to exchange experiences regarding their implementation.

3. Conclusions

Water challenges are becoming ever more urgent in a world of unprecedented urbanisation and population growth, depleting resources and increasing climate change impacts. Combined impacts of sea-level rise, river flooding, increased frequency and magnitude of extreme rainfall, heatwaves, water scarcity, water pollution, ageing or lacking infrastructures for water, wastewater and solid waste in rapidly urbanising regions in the world call for improved water management and governance capacity in cities to accelerate the transition to water-wise cities.

Cities and their hinterlands face many challenges. In many places, good water governance is the main bottleneck. Cities require a long-term strategy and a multilevel water governance approach. Research has shown how important it is to involve the civil society and private parties early on in this process to create success. Collaboration among cities and regions by sharing best practices for rapid implementation is crucial not only to cope with SDG6 but also with many of the other SDGs.

Integrated solutions are needed, such as water-sensitive design, including rainwater harvesting, recycling, reuse, pollution prevention and other innovative urban water approaches. The contributors to this Special Issue provide a series of papers to create further awareness and solutions by presenting examples of integrated approaches, advanced water management practices and water governance strategies.

Author Contributions: K.v.L. conceived and led the development of the Special Issue and this paper; J.H., P.P.J.D. and J.F. each contributed substantially to the writing of this paper.

Acknowledgments: The authors of this paper, who served as guest editors of this Special Issue, wish to thank the journal editors, all authors submitting papers to this Special Issue, and the many referees who contributed to paper revision and improvement of all published papers.

Conflicts of Interest: The authors declare no conflict of interest.

References

1. Makarigakis, A.; Jimenez-Cisneros, B. UNESCO's Contribution to Face Global Water Challenges. *Water* **2019**, *11*, 388. [CrossRef]
2. Romano, O.; Akhmouch, A. Water Governance in Cities: Current Trends and Future Challenges. *Water* **2019**, *11*, 500. [CrossRef]
3. Kim, H.; Son, J.; Lee, S.; Koop, S.; Van Leeuwen, K.; Choi, Y.; Park, J. Assessing Urban Water Management Sustainability of a Megacity: Case Study of Seoul, South Korea. *Water* **2018**, *10*, 682. [CrossRef]
4. Madonsela, B.; Koop, S.; van Leeuwen, K.; Carden, K. Evaluation of Water Governance Processes Required to Transition towards Water Sensitive Urban Design—An Indicator Assessment Approach for the City of Cape Town. *Water* **2019**, *11*, 292. [CrossRef]
5. Šteflová, M.; Koop, S.; Elelman, R.; Vinyoles, J.; Van Leeuwen, K. Governing Non-Potable Water-Reuse to Alleviate Water Stress: The Case of Sabadell, Spain. *Water* **2018**, *10*, 739. [CrossRef]
6. Benavides, L.; Avellán, T.; Caucci, S.; Hahn, A.; Kirschke, S.; Müller, A. Assessing Sustainability of Wastewater Management Systems in a Multi-Scalar, Transdisciplinary Manner in Latin America. *Water* **2019**, *11*, 249. [CrossRef]
7. Lahmouri, M.; Drewes, J.E.; Gondhalekar, D. Analysis of Greenhouse Gas Emissions in Centralized and Decentralized Water Reclamation with Resource Recovery Strategies in Leh Town, Ladakh, India, and Potential for Their Reduction in Context of the Water–Energy–Food Nexus. *Water* **2019**, *11*, 906.
8. Zhang, S.; Li, Y.; Ma, M.; Song, T.; Song, R. Storm Water Management and Flood Control in Sponge City Construction of Beijing. *Water* **2018**, *10*, 1040. [CrossRef]
9. Szeląg, B.; Kiczko, A.; Dąbek, L. Stormwater Reservoir Sizing in Respect of Uncertainty. *Water* **2019**, *11*, 321. [CrossRef]
10. Haghighatafshar, S.; La Cour Jansen, J.; Aspegren, H.; Jönsson, K. Conceptualization and Schematization of Mesoscale Sustainable Drainage Systems: A Full-Scale Study. *Water* **2018**, *10*, 1041. [CrossRef]

11. Hofman-Caris, R.; Bertelkamp, C.; de Waal, L.; van den Brand, T.; Hofman, J.; van der Aa, R.; van der Hoek, J.P. Rainwater Harvesting for Drinking Water Production: A Sustainable and Cost-Effective Solution in The Netherlands? *Water* **2019**, *11*, 511. [CrossRef]

12. Nicklin, H.; Leicher, A.M.; Dieperink, C.; van Leeuwen, K. Understanding the costs of inaction—An assessment of pluvial flood damages in two European cities. *Water* **2019**, *11*, 801. [CrossRef]

13. Driessen, P.; Hegger, D.; Kundzewicz, Z.; Van Rijswick, H.; Crabbé, A.; Larrue, C.; Matczak, P.; Pettersson, M.; Priest, S.; Suykens, C.; et al. Governance Strategies for Improving Flood Resilience in the Face of Climate Change. *Water* **2018**, *10*, 1595. [CrossRef]

14. Shifflett, S.D.; Newcomer-Johnson, T.; Yess, T.; Jacobs, S. Interdisciplinary Collaboration on Green Infrastructure for Urban Watershed Management: An Ohio Case Study. *Water* **2019**, *11*, 738. [CrossRef] [PubMed]

15. Lieberherr, E.; Ingold, K. Actors in Water Governance: Barriers and Bridges for Coordination. *Water* **2019**, *11*, 326. [CrossRef]

16. Tosun, J.; Leopold, L. Aligning Climate Governance with Urban Water Management: Insights from Transnational City Networks. *Water* **2019**, *11*, 701. [CrossRef]

Perspective

UNESCO's Contribution to Face Global Water Challenges

Alexandros K. Makarigakis * and Blanca Elena Jimenez-Cisneros

International Hydrological Programme (IHP), Natural Sciences Sector, UNESCO, 7 Place de Fontenoy, 75007 Paris, France; bjimenezc@iingen.unam.mx
* Correspondence: a.makarigakis@unesco.org; Tel.: +33-1-456-80806

Received: 21 November 2018; Accepted: 10 February 2019; Published: 23 February 2019

Abstract: The current world population of 7.6 billion is expected to reach 8.6 billion in 2030, 9.8 billion in 2050 and 11.2 billion in 210, with roughly 83 million people being added every year. The upward trend in population size along with an improved quality of life are expected to continue, and with them the demand for water. Available water for human consumption and development remains virtually the same. Additional to the different pressures of the demand side on the available resources (offer side), climate variability and change apply further pressures to the management of the resource. Additional to the increase in evaporation due to temperature rise, climate change is responsible for more frequent and intense water related extreme events, such as floods and droughts. Anthropogenic activities often result in the contamination of the few pristine water resources and exacerbate the effects of climate change. Furthermore, they are responsible for altering the state of the environment and minimizing the ecosystem services provided. Thus, the water security of countries is compromised posing harder challenges to poor countries to address it. This compromise is taking place in a complex context of scarce and shared resources. Across the world, 153 countries share rivers, lakes and aquifers, home to 40% of the world's current population. The United Nations Educational, Scientific and Cultural Organization (UNESCO) is the scientific arm of the United Nations and its International Hydrological Programme (IHP) is the main vehicle for work in water sciences at an intergovernmental level. IHP VIII, IHP's medium term strategy, aims to assist UNESCO's Member States (MS) in achieving water security by mobilizing international cooperation to improve knowledge and innovation, strengthening the science-policy interface, and facilitating education and capacity development in order to enhance water resource management and governance. Furthermore, the organization has established an Urban Water Management Programme (UWMP) aiming at promoting sustainable water resource management in urban areas.

Keywords: climate change; IHP; intergovernmental; science and technology; sustainability; UNESCO; water management; water security; Urban Water Management Programme

---

1. Introduction

The International Hydrological Programme (IHP) is the only intergovernmental programme of the UN system devoted to the scientific, educational, cultural and capacity building aspects of hydrology for the better management of water resources. Drawing on more than four decades of experience, UNESCO-IHP fosters and consolidates cross-disciplinary and cross sectoral networks that facilitate cooperation within research and capacity building, and development of analytical tools and data sharing, primarily across national boundaries. UNESCO-IHP also enhances awareness raising of policy-makers at the national, regional and international level on the predictions and risks related to global change, including climate change and human impact.

IHP is a truly intergovernmental programme, having its planning, definition of priorities, and supervision of the execution to be decided by the Intergovernmental Council. The Council is composed of 36 UNESCO Member States elected by the General Conference of UNESCO at its ordinary sessions held every two years. Equitable geographical distribution and appropriate rotation of the representatives of the Member States are ensured in the composition of the Council. Each of UNESCO's six electoral regions elects Member States for membership in the Council.

Consequently, the Council elects a chairperson and four vice-chairpersons. These, with the chairperson of the previous Bureau as ex-officio member, constitute the Council's Bureau. The composition of the Bureau so formed reflects an equitable geographical distribution, each representing UNESCO's six electoral regions. The members of the Bureau remain in office until a new Bureau has been elected (It needs to be noted that following the 23rd session of IHP's Intergovernmental Council, the role of the ex-officio member will no longer apply and Member States will elect a chairperson, a rapporteur and four vice-chairpersons).

Responding to the need to have an impact on the practical management of water resources, IHP networks comprise not only the scientists but also professionals, different sectors, and the society at large, including youth, gender and children groups. There is no other international Member States' water network with such a wide range of disciplines, sectors and stakeholders.

2. Intergovernmental Hydrological Programme: Origin and Strategy

At the end of the first International Hydrological Decade (IHD, 1965–1974) the international scientific community together with governments realized that water resources often were one of the primary limiting factors for harmonious socio-economic developments in many regions of the world. Moreover, they realized that to solve problems, internationally coordinated cooperation mechanisms were necessary to enhance the knowledge base, capacity and rational management. This gave birth to the UNESCO's IHP.

IHP facilitates an inter- and transdisciplinary integrated approach to watershed and aquifer management, incorporating the social, economic and human dimensions of water resources. To advance knowledge development and dissemination, IHP uses all available experience and promotes and develops international cooperative research in hydrological and freshwater sciences. IHP was planned and implemented in six-year phases, covering themes reflecting the current priorities decided by Member States; as of 2014, the planning exercise has shifted to an eight-year cycle.

The core themes of the first three phases of IHP (1971–1989) followed the same directions of the International Hydrological Decade, focusing on research and capacity building in hydrological science in its strict sense. Since then, the different phases of IHP (Figure 1) were always in advance of the major challenges the world had to face concerning water.

In the nineties, more than 25 years prior to the Agenda 2030 and the Sustainable Development Goals, the programme, being in its fourth phase, IHP-IV (1990–1995), identified sustainability and water resource development and management as key elements, adopting "Hydrology and Water Resources for Sustainable Development" as a core theme. Similarly, the work in the fifth phase, IHP-V (1996–2001), had "Hydrology and Water Resources Development in a Vulnerable Environment" as a core theme.

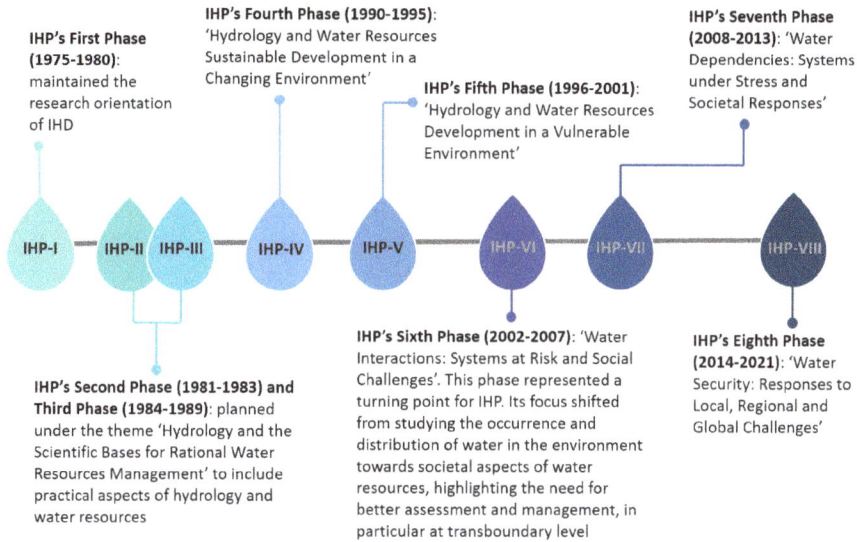

Figure 1. IHP Phases.

UNESCO, being the scientific arm of the UN family is required to lead the work in water breaking scientific barriers using an out of the box approach.

Recognizing the need for a paradigm shift in thinking on water from fragmented compartments of scientific inquiry to a more holistic, integrated approach, the core theme for IHP-VI (2002-2007) was defined as "Water Interactions: Systems at Risk and Social Challenges". The same trend continued in the formulation of the IHP-VII (2008–2013), which adopted 'Water Dependencies: Systems under Stress and Societal Responses' as a core theme, further emphasizing the interacting dependencies of the system components and the important role of society. All these themes were well in advance of the national research agendas setting new trends on the need to develop knowledge.

Following the 2000–2015 and the Millennium Development Goals, Member States came to an agreement for establishing an ambitious and interconnected development agenda of 17 Sustainable Development Goals (SDGs), Agenda 2030. Sustainable Development Goal 6 aims at ensuring availability and sustainable management of water and sanitation for all. A closer look at the SDGs reveals that many of the SDGs have a strong relationship with sustainable water use and consumption, e.g., SDGs 2 (zero hunger), 3 (good health and well-being), 11 (sustainable cities and communities), 12 (responsible production and consumption), 13 (climate action), and 14 (life below water).

UNESCO's focus of its eighth programmatic phase IHP VIII (2014–2021), has adopted 'Water Security: Responses to Local, Regional and Global Challenges' as its overarching idea. Given population growth, deteriorating water quality, the growing impact of floods and droughts and the other hydrological effects of global change, water security is a growing concern. It touches upon all aspects of life and requires a holistic approach, which actively integrates social, cultural and economic perspectives, scientific and technical solutions and attention to societal dynamics. In 2016, the World Economic Forum identified the water crisis as the global risk of highest concern for people and economies over the next ten years.

Water security has been defined by UNESCO's Member States as the capacity of a population to safeguard access to adequate quantities of water of an acceptable quality for sustaining human and ecosystem health on a watershed basis, and to ensure efficient protection of life and property against water-related hazards such as floods, landslides, land subsidence, and droughts. To date this is the only intergovernmentally approved definition. Although other better-rounded definitions have been developed, their use often causes political challenges and thus are frequently avoided.

The activities of the eighth phase of IHP (IHP-VIII) are being conducted along three strategic axes: (a) mobilizing international cooperation to improve knowledge and innovation to address water security challenges; (b) strengthening the science-policy interface to achieve water security at all levels; and (c) developing institutional and human capacities for water security and sustainability.

The work of IHP-VIII focusing on capacity building and awareness raising on six thematic areas to assist Member States in their challenging endeavor to better manage and secure water and to ensure the necessary human and institutional capacities. These are:

- Theme 1: Water-related Disasters and Hydrological Changes
- Theme 2: Groundwater in a Changing Environment
- Theme 3: Addressing Water Scarcity and Quality
- Theme 4: Water and Human Settlements of the Future
- Theme 5: Ecohydrology, Engineering Harmony for a Sustainable World
- Theme 6: Water Education, Key to Water Security

IHP-VIII on water security aims to address challenges identified in Agenda 2030, Sendai Framework for Disaster Risk Reduction and the New Urban Agenda and the Paris Agreement.

Within the framework of water security, IHP builds capacity of member states by synergistically integrating the experience and tools available within the activities implemented in all six thematic areas. The goal is to provide the scientific knowledge base for sound policy advice, in order to manage and cope with challenges to water resources in the practice, and to increase the resilience of natural and human systems with an emphasis on vulnerable communities.

3. IHP's Urban Water Management Programme (UWMP)

IHP's UWMP aims to promote sustainable water resource management in urban areas by helping countries develop and implement effective strategies and policies for urban water management through the dissemination of scientifically-sound policy guidelines, scientific knowledge and information on new and innovative approaches, solutions and tools for sustainable urban water management, as well as by providing capacity building support on key urban water issues.

4. Global Water Challenges

A frequent expression used by numerous professionals to describe water related challenges is that "there is too much, too little or too polluted". UNESCO, within the concept of water security, is working to ensure that all three challenges are addressed. As it is quite difficult to capture the results of the work of more than 3000 professionals comprising UNESCO's Water Family, a few selected issues are presented below.

4.1. Pressures on Water Availability

4.1.1. Population Growth

We currently live in the Great Acceleration period of the Anthropocene. In tracking the effects of human activity upon the Earth, a number of socioeconomic and earth system parameters are utilized including population, economics, water usage, food production, transportation, technology, green-house gases, surface temperature, and natural resource usage. Since 1950, these trends are increasing significantly if not exponentially. The current world population of 7.6 billion is expected to reach 8.6 billion in 2030, 9.8 billion in 2050 and 11.2 billion in 2100 [1].

In 1990, 43% (2.3 billion) of the world's population lived in urban areas. In 2015, the urban population had grown to 54% (4 billion) and it is expected to increase to 66% by 2050. It is projected that 2.5 billion people will be added to urban populations by 2050, 90% of which will be in Asia and Africa [2,3]. The urbanization trend has experienced a remarkable increase in the absolute numbers of urban dwellers, from a yearly average of 57 million between 1990–2000 to 77 million between

2010–2015. It has to be noted that, although cities impose challenges on the environment, natural resources and the hydrological cycle, no questioning to this model for development has been made.

As the world's population has increased by around 4-fold in the 20th century, the human water consumption has increased around 5, 18 and 10 times for agricultural, industrial and municipal use, respectively [4,5]. A rise in the world population and its standards of living, along with unsustainable practices, has put water resources under ever-increasing pressure globally.

4.1.2. Agriculture

Agriculture is the world's largest user of water. Considering water abstraction, agricultural use of water represents near 70% of the global use [6] with clear differences among developed countries and developing countries where rain-fed irrigation accounts for 60% of their production. Irrigation water withdrawal in developing countries is expected to grow by about 14 percent from the current 2130 km^3 per year to 2420 km^3 in 2030 [7].

In addition to this use of water to produce crops, water is also used to manufacture food. In Europe, for example, the manufacturing of food products consumes on average about 5 m^3 of water per person per day [8]. At the same time, with as much as 1.3 billion tons of food wasted annually [9], 250 km^3 of water is being "lost" per year due to food waste (food waste can be defined as the discarding of food that was fit for human consumption but has become spoiled, expired or otherwise unwanted) worldwide [9]. At the global level, meat and cereals clearly stand out in the global proportion of food waste by 21.7% and 13.4%, respectively [10].

4.1.3. Water Scarcity

By 2050, it is estimated that 40 per cent of the global population will be living in river basins that experience severe water stress, particularly in Africa and Asia. Approximately 450 million people in 29 countries face severe water shortages [11]; about 20% more water than is now available will be needed to feed the additional three billion people by 2025; as much as two-thirds of the world population could be water-stressed by 2025 [12]; Water scarcity is projected to become a more important determinant of food scarcity than land scarcity, according to the view held by the UN [13].

4.1.4. Climate Variability and Change

Climate variability and change intensifies in a significant manner such water-related threats [14]. A recent model intercomparison study reveals that 2 °C of global warming will result in a severe decrease in the available water resources for 15% of the global population and will increase the number of people living under absolute water scarcity by another 40% compared to the effect of population growth alone [15]. Furthermore, numerous studies show that warming weather can trigger more water use and aggressive extraction from water resources [16–18], which together with changes in operation patterns [19–23] pose additional pressure on the already scarce water resources.

Changes in global climate are also expected to reduce groundwater recharge to aquifers, storage and discharge [24]. These reductions will have significant negative effects on available groundwater for development as well as for groundwater dependent ecosystems and the services they provide to both humans and the environment. Moreover, in the case of coastal aquifers, the combination of groundwater level drop and sea level rise due to the direct and/or indirect effects of climate change will cause an increase in saltwater intrusion, which in turn will pose serious threats to the livelihoods of one of the most vulnerable populations to climate change: islanders.

The role public and non-governmental organizations, including research and academic organisations, play in enabling adaptation at multiple scales has been shown to be crucial by recent studies [25].

4.2. Water Quality

Further to the pressure for additional water sources due to population growth and increase of the standard of living, human activities have increased the release of various contaminants in ground and surface water resources, resulting in a wide range of consequences from major decline in water availability and water quality to massive environmental changes. Half of the world's rivers and lakes are polluted; and major rivers, such as the Yellow river, Ganges river, and Colorado river, do not flow to the sea for much of the year because of upstream withdrawals [26]. Inefficient and ineffective use of fertilizers and pesticides in agriculture is a major contributor to ground and surface water contamination. As an example, groundwater pollution in Greece is often related to the use or abuse of fertilizers, which diffuse into soils and contaminate the aquifers. Additionally, coastal aquifers are subject to a negative water balance, owing to their overexploitation that triggers saline water intrusions [27]. Ineffective waste management often results in the release of contaminants related to noxious compounds whereas fertilized agricultural fields and wastewater treatment plants often discharge nutrients in significant quantities [28].

Such unfolding issues result in an increasing cost for water treatment and have exacerbated sociopolitical tensions over decreasing water availability, which have made water management and controlling the competition over water allocation extremely complex and sensitive [29].

Finally, water quality can also be largely degraded by climate change [6,30,31], although a comprehensive global understanding of water quality consequences of climate change is currently lacking.

There is a consensus however, that there could be significant water quality issues resulting from planned and unintended responses to climate change. Thus, any plan to address undesirable water quality impacts will require a holistic approach integrating activities of institutions responsible for managing air, land and water resources.

UNESCO [32] estimates that around 2 billion people currently live in water-stressed areas and over 800 million people have inadequate access to safe drinking water, which is supported by the findings of the latest Joint Monitoring report [33], stating that 844 million people still lacked even a basic drinking water service.

4.3. Water Related Hazards

Water-related hazards account for around 90% of all natural hazards globally, marking floods and droughts as the two most destructive natural threats to human societies. Climate change is expected to cause a rise in their intensity and frequency of extreme water events. Only throughout 2010, water-related disasters killed nearly 300,000 people, affected around 208 million others and cost nearly $110 billion [32,34]. Hence, recent years have seen increased attention for strategic flood risk assessments, and their inclusion in global integrated assessments [35].

In addition, the impact of extreme events on water-related hazards is expected to also become more intense and more geographically spread under climate change conditions causing an increase on social vulnerability. Recent multi-model studies highlight a likely increase in the global severity of drought by the end of the 21st century, in which the frequency of drought increases by more than 20% in highly populated regions, such as South America and Central and Western Europe [36]. Drought due to reduced rainfall has been the cause of a 95% reduction in Lake Chad's area between 1960 and 1985 [37]. Albeit water levels have risen since the 2000s, ecosystems have been significantly imbalanced and weakened, unable to provide livelihood related services. Furthermore, conflicts plagued the area over access to the diminished resource, and significant waves of migration have occurred through the years [38].

Water-related hazards are continuously present in the local and global news. While California is currently recovering from a major 5-year drought [39], southern Quebec was in a state of emergency due to major flooding in the area, including parts of Montreal [34,40]. On 18 January 2018, Cape Town residents, South Africa's second-largest city, woke up to their Mayor's, Ms Patricia De Lille, proclamation that "Day Zero," the day the city would run out of water, was fast approaching. On February 1—the height of summer in the southern hemisphere, when water demand is greatest—the city clamped down, harder than any city in the world with its living standards. Officials set a target of 50 Liters (13 gallons) per person, per day, for all domestic uses: cooking, bathing, toilet flushing, washing clothes. Watering lawns and scrubbing cars with city water had already been banned for months. "The abuse of water means that we will all suffer," De Lille had warned.

On the other hand, an increase in flooding frequency is projected in more than half of the world, particularly in the non-snow dominated regions, which naturally have a greater population [41]. The recent flood in Houston, Texas, during the course of hurricane Harvey, resulted in more than 80 fatalities and an estimated economic cost that exceeded $150 billion USD [42]. Globally, economic losses from flooding exceeded $19 billion in 2012 [21], and have risen over the past half century [30,43]. Hence, recent years have seen increased attention for strategic flood risk assessments, and their inclusion in global integrated assessments [35].

Only throughout 2010, water-related disasters killed nearly 300,000 people, affected around 208 million others and cost nearly $110 billion [32]. Investing in disaster risk reduction is thus a precondition for developing sustainably in a changing climate. It is a precondition that can be achieved and that makes good financial sense.

4.4. Servicing the Most Vulnerable: Slum Populations

Even though significant progress has been made globally towards improving access to water, almost 700 million people still lack access to clean drinking water globally [43]. Informal settlements (slums) constitute a significant percentage of the urban population. There were more slum dwellers in 2012 than in 2000, a trend that will likely continue in the future [44].

Slum dwellers most often lack water and sanitation related services, as well as many other public services. For instance, in India, 56% of the population in the top 20% (household income group) has access to piped water, compared to 6% of the bottom 20% [32]. Furthermore, they often have to pay higher rates to receive water than citizens covered by the piped network do; water price can be up to thirty three times higher than the one charged by the operators [45]. Sanitation facilities are usually non-existent, having people frequently relying on communal toilets or open defecation.

Slum populations constitute the most vulnerable in an urban or peri-urban setting, having high exposure to natural hazards, often settled in areas that are not in the city plan and which are not suitable for human settlements, such as flood plains or hill sides prone to landslides. Rapid urban expansion aggravates these challenges and the people are also disproportionately affected by the impacts of climate hazards [2,33].

5. UNESCO's Contribution to Global Water Challenges

UNESCO is the scientific organization of the United Nations, whose purpose is advancing, through the educational, scientific, and cultural relations of the peoples of the world and the free exchange of ideas and knowledge, the objectives of international peace and of the common welfare of mankind.

Addressing the increasing global water challenges will be achieved by narrowing the problem within the broader framework, including the UNESCO mandate to go beyond the Integrated Water Resources Management (IWRM), water-energy-food-environment nexus or new holistic frameworks. These frameworks should aim to open up a dialogue with practitioning, management and policy communities.

As presented in Figure 2, IHP works on the basis of three axes to address global water challenges: networking, science-policy interface and building/strengthening both institutional and human capacity. In the text below, a few examples of the type of activities being implemented is provided.

Facing these global challenges requires pushing the boundary of current advancements within the water security domain. First, a common acceptance and international recognition of the water security concept and its political acceptance should be made to facilitate progress in this field. Secondly, new insights, tools and methodologies are needed for better representation of complex interactions within coupled human and natural systems, especially in urban regions across a range of temporal and spatial scales. Such attempts should be made with the greater goal of diagnosing water-related threats as a result of extreme or gradual changes in natural and anthropogenic conditions, in light of current limitations in future projections [46]. Thirdly, an entire change in the general mindset society has towards water and water related issues is needed to effectively minimize the increasing challenges and to eliminate new ones. At a later stage, scientific knowledge needs to be acquired and used to change cultural aspects through education.

Scientific contributions, addressing the above-mentioned challenges, are emerging [46]; however much more needs to be done. First, new technologies are required to implement the scientific solutions, particularly with respect to water conservation, treatment, and reuse. While there are practicable water conservation technologies around, much more is needed in the water quality domain, particularly with respect to operational regulation and exotic contaminants.

Figure 2. The water centric 17 Sustainable Development Goals.

5.1. Science & Technology (Tools and Methodologies)

5.1.1. Water Availability

IHP's "Water for Human Settlements"and its UWMP, have been focusing their efforts in disseminating and promoting the use of artificial intelligence (AI) and of the internet of things (IoT) in urban water management to address both the issue of water availability in an urban setting, as well as water quality. The use of sensors, transmitters, Supervisory Control and Data Acquisition (SCADA) systems, modelling and other tools, can effectively reduce the non-revenue water (NRW) into single digits and ensure enough water of good quality for people. Furthermore, the Programme and its urban based initiative (UWMP) have been focusing on intermittent water supply in order to better understand how this could be avoided or done in a secure and safe manner, to the extent possible.

5.1.2. Water Quality

Water quality can be remotely monitored via the use of satellite information. The International Initiative on Water Quality [47] has initiated a project that supports monitoring of Sustainable Development Goal 6's targets 6.3.2 (Proportion of bodies of water with good ambient water quality) and 6.6.1 (Change in the extent of water-related ecosystems over time), where remote sensing technology is used to provide a time series of information related to the pollution (or absence) of the surface water bodies. The UWMP is currently studying the effects of flooding episodes on water quality and a publication to this extent is expected in late 2019.

5.1.3. Water Related Hazards

In order to enhance the resilience of communities to floods and droughts, IHP has been developing the Flood and Drought Monitor [48] and has been providing the technology to various regions around the globe. The monitor permits the forecasting of extreme events (flooding, drought) well before they take place. In order to provide an end to end solution, similar tools have been developed, such as the drought atlas [49], which is coupled with a forecasting system and telephone apps that provide information to farmers in Latin American countries related to their crops' irrigation.

5.1.4. Nature Based Solutions: Ecohydrology

The fifth annual theme-oriented report of the United Nations World Development Report (WWDR) produced by UNESCO's World Water Assessment Programme focuses on opportunities to harness the natural processes that regulate various elements of the water cycle, which have become collectively known as nature-based solutions (NBS) for water [49]. IHP's work on ecohydrology [50] promotes the use of the interactions between biota and hydrology to regulate, remediate and conserve ecosystems to stabilize and improve the quality of water resources. Implementation of ecohydrology is undertaken through "harmonization" with existing and planned hydrotechnical infrastructures. Twenty-three pilot projects have been established worldwide to validate and quantify the effectiveness of ecohydrological solutions [50]. It needs to be stressed that the application of ecohydrology principles can be utilized to provide clean potable water, as well as to minimize the effects of water related hazards to communities and the environment.

UWMP has produced knowledge on this topic focused on an urban setting with two publications: "Capacity building for ecological sanitation: concepts for ecologically sustainable sanitation in formal and continuing education" and "Aquatic Habitats in Sustainable Urban Water Management".

5.1.5. Data Management

Data need to be stored in a safe environment that allows its analysis with a view to produce information for improved water resource management and decision making. Various databases and platforms have been developed during the past decades, usually at an inhibitory cost to developing states; especially when multiple licenses are required. IHP's Water Information Network System (WINS) is an open access, open source platform for sharing, accessing and visualizing water-related information, as well as for connecting water stakeholders [51]. WINS allows access to various types of information (maps, reports, graphs, etc.) covering the entire water cycle, ranging from groundwater to urban water through gender issues, from a local to a global scale. Information provided in the form of maps can be combined directly on the platform in order to create new information, and generate customized maps that can be shared with a large panel of stakeholders such as policy makers, institutions, researchers, or the civil society.

5.2. International Cooperation

Transboundary basins cover more than half of the Earth's land surface, account for an estimated 60% of global freshwater flow and are home to more than 40% of the world's population. Across the world, 153 countries share rivers, lakes and aquifers, and 592 transboundary aquifers have been inventoried by UNESCO's International Hydrological Programme to date. Transboundary water cooperation is thus critical for ensuring sustainable management of water resources.

IHP's PCCP (From Potential Conflict to Cooperation Potential) project facilitates multi-level and interdisciplinary dialogues in order to foster peace, cooperation and development related to the management of transboundary water resources. The project follows the idea that although transboundary water resources can be a source of conflict their joint management can be strengthened and even used as a means for further cooperation, contributing to UNESCO's mandate: to nurture the idea of peace in human minds.

Further to initiatives such as PCCP, UNESCO and UNECE are co-custodians of the SDG 6 indicator 6.5.2 on water cooperation, who provided in July 2018 the first global baseline. The work to date reveals that although significant progress has been made, arrangements for transboundary water cooperation are often absent.

UNESCO was designated as the agency to lead the United Nations International Year of Water Cooperation (IYWC) in 2013. The organization mobilised an estimated 25 million people around the world that year, positioning the idea to cooperate instead of compete/fight among countries, regions and different stakeholders to manage water.

At an urban level, UNESCO via its publications has examined and analysed urban water conflicts, their origins and nature, and have presented several historical urban water conflict cases and illustrations of changing conflict nature, including a theoretical analysis of ecological–economic factors to provide a basis for urban water conflict solution guidelines [52].

5.3. Science-Policy Interface

A collaborative, two-way interaction between science and policy spheres is the key to achieving practicable water security solutions.

As an intergovernmental organization, UNESCO's efforts are mainly focusing on decision makers. Tools developed are designed to be simple to use and contain the information that is required to make a science-based decision.

Recently, IHP established a Science Policy Interface Colloquium in Water (SPIC Water) as part of its Water Dialogues framework. The 1st SPIC Water took place on 14 June 2018 at UNESCO's Headquarters in Paris, France, and brought together ministers responsible for water resource management in 13 countries, along with experts and representatives of Member States [53]. The Colloquium was an opportunity to take stock of the progress made towards achieving the Sustainable Development Goal on Water and Sanitation (SDG6). It was organized at the request of Member States to discuss how UNESCO's International Hydrological Programme (IHP) can help to identify science-based solutions, effective policies and practices on water and sanitation, and support countries in their efforts to implement the 2030 Agenda.

The ministerial messages highlighted that the 2030 Agenda is promoting local action and positive changes in institutions at the country level. However, the sustainability of actions remains a challenge. They also noted the need to harmonize activities and policies at the global, regional and local level and to adapt targets to the local context. All underlined the need for reinforced human capacity if the 2030 Agenda was to be implemented in the domain of water. They welcomed the existence of a forum like SPIC Water, where policy-makers could exchange viewpoints with experts, who provide the knowledge and information needed to adapt policies based on available knowledge.

The Science Policy Interface Colloquiums on Water will play a significant role in the implementation of the 2030 Agenda. SDG 6 provides the platform for decision makers at the highest political level in water resource management to express the challenges they face and for scientists to adjust their work to cater for their needs. It will thus, guide future research and scientific work to pursue solutions that can be applied by countries.

SPIC water is designed to complement existing international fora, such as the World Water Forum, Dushanbe Conference, International Water Weeks, etc., and feed into the discussions during the High Level Political Forum in New York, when SDG 6 is examined.

5.4. Human Capital

The uptake of scientific and technological solutions requires particular attention to the socio-economic drivers at the managerial and public levels. The importance of social capital cannot and should not be underestimated in achieving water security.

The availability of sound scientific and technological tools cannot provide a solution to water resource management alone; it requires trained professionals to use them, sensitized decision makers to understand their importance and informed citizens to accept their results. Capacitating the human capital is thus the main focus of UNESCO IHP's investment.

An average of 10,000 experts, decision makers and communities have been trained and/or been made aware of various issues related to water security over the past two years (2016–2017) in a wide range of themes (Figure 3) by the efforts of UNESCO's Water Family.

Training on issues of water security in an urban context have been spearheaded by UNESCO's water related Chairs and Category 2 Centres.

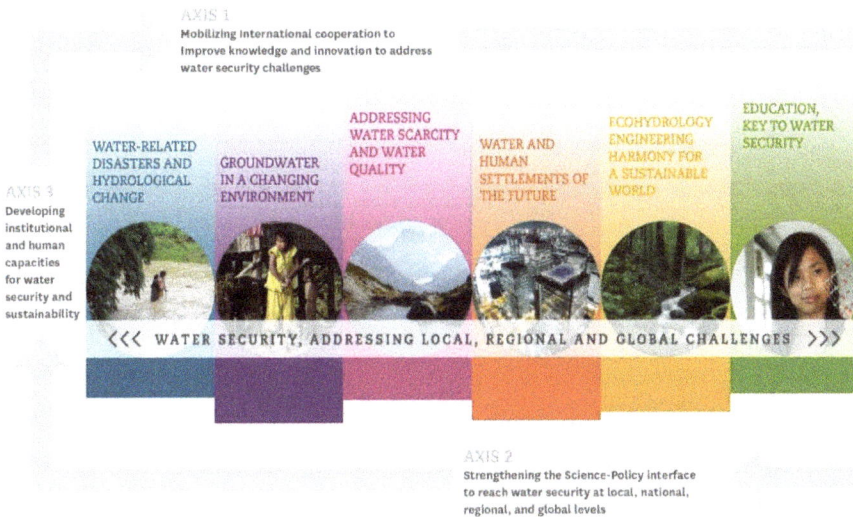

Figure 3. IHP VIII (2014–2021); Water Security: Responses to Local, Regional and Global Challenges.

5.5. Networking

UNESCO's Water Family is a network of networks comprising of 169 National IHP Committees or focal points, 37 Category 2 Centres and 50 UNESCO Chairs, along with the Secretariat of the UNESCO Water Science Division, IHP, WWAP and regional offices that surpass in numbers the staggering amount of 3000 experts.

Working quite often within the framework of IHP's International Initiatives [54] and/or within the implementation of projects and activities in the framework of IHP VIII, UNESCO's Water Family provides technical support to Member States in achieving water security and through this, internationally agreed goals, such as SDGs 6, 11 and 13, and agreement such as the Sendai Framework, Paris Agreement and the New Urban Agenda.

The framework of IHP's 17 international initiatives (see Table 1), provides yet an additional network of experts, who do not necessarily belong to UNESCO's Water Family institutions but who contribute to the strategic goals of the organization.

Table 1. IHP's 17 international initiatives.

Initiative	Description	Contact Officer
	Flow Regimes from International Experimental and Network Data, an international research initiative that helps to set up regional networks for analyzing hydrological data through the exchange of data, knowledge and techniques at the regional level	Mr Abou Amani a.amani@unesco.org
	Global Network on Water and Development Information in Arid Lands, a global network on water resources management in arid and semi-arid zones whose primary aim is to build an effective global community to promote international and regional cooperation in the arid and semiarid areas	Mr Anil Mishra a.mishra@unesco.org
GLOBAL NETWORK OF WATER MUSEUMS	Global Network of Water Museums, is an IHP initiative to create synergies within UNESCO with the aim of better using water museums to improve water management via communication and educational activities	Mr Alexander Otte a.otte@unesco.org
	Groundwater Resources Assessment under the Pressures of Humanity and Climate Change, a UNESCO-led project seeking to improve our understanding of how groundwater interacts within the global water cycle, how it supports human activity and ecosystems, and how it responds to the complex dual pressures of human activity and climate change	Ms. Alice Aureli a.aureli@unesco.org
	Hydrology for the Environment, Life and Policy, a new approach to integrated catchment management by building a framework for water law and policy experts, water resource managers and water scientists to work together on water-related problems	Mr Abou Amani a.amani@unesco.org
IWRM	Integrated Water Resources Management, an initiative implementing IWRM at the river basin level as an essential element to managing water resources more sustainably, leading to long-term social, economic and environmental benefits	Mr Alexandros Makarigakis a.makarigakis@unesco.org
United Nations Educational, Scientific and Cultural Organization . Regional Centre on Urban Water Management (under the auspices of UNESCO)	International Drought Initiative, an initiative aiming at providing a platform for networking and dissemination of knowledge and information between international entities that are actively working on droughts	Mr Abou Amani a.amani@unesco.org
	International Flood Initiative, an interagency initiative promoting an integrated approach to flood management which takes advantage of the benefits of floods and the use of flood plains, while reducing social, environmental and economic risks. Partners include the World Metereological Organization (WMO), the United Nations University (UNU), the International Association of Hydrological Sciences (IAHS) and the International Strategy for Disaster Reduction (ISDR).	Mr Abou Amani a.amani@unesco.org
IIWQ	International Initiative on Water Quality, an international platform to strengthen knowledge, research and policy, and develop innovative approaches to tackle water quality challenges	Ms. Sarantuyaa Zandaryaa s.zandaryaa@unesco.org

Table 1. *Cont.*

Initiative	Description	Contact Officer
International Sediment Initiative	International Sediment Initiative, an initiative to assess erosion and sediment transport to marine, lake or reservoir environments aimed at the creation of a holistic approach for the remediation and conservation of surface waters, closely linking science with policy and management need	Mr Anil Mishra a.mishra@unesco.org
isarm	Internationally Shared Aquifer Resources Management, an initiative to set up a network of specialists and experts to compile a world inventory of transboundary aquifers and to develop wise practices and guidance tools concerning shared groundwater resources management	Ms. Alice Aureli a.aureli@unesco.org
LAND SUBSIDENCE INTERNATIONAL INITIATIVE	Land Subsidence International Initiative is a global IHP platform for scientific researchers and institutions, aimed at voluntarily creating the knowledge base to build, facilitate and foster cooperation concerning planning, hydrogeological sciences and water security in urban and coastal areas, by exchanging expertise and good practices for a better transfer of knowledge to public policies;	Ms. Alice Aureli a.aureli@unesco.org
MAR	Managing Aquifer Recharge, an initiative that aims to expand water resources and improve water quality with the adoption of improved practices for management of aquifer recharge (storage and recovery)	Ms. Alice Aureli a.aureli@unesco.org
P C → C P	From Potential Conflict to Cooperation Potential, a project facilitating multi-level and interdisciplinary dialogues in order to foster peace, cooperation and development related to the management of shared water resources	Ms. Renee Gift r.gift@unesco.org
UWMP	Urban Water Management Programme, an initiative that generates approaches, tools and guidelines which will allow cities to improve their knowledge, as well as analysis of the urban water situation to draw up more effective urban water management strategies	Mr Alexandros Makarigakis a.makarigakis@unesco.org
WHYMAP	World Hydrogeological Map, an initiative to collect, collate and visualize hydrogeological information at the global scale to convey groundwater-related information in a way appropriate for global discussion on water issues	Ms. Alice Aureli a.aureli@unesco.org
	World Large Rivers Initiative, while excluding operational management, aims to establish a purely scientific global platform of researchers and institutions to develop, on a voluntary basis, the scientific foundation for integrated river research by exchanging expertise and good practice	Mr Abou Amani a.amani@unesco.org

5.6. More than Science & Technology

The role of social and cultural processes in water security, and social processes ultimately should be embedded in IWRM models and the nexus as new algorithms. This can lead into new understanding of the complex dynamics between human and natural systems and can pave the way to extending the scope of risk management.

UNESCO's multidisciplinary mandate allows the organization to bring solutions that include various social and human elements in tandem with scientific and technological opportunities. It ensures that the education element stands on the top of the agenda and that cultural beliefs and customs are taken into consideration when one designs a training or a way to manage the valuable resource.

Principles of ethics in the use of technology need to be examined ensuring that services will be provided in an inclusive manner.

6. Conclusion

The importance of water has, at last, been receiving considerable attention at various fora (e.g., Davos World Economic Forum [55] and has been identified as an important element for development in the post 2015 agenda, receiving the sixth goal of the 2030 Agenda and having a fundamental role in the Sendai Framework and the New Urban Agenda.

UNESCO's International Hydrological Programme's role is to raise awareness of communities and decision makers alike on the importance of water in human development and the environment; to do so in an inclusive and culturally sensitive manner and to assure that a critical mass of experts exist with geographical and gender balance to support activities and policies geared towards the solution of the identified challenges.

Water resources management and service delivery face multiple challenges at the local, regional and global level. When sustainable development is thought of in combination with the conservation of the environment and the protection of people and from water related natural hazards, the principle of water security is formed.

UNESCO's International Hydrological Programme has water security as the core of its work during its current medium term strategy, IHP VIII (2014–2021) and supports Member States in their efforts to achieve it. Within this framework, IHP is developing scientific and technological tools for science based decision making, promotes international cooperation through networking, enhances the science policy interface and focuses its efforts in the education and training of the human capital at local, regional and global levels.

Furthermore, the programme operating as the scientific arm of the United Nations on issues related to Water, plays a forecasting role to ensure the identification of future challenges and that enough scientific research will be conducted towards the provision of solutions to these challenges, as they will be the center of development in the near future.

In an urban context, UNESCO's IHP-VIII fourth theme is dedicated to water for human settlements and together with the initiative on Urban Water Management they provide a platform where new technologies, methodologies and techniques can be identified and tested to achieve a holistic way of managing water resources and providing sustainable services in the face of water scarcity and other pressures (such as climate change, pollution and population growth).

An open call is made through this paper for the scientific community active in water to participate actively in UNESCO's water related activity and to become a member of the UNESCO Water Family.

Author Contributions: Conceptualization and methodology, A.K.M. and B.E.J.-C.; writing—original draft preparation, A.K.M. and B.E.J.-C; writing—review and editing, A.K.M.

Funding: The APC was funded by UNESCO's Natural Science Sector, Water Science Division.

Conflicts of Interest: The authors declare no conflict of interest. The funders had no role in the design of the study; in the collection, analyses, or interpretation of data; in the writing of the manuscript, or in the decision to publish the results.

References

1. UNDP (United Nations Development Programme). *Human Development Report. Beyond scarcity: Power, Poverty and the Global Water Crisis*; United Nations Development Programme: New York, NY, USA, 2006.
2. UNISDR. Making Development Sustainable: The Future of Disaster Risk Management. In *Global Assessment Report on Disaster Risk Reduction*; United Nations Office for Disaster Risk Reduction (UNISDR): Geneva, Switzerland, 2015.
3. *Water, Megacities and Global Change*; UNESCO/ARCEAU IdF: Paris, France, 2016; ISBN 978-92-3-100161-1. Available online: https://unesdoc.unesco.org/ark:/48223/pf0000245419 (accessed on 17 December 2018).
4. Shiklomanov, I.A. Appraisal and Assessment of World Water Resources. *Water Int.* **2000**, *25*, 11–32. [CrossRef]

5. United Nations, Department of Economic and Social Affairs, Population Division. *World Population Prospects: The 2017 Revision, Key Findings and Advance Tables*; Working Paper No. ESA/P/WP/248; United Nations, Department of Economic and Social Affairs, Population Division: New York, NY, USA, 2017.

6. WWAP (World Water Assessment Programme). *The United Nations World Water Development Report 4: Managing Water under Uncertainty and Risk*; UNESCO: Paris, France, 2012.

7. Agriculture, Food and Water, 2003. Available online: http://www.fao.org/docrep/006/Y4683E/y4683e07.htm (accessed on 16 October 2018).

8. Förster, J. *Eurostat Statistics Explained. Water Use in Industry*; Eurostat: Luxembourg, 2014; Available online: https://ec.europa.eu/eurostat/statistics-explained/index.php?title=Archive:Water_use_in_industry&oldid=197333 (accessed on 29 September 2018).

9. FAO (Food and Agriculture Organization of the United Nations). *Food Wastage Footprints. Sustainable Pathways*; FAO: Rome, Italy, 2013.

10. Madani, K.; Lund, J.R. Strategic multi-criteria decision making under uncertainty. In *World Environmental and Water Resources Congress 2011: Bearing Knowledge for Sustainability—Proceedings of the 2011 World Environmental and Water Resources Congress, Palm Springs, CA, USA, May 2011*; pp. 3096–3102. [CrossRef]

11. Serageldin, I. Assuring water for food: The challenge for the coming generation. *Water Resour. Dev.* **2001**, *17*, 521–525. [CrossRef]

12. Shiklomanov, I. World water resources. water in crisis, Oxford, New York. In *Comprehensive Assessment of the Freshwater Resources of the World*; Gleick, P., Ed.; World Meteorological Organization: Stockholm, Sweden, 1997; Volume 88.

13. UNDP (United Nations Development Programme). Millenium Development Goals Report for Uganda 2015: Special Theme; Results, Reflections and the Way Forward. Available online: https://ug.usembassy.gov/wp-content/uploads/sites/42/2016/07/UGANDA-MDG_2015-FINAL.-REPORT.pdf (accessed on 29 September 2018).

14. Seckler, D.; Barker, R.; Amarasinghe, U. Water scarcity in the twenty-first century. *Int. J. Water Resour. Dev.* **1999**, *15*, 29–42. [CrossRef]

15. Schiermeier, Q. IPCC report under fire. *Nature* **2014**, *508*, 298. [CrossRef] [PubMed]

16. Elliott, J.D.; Deryng, C.; Müller, K.; Frieler, M.; Konzmann, D.; Gerten, M.; Glotter, M.; Flörke, Y.; Wada, N.; Best, S.; et al. Constraints and potentials of future irrigation water availability on agricultural production under climate change. *Proc. Natl. Acad. Sci.* **2014**, *111*, 3239–3244. [CrossRef] [PubMed]

17. Haddeland, I.; Heinke, J.; Biemans, H.; Stephanie, E.; Martina, F.; Naota, H.; Markus, K.; Fulco, L.; Yoshimitsu, M.; Jacob, S.; et al. Global water resources affected by human interventions and climate change. *PNAS* **2014**, *111*, 3251–3256. [CrossRef] [PubMed]

18. Schewe, J.; Heinke, J.; Gerten, D.; Ingjerd, H.; Nigel, W.A.; Douglas, B.C.; Rutger, D.; Stephanie, E.; Balázs, M.F.; Felipe, J.C.-G.; et al. Multi-model assessment of water scarcity under climate change. *PNAS* **2014**, *111*, 3245–3250. [CrossRef] [PubMed]

19. Mirchi, A.; Madani, K.; Roos, M.; Watkins, D.W. Climate Change Impacts on California's Water Resources. 2013. Available online: https://www.researchgate.net/publication/273378884_Climate_Change_Impacts_on_California\T1\textquoterights_Water_Resources (accessed on 29 September 2018). [CrossRef]

20. Gohari, A.; Bozorgi, A.; Madani, K.; Elledge, J.; Berndtsson, R. Adaptation of surface water supply to climate change in central Iran. *J. Water Clim. Chang.* **2014**, *5*, 391–407. [CrossRef]

21. Nazemi, A.; Wheater, H.S.; Chun, K.P.; Bonsal, B.; Mekonnen, M. Forms and drivers of annual streamflow variability in the headwaters of Canadian Prairies during the 20th century. *Hydrol. Process.* **2017**. [CrossRef]

22. Nazemi, A.; Wheater, H.S. On inclusion of water resource management in Earth system models—Part 1: Problem definition and representation of water demand. *Hydrol. Earth Syst. Sci.* **2015**, *19*, 33–61. [CrossRef]

23. Nazemi, A.; Wheater, H.S. On inclusion of water resource management in Earth system models—Part 2: Representation of water supply and allocation and opportunities for improved modeling. *Hydrol. Earth Syst. Sci.* **2015**, *19*, 63–90. [CrossRef]

24. Holger, T.; Martin-Bordes, J.L.; Jasonm, G. *Climate Change Effects on Groundwater Resources: A Global Synthesis of Findings and Recommendations*; Taylor & Francis Group: London, UK, 2012; ISBN 978-0-415-68936-6.

25. Agrawal, A. Local institutions and adaptation to climate change. In *Social Dimensions of Climate Change: Equity and Vulnerability in a Warming World, Mearns*; Mearns, R., Norton, A., Eds.; The World Bank: Washington, DC, USA, 2010; pp. 173–198.

26. Rosenzweig, C.J.; Elliott, D.; Deryng, A.C.; Ruane, C.; Müller, A.; Arneth, K.J.; Boote, C.; Folberth, M.; Glotter, N.; Khabarov, K.; et al. Assessing agricultural risks of climate change in the 21st century in a global gridded crop model intercomparison. *Proc. Nat. Acad. Sci.* **2014**, *111*, 3268–3273. [CrossRef] [PubMed]

27. Daskalaki, P.; Voudouris, K. Groundwater quality of porous aquifers in Greece: A synoptic review. *Environm. Geol.* **2007**, *54*, 505–513. [CrossRef]

28. Carey, R.O.; Migliaccio, K.W. Contribution of wastewater treatment plant effluents to nutrient dynamics in aquatic systems: A review. *Environ. Manag.* **2009**, *44*, 205–217. [CrossRef] [PubMed]

29. Managing Water Resources in Arid and Semi Arid Regions of Latin America and the Caribbean, MWAR-LAC. Available online: http://www.cazalac.org/mwar_lac/index.php?id=12 (accessed on 16 October 2018).

30. IPCC (Intergovernmental Panel on Climate Change). *Climate Change 2014: Synthesis Report. Contribution of Working Groups I, II and III to the Fifth Assessment Report of the Intergovernmental Panel on Climate Change*; Pachauri, R.K., Meyer, L.A., Eds.; IPCC: Geneva, Switzerland, 2014; p. 151.

31. Jiménez Cisneros, B.E.; Oki, T.; Arnell, N.W. Freshwater resources. In *Climate Change 2014: Impacts, Adaptation, and Vulnerability. Part A: Global and Sectoral Aspects. Contribution of Working Group II to the Fifth Assessment Report of the Intergovernmental Panel on Climate Change*; Field, C.B., Barros, V.R., Dokken, D.J., Eds.; Cambridge University Press: Cambridge, UK, 2014.

32. World Bank. *Reducing Inequalities in Water Supply, Sanitation, and Hygiene in the Era of the Sustainable Development Goals: Synthesis Report of the WASH Poverty Diagnostic Initiative*; WASH Poverty Diagnostic; World Bank: Washington, DC, USA, 2017.

33. WWF (World Wide Fund for Nature). *Food Loss and Waste: Facts and Futures*; WWF South Africa: Gland, Switzerland, 2017.

34. Munich Re 2013 NatCatSERVICE Database (Munich: Munich Reinsurance Company). 2013. Available online: https://www.munichre.com/en/reinsurance/business/non-life/natcatservice/index.html (accessed on 29 September 2018).

35. Prudhomme, C.; Giuntoli, I.; Robinson, E.L.; Douglas, B.C.; Nigel, W.A.; Rutger, D.; Balázs, M.F.; Wietse, F.; Dieter, G.; Simon, N.G.; et al. Hydrological droughts in the 21st century hotspots and uncertainties from a global multimodel ensemble experiment. *PNAS* **2014**, *111*, 3262–3267. [CrossRef] [PubMed]

36. Richter, B.D.; Mathews, R.; Harisson, D.L.; Wigington, R. Ecologically sustainable water management: Managing river flows for ecological integrity. *Ecol. Appl.* **2003**, *13*, 206–224. [CrossRef]

37. UNESCO, 1st Water Science-Policy Interface Colloquium. Available online: https://en.unesco.org/news/1st-water-science-policy-interface-colloquium (accessed on 29 September 2018).

38. UNESCO, International Initiative on Water Quality (IIWQ). Available online: https://en.unesco.org/waterquality-IIWQ (accessed on 29 September 2018).

39. Guégan, M.; Uvo, C.B.; Madani, K. *"Developing a Module for Estimating Climate Warming Effects on Hydropower Pricing in California" Energy Policy*; Elsevier: Amsterdam, Netherlands, 2012; Volume 41, pp. 261–271.

40. Montreal, Laval Declare States of Emergency over Flooding; Benjamin Shingler, Kalina Laframbois, CBC News. Available online: https://www.cbc.ca/news/canada/montreal/quebec-flooding-province-army-1.4103506 (accessed on 17 September 2018.).

41. Dankers, R.; Arnell, N.W.; Clark, D.B.; Pete, D.F.; Balázs, M.F.; Simon, N.G.; Jens, H.; Hyungjun, K.; Yoshimitsu, M.; Yusuke, S.; et al. First look at changes in flood hazard in the Inter-Sectoral Impact Model Intercomparison Project ensemble. *PNAS* **2014**, *111*, 3257–3261. [CrossRef] [PubMed]

42. Balaguru, K.; Foltz, G.R.; Ruby, L.L.; Hagos, S.M.; Judi, D.R. On the Use of Ocean Dynamic Temperature for Hurricane Intensity Forecasting. *Am. Meteorol. Soc.* **2018**. [CrossRef]

43. WHO and UNICEF. *Progress on Sanitation and Drinking Water—2015 Update and MDG Assessment*; World Health Organization (WHO) and the United Nations Children's Fund (UNICEF): Geneva, Switzerland, 2015; ISBN 9 789241 509145.

44. WWAP (World Water Assessment Programme). *The United Nations World Water Development Report 2017. Wastewater: The Untapped Resource*; UNESCO: Paris, France, 2017.

45. UNESCO, Ecohydrology. Available online: https://en.unesco.org/themes/water-security/hydrology/ecohydrology/ (accessed on 29 September 2018).

46. OECD. *OECD Environmental Outlook to 2050: The Consequences of Inaction*; OECD Publishing: Paris, France, 2012. [CrossRef]

47. UNESCO, The Water Information Network System (WINS). Available online: https://en.unesco.org/ihp-wins (accessed on 29 September 2018).
48. G-WADI, UNESCO IHP's "Global Network on Water and Development Information for Arid Lands. Available online: http://gwadi.org/tools-resources/hydrologic-monitoring/african-monitoring (accessed on 16 October 2018).
49. Mirchi, A.; Madani, K.; Watkins, J.D.; Ahmad, S. Synthesis of System Dynamics Tools for Holistic Conceptualization of Water Resources Problems. *Water Resour. Manag.* **2012**, *26*, 2421–2442. [CrossRef]
50. UNESCO, Hydrology Initiatives. Available online: https://en.unesco.org/themes/water-security/hydrology/programmes (accessed on 29 September 2018).
51. UN HABITAT, Planning and Design for Sustainable Urban Mobility: Global Report on Human Settlements 2013. ISBN Series Number: 978-92-1-131929-3. Available online: https://unhabitat.org/planning-and-design-for-sustainable-urban-mobility-global-report-on-human-settlements-2013/ (accessed on 29 September 2018).
52. *UN HABITAT, World Cities Report 2016*; UN-Habitat: Nairobi, Kenya; ISBN 978-92-1-132708-3. Available online: https://unhabitat.org/books/world-cities-report/ (accessed on 29 September 2018).
53. UNESCO BIOPALT project. Available online: https://en.unesco.org/biopalt?language=en. (accessed on 4 January 2019).
54. UNESCO, Urban Water Management Programme. Available online: https://en.unesco.org/uwmp (accessed on 29 September 2018).
55. Davos World Economic Forum 2015. Available online: https://www.weforum.org/agenda/2015/12/why-we-need-to-address-global-water-security-now/ (accessed on 16 October 2018).

water MDPI

Perspective

Water Governance in Cities: Current Trends and Future Challenges

Oriana Romano * and Aziza Akhmouch

Unit for Climate, Water, and SDGs, Cities, Urban Policies, and Sustainable Development Division, Centre for Entrepreneurship, SMEs, Regions and Cities, Organisation for Economic Co-operation and Development, 2 Rue André Pascal, 75775 Paris, France; aziza.akhmouch@oecd.org
* Correspondence: oriana.romano@oecd.org; Tel.: +33-145247686

Received: 19 December 2018; Accepted: 21 February 2019; Published: 10 March 2019

Abstract: Adapting water governance to changing needs, while coping with the uncertainties caused by climate change and the consequences of urbanisation and demographic growth, is key for inclusive, safe and resilient cities. The urgency of the challenges calls for innovative practices to enhance water security and provide better services to citizens, as foreseen by the Sustainable Development Goal (SDG) 6. The key question is: how to accomplish these objectives? While there is no doubt that technical solutions are available and play a fundamental role, they represent only part of the solution. Cities must ensure that the institutional frameworks in place are "fit to fix the pipes", from accessible information to adequate capacity, from sufficient funding to transparency and integrity, and from meaningful stakeholder engagement to coherence across sectoral policies. Building mainly on recent studies on water governance carried out by The Organisation for Economic Co-operation and Development (OECD) and specifically on urban water governance, this paper will discuss current trends and provide a set of tools for policy solutions based on OECD's 3Ps framework: people, policies and places. It will conclude by highlighting the importance of improving monitoring and evaluation for better design and implementation of urban water governance.

Keywords: water governance; infrastructure; urban water management; indicators; SDGs; stakeholder participation; water policy

---

1. Water and Cities: A Challenging Future Ahead

People's well-being and economic activities unquestionably hinge upon a critical component: water. In cities, water represents both an *opportunity* to carry out economic and social functions, and a *threat*, when consequences of disastrous events hit local economies and ecosystems. Yet, whether water is a challenge or an opportunity for cities largely depends on how well and efficiently it is governed. Indeed, urban water governance is about "doing things right" when managing too much, too little and too polluted water in cities and their hinterlands and providing adequate services.

Megatrends such as demographic growth, urbanisation and climate change increasingly affect water availability and quality in cities, where most people live and will be living in the future (70% by 2050 [1]). By 2050, water demand will increase by 55% compared to the year 2000, while four billion people will be living in water-stressed areas. Moreover, 240 million people will lack access to improved water sources, and almost 1.4 billion people are projected to lack access to basic sanitation [2]. In some rural and peri-urban areas of Mexico, Greece, Italy, and Portugal, amongst others, fractions of the population are not connected to water systems or have irregular access to water due to water scarcity [3].

Extreme water-related events are becoming increasingly frequent in cities all around the world. This is a challenge for over 80% of surveyed cities (Acapulco de Juarez, Amsterdam, Athens, Barcelona, Belo Horizonte, Bologna, Budapest, Calgary, Chihuahua, Cologne, Copenhagen, Culiacan,

Daegu, Edinburgh, Glasgow, Grenoble, Hermosillo, Hong Kong, China; Kitakyushu, Krakow, Lisbon, Liverpool, Malaga, Marseille, Mexico City, Milano, Montreal, Nantes, Naples, New York City, Okayama, Oslo, Paris, Phoenix, Prague, Queretaro, Rio de Janeiro, Rome, San Luis Potosi, Singapore, Stockholm, Suzhou, Toluca, Turin, Tuxtla, Veracruz, Zaragoza and Zibo) from OECD and non-OECD countries [3]. Projections show that more people will be at risk from floods by 2050 (from 1.2 billion today to 1.6 billion), especially in coastal cities [2]. At the same time, cities are facing or are at high risk of drought. In 2015, in Brazil, for example, a country where 12% of the world's freshwater resources are concentrated, Rio de Janeiro and São Paulo were hit by the worst drought in 84 years, while other areas in the country were experiencing flooding [4]. Moreover, these extreme events are incredibly costly: the severe flooding that hit Copenhagen in 2011 caused about EUR 700 million of damages; hurricane Sandy in New York City generated USD 19 billion of economic losses in 2012. In October 2018, hurricane Michael in Florida may have caused USD 25 billion in economic losses. Overall, between 2010 and 2050 the economic value of assets at risk of flood is projected to grow by 340%, reaching USD 45 trillion [2].

Significant investment is required to renew and upgrade infrastructure. Investment in water supply and sanitation alone will require USD 6.7 trillion by 2050 and this bill could triple by 2030 if investment is extended to a wider range of water-related infrastructure [5]. For a total of 92% of surveyed cities obsolete or lacking infrastructure represents the most important challenge for the future of water management [3].

Current levels of service delivery and water security in OECD and emerging economies should not be taken for granted. Although cities in the OECD area can provide high quality water services, they cannot rely on current infrastructure and procedures to maintain acceptable levels of water supply and sanitation. Global agreements and frameworks, such as the 2030 Agenda for Sustainable Development, the Sendai Framework and the New Urban Agenda call upon cities to be better prepared for water-related disasters, and be more resilient and inclusive when providing water services. New socio-economic paradigms such as the circular economy are calling upon better use and re-use of natural resources, including water. The key question is how to accomplish these objectives? While technical solutions are well-known and available, they represent only part of the solution for cities to manage water in a sustainable, integrated and inclusive way, at an acceptable cost, and in a reasonable timeframe. Therefore, beyond determining "what-to-do", it is important to know "who does what", "at which level of government" and "how" [5]. In other words, it is essential to implement governance frameworks that can help cities to adapt to changing circumstances, while maintaining their central role in local, national and global contexts.

2. Water Governance as a Means to an End

Often water crises are water governance crises: managing water risks of too much, too little, and too polluted water is all the more challenging if the roles and responsibilities are not clearly allocated, stakeholders are not engaged, information is not shared and the capacities are not adequate to anticipate and tackle the risks [6].

The OECD (Organisation for Economic Co-operation and Development) defines water governance as "the set of administrative systems, with a core focus on formal institutions (laws, official policies) and informal institutions (power relations and practices) as well as organisational structures and their efficiency" [6] (p. 28). As such, *governance* is not synonymous with *government*, and is distinct from water *management*, which refers to operational activities, for instance delivery and recycling [6]. As a means to an end, governance is "good" if it can solve water challenges; it is "bad" if does not respond to place-based needs [5].

At urban level, three models of water governance can be distinguished [7]: Hierarchical, Market and Network governance. The hierarchical model relies on top-down approaches in decision-making and implementation for water supply and sanitation with centralised public authorities, vertical accountability and poor stakeholder engagement; the market model is based on a greater empowerment of stakeholders for water management and ownership of water assets. It began developing in the nineties through different forms (e.g., privatisation, corporatisation, contracts between private operators and municipalities). Finally, the network model builds on the co-operation of private, civil and public actors and decentralised management approaches [8]. Beyond the theoretical distinctions, in practice, governance models are hybrid. Market signals, public policies and collective action can reinforce each other in complex polycentric social systems, where actors at different scales adapt their rules over time according to the problems they are addressing [9]. In order to do so, a number of principles and requirements are important, including information provision (e.g., state of the environment, uncertainty and values); compliance with rules; institutional infrastructure (e.g., research, social capital, and rules), coordination across levels of government [10].

As a matter of fact, cities are unable to address the complexity of water challenges on their own, but need to work with lower and higher levels of governments [3] and put in place meaningful mechanisms for participation. "System thinking" can reduce institutional fragmentation, while improving co-ordination and coherence across different policies [11].

To provide better understanding and policy guidance on water governance to public, private and non–profit actors, the OECD together with member states and water experts gathered in the OECD Water Governance Initiative developed 12 Principles on Water Governance [5]. The Principles are structured around three pillars: effectiveness, efficiency, and trust and engagement. Governance should contribute to the definition and implementation of policy goals (effectiveness), at the lowest possible cost to society (efficiency), while ensuring inclusiveness of stakeholders (trust and engagement) (Figure 1).

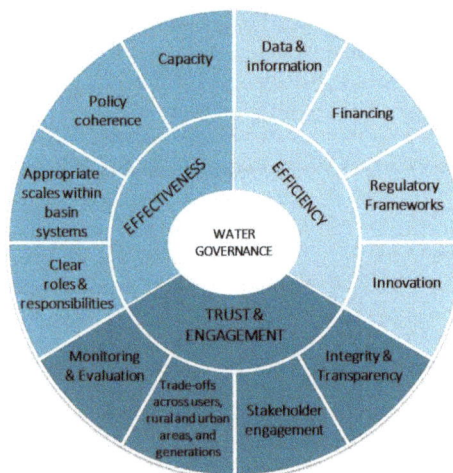

Figure 1. The Organisation for Economic Co-operation and Development (OECD) Principles on Water Governance. Source: OECD (2015), OECD Principles on Water Governance [5].

The 12 Principles refer to the water policy cycle, from the clear allocation of roles and responsibilities for water policy making, policy implementation, operational management and regulation (Principle 1) to regular monitoring and evaluation of water policy and governance (Principle 12).

3. Water Governance in Cities

In order to identify challenges and responses, the OECD employed an analytical framework that combined: (i) an assessment of the key factors affecting the effectiveness of urban water governance; (ii) a mapping of the roles and responsibilities at different levels of government; (iii) an appraisal of the main multi-level governance gaps to urban water management; and (iv) a focus on the policy responses to mitigate fragmentation and to foster integrated urban water management in cities and their hinterlands [3] (Figure 2).

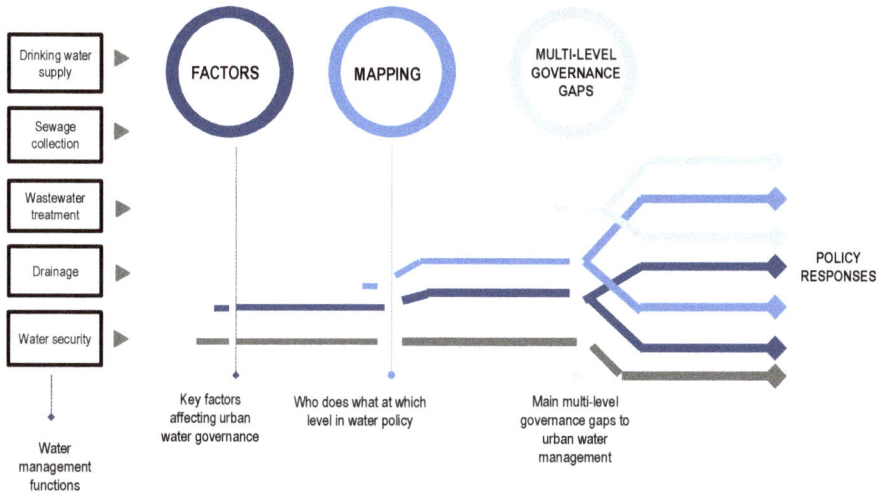

Figure 2. The analytical framework for assessing water governance in cities. Source: OECD (2016), Water Governance in Cities. OECD Publishing, Paris.

3.1. Key Factors Affecting the Effectiveness of Urban Water Governance

Several factors are shaping water governance in cities. According to the main results of the OECD survey carried out across 48 cities from OECD and non-OECD countries, water decisions in cities are affected by internal factors as well as by factors external to the water sector. The water sector is typically capital-intensive, requiring huge investment for infrastructure development and maintenance [12]. Water infrastructure is ageing, with negative impacts on efficiency and increasing operative costs due to leakages. This represents one of the greatest challenges for almost all surveyed cities (92%). Cities like Liverpool, Lisbon and Zaragoza, amongst others, have heavily invested to reduce leakages and to rehabilitate the pipeline network. In Zaragoza, for example, water losses from the distribution network have been reduced by more than 40% over a period of ten years (1997–2007). However, beyond technical solutions, improving the information system, flow monitoring and the use of performance indicators related to water losses can help reduce both inefficiencies and environmental- and financial-related costs.

Institutional factors, external to the water sector, highly influence urban water governance. Amongst them, territorial reforms are affecting the water governance system in 52% of surveyed cities. For example, in terms of re-organisation of water services delivery, information sharing across actors initiating new horizontal and vertical interactions, stakeholder engagement, and policy complementarities across different sectors and between cities and surrounding areas are all crucial. This has been the case in France, where in 2015, the territorial reform (Nouvelle Organisation Territoriale de la République, NOTRe) had implications for the transfer of responsibilities on water and sanitation to communities of municipalities.

Inevitably, water governance is also affected by megatrends such as climate change and urban growth (79% and 63% of surveyed cities, respectively [3]). Climate change is likely to increasingly affect the risks of "too much", "too little" or "too polluted" water. This can exacerbate the competition between water users. To cope with these challenges, cities would need to combine regulatory and economic instruments and to remove governance obstacles to long-term planning for climate change adaptation.

3.2. Mapping of Roles and Responsibilities

Urban water governance is a shared responsibility across different levels of government. While central governments have a prominent role in policy-making and regulatory functions, local governments have a more operative role in water functions, such as drinking water supply and drainage. Central governments tend to play an important role in water security policy-making and implementation and are also heavily involved in the regulation of water services [3]. In general, there is a trend towards the establishment of dedicated water regulatory bodies dealing with tariff regulation and performance monitoring, amongst other things. In general, this trend accompanies a reform of the water industry, which might imply a reorganisation of water provisions around fewer but bigger operators. For instance, this has occurred in Italy, Portugal, England and Wales, where regulators work with both national and sub-national actors [13].

In most cities surveyed in the 2016 OECD report [3], local governments (municipalities) are the primary sub-national authorities in charge of designing and/or implementing policies for drinking water supply and wastewater services. Metropolitan authorities may deal with water supply and sanitation. For example, the metropolitan area of Barcelona, which is formed by 36 municipalities, promotes integrated management of water supply and sanitation in the metropolitan area.

There are also a series of co-ordination mechanisms at vertical and horizontal levels to enhance water security: in Glasgow (United Kingdom) the Metropolitan Glasgow Strategic Drainage Partnership (MGSDP) is a collaborative venture between local authorities, the Scottish Environment Protection Agency (SEPA), Scottish Water, Scottish Enterprise, Clyde Gateway and Scottish Canals. The scope of its responsibilities includes flood reduction and improved water quality. In Italy, the authorities of the optimal territorial areas (Ambiti Territoriali Ottimali, ATO) ensure local stakeholder participation in order to manage water services in an integrated manner. The advantage of the coordination mechanisms is to gather several authorities and stakeholders for a concerted action towards greater water security and coherent water management, while avoiding overlaps and duplications. The accomplishment of expected results depends on internal and external circumstances, including political willingness.

3.3. Multi-Level Governance Gaps to Urban Water Management

Cities face several multi-level governance gaps. In particular, cities may suffer from unstable or insufficient revenues undermining effective implementation of water responsibilities. A total of 69% of surveyed cities in Reference [3] reported difficulties in raising tariffs for water services. At the same time, many cities have introduced affordability measures for low-income groups. They consist of using progressive social tariffs (e.g., Grenoble, Hermosillo, Lisbon); and implementing pro-poor policies (e.g., Budapest, Calgary, Hong Kong, China) and providing assistance to rural communities (e.g., Veracruz); grants for low-income families (e.g., Singapore); or social funds for people living in disadvantaged areas (e.g., Grenoble and Malaga).

Capacity is often the "Achilles' heel" of sub-national governments: many cities are facing technical and human resources gaps to efficiently manage water. The former relates to planning, quality information, monitoring and evaluation. The latter covers issues regarding staff, expertise and managerial capabilities. Water management in cities involves expertise from different fields and requires the capacity to respond to emergencies (such as in cases of water-related extreme events), to set up measures for disaster prevention, as well as to carry out ordinary duties, which must all be implemented in coherence with citizens' needs and in co-ordination with other policies and sectors. The cities involved in the OECD 2016 [3] survey reported the lack of staff and managerial competencies (65%) as the main source of their capacity gap.

Other multi-level governance gaps are the following: weak articulation between institutional, functional and hydrological boundaries, which can hinder integrated water management that would optimise the opportunity cost of investments and the efficient use of water (administrative gap); fragmentation of tasks and lack of strategic vision across water-related sectors (policy gap); lack of institutional incentives for co-operation and contradictions between legal and regulatory instruments at different levels of government (objective gap), as obstacles to long-term and co-ordinated urban water governance. Moreover, weak stakeholder engagement (accountability gap) also represents a challenge. Data production (e.g., on the state of environment) can be incomplete or collected irregularly (information gap).

4. The "3Ps" Framework

OECD (2016) developed the "3Ps" framework (policy, people and places) in response to the above challenges [3] (Figure 3):

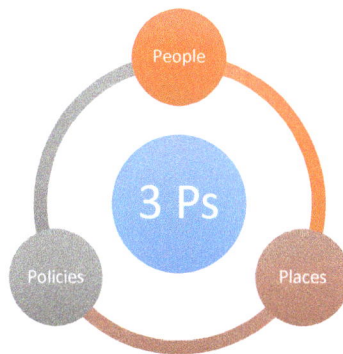

Figure 3. The "3Ps"Framework Source: OECD (2016), Water Governance in Cities. OECD Publishing, Paris.

Policy: Water governance has consequences for, and can be affected by a number of intrinsically related policies, such as land use, spatial planning, transport, energy, solid waste, environment, and agriculture, with impacts on water resource consumption, quality and security. Co-ordination across policies favours inter-sectoral complementarities while efficiently allocating resources. In the Netherlands, municipalities carry out "water assessments" to factor in water-related consequences and costs in spatial planning decisions. Building codes and housing regulation increasingly aim to reduce water consumption and to protect from water-related risks. In Germany, the City of Cologne co-ordinates water and spatial planning for new building areas to prevent flood damages from heavy rainfalls. Beyond planning and legal instruments, policy co-ordination can also take the form of financial incentives, as in the case of the City of Paris, which defined incentives for farmers to reduce their use of pesticides in order to protect water and natural resources.

People: A plethora of people from public, private, non-profit sectors to water users themselves have a stake or play a role in urban water management: urban planners, water service providers, regulators, advisors and civil society. They all contribute to dynamic and integrated approaches for water management. Stakeholder engagement can help build trust and ownership, secure willingness to pay for water services, ensure the accountability of city managers and service providers to end-users and citizens, set convergent objectives across policy areas and prevent and manage conflicts over water allocation [14]. Stakeholder engagement is important to raise awareness about current and future water risks and to build the social and political acceptability of reforms. For example, within the Local Urban Environment Adaptation Plan for a Resilient City (BLUE AP), the City of Bologna (Italy) engaged 150 stakeholders during a year of consultations to set climate change adaptation measures, including specific measures for water management to cope with water scarcity and floods. Overall, 70 project ideas and six pilot actions were presented.

Places: Water cuts across boundaries. As such, developing a place-based understanding is required to overcome territorial mismatches and to favour co-operation between cities and their surroundings (rural and watersheds). In this case, rural-urban partnerships represent win-win-win solutions benefiting cities, upstream and downstream communities and ecosystems. They consist in cross-sectoral and holistic sets of initiatives (e.g., within a wider package of environmental policy initiatives) or are focused on single objectives/projects (i.e., management of water resources) [15]. In Montreal, a multi-stakeholder committee helps improve the quality of discharged water in catchment areas; in New York City, contracts between the utility and watershed communities help preserve both water quality and the economic dynamism of the area.

5. Ways Forward: Improving Water Governance in Cities

Urban water governance is critical for managing water-related risks at an acceptable cost and in a proper time, so that the next generation does not inherit liabilities and costs. Clearly, there is no "one size fits all" solution. Moreover, cities have different capacities to respond to challenges. Correctly addressing these challenges requires assessing the range of political, institutional and administrative rules, practices and processes (formal and informal) through which decisions are taken and implemented. To support governments in this endeavour, the OECD together with experts from the public, private and non-for-profit sectors gathered within the OECD Water Governance Initiative, developed a Water Governance Indicator Framework that cities can use to identify whether water governance conditions are in place and function or need improvement [16]. The OECD Water Governance Indicator Framework is a self-assessment tool, which is part of the implementation strategy of the OECD Principles on Water Governance. The framework is composed of 36 indicators, measuring the *what* (policy framework), the *who* (institutions in charge) and the *how* (co-ordination tools for water policies) for each Principle. Differently from other measurement frameworks, the OECD Water Governance Indicator Framework can provide a global picture on the water governance system, rather than focusing on specific dimensions (e.g., transparency) or specific functions (e.g., water supply and sanitation). Information should be gathered through multi-stakeholder dialogues, which can lead to greater accountability of governments and stakeholders on how they share responsibilities and deliver the intended outcomes.

Cities can greatly benefit from the multi-stakeholder process to acquire information for each indicator. In fact, the self-assessment should be carried out in a participative manner in order to ensure that the process is transparent, neutral and open, as indicated by the 10-step methodology for self-assessment [16]. An open dialogue across stakeholders can inform policy makers in cities on the state-of-the-art of water governance and future expectations. This assessment in fact takes into account the diversity of opinions during the discussion. For each indicator, the assessment is complemented by the level of consensus amongst stakeholders. This is visually represented by water drops from 1 (low level of consensus) to 3 (high level of consensus) (Figure 4).

Identifying challenges is the first step towards effective responses for improved water governance in cities. Notably, cities will be increasingly facing challenges due to climate change and urbanisation, with consequences on competition for water resources between different users, on water quality and water security. However, cities are also laboratories for innovation, where experiments and pilots can take place. As such, fostering dialogue at a local level and building a consensus across a range of public authorities and stakeholders can help identify innovative, effective and efficient ways forward to better manage too much, too little and too polluted water now and in the future.

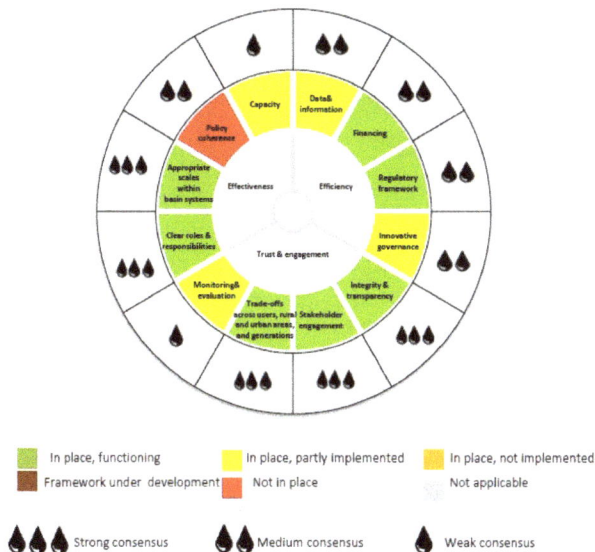

Figure 4. Visual representation of the Traffic Light System. Source: OECD (2018), Implementing the OECD Principles on Water Governance: indicator Framework and evolving practices, OECD Publishing.

Author Contributions: Conceptualization A.A. and writing—original draft preparation, review and editing, O.R.

Conflicts of Interest: The authors declare no conflict of interest.

References

1. UNDESA. Revision of World Population Prospects. Available online: www.un.org/development/desa/en/news/population/2015-report.html (accessed on 21 February 2019).
2. OECD. *OECD Environmental Outlook to 2050*; OECD Publishing: Paris, France, 2012.
3. OECD. *Water Governance in Cities, OECD Studies on Water*; OECD Publishing: Paris, France, 2016.
4. OECD. *Water Resources Governance in Brazil*; OECD Publishing: Paris, France, 2015.
5. OECD. OECD Principles on Water Governance. OECD: Paris, France, 2015. Available online: www.oecd.org/governance/oecd-principles-on-water-governance.htm (accessed on 9 March 2019).
6. OECD. *Water Governance in OECD Countries: A Multi-level Approach*; OECD Publishing: Paris, France, 2011.
7. Van de Meene, S.J.; Brown, R.R.; Farrelly, M.A. 2011 Towards understanding governance for sustainable urban water management. *Glob. Environ. Chang.* **2011**, *21*, 1117–1127. [CrossRef]
8. Ostrom, E. Beyond markets and states: polycentric governance of complex economic systems. *Am. Econ. Rev.* **2010**, *100*, 641–672. [CrossRef]
9. Andersson, K.P.; Ostrom, E. Analyzing Decentralized Resource Regimes from a Polycentric Perspective. *Policy Sci.* **2008**, *41*, 71–93. [CrossRef]
10. Dietz, T.; Ostrom, E.; Stern, P.C. The Struggle to Govern the Commons. *Science* **2003**, *302*, 1907–1912. [CrossRef] [PubMed]
11. Tortajada, C. Water governance: Some critical issues. *Int. J. Water Resour. Dev.* **2010**, *26*, 297–307. [CrossRef]
12. Winpenny, J. Water: Fit to Finance? Catalyzing National Growth through Investment in Water Security, Report of the High Level Panel on Financing Infrastructure for a Water-Secure World. World Water Council: Marseille, 2015. Available online: http://www.worldwatercouncil.org/sites/default/files/2017-10/WWC_OECD_Water-fit-to-finance_Report.pdf (accessed on 9 March 2019).
13. OECD. *The Governance of Water Regulators, OECD Studies on Water*; OECD Publishing: Paris, France, 2015.
14. OECD. *Stakeholder Engagement for Inclusive Water Governance*; OECD Publishing: Paris, France, 2015.

15. OECD. *Rural-Urban Partnerships: An Integrated Approach to Economic Development, OECD Rural Policy Reviews*; OECD Publishing: Paris, France, 2013.
16. OECD. *Implementing the OECD Principles on Water Governance: Indicator Framework and Evolving Practices, OECD Studies on Water*; OECD Publishing: Paris, France, 2018.

water

MDPI

Article

Assessing Urban Water Management Sustainability of a Megacity: Case Study of Seoul, South Korea

Hyowon Kim [1], Jaewoo Son [1], Seockheon Lee [2], Stef Koop [3], Kees van Leeuwen [3], Young June Choi [4] and Jeryang Park [1,*]

[1] School of Urban and Civil Engineering, Hongik University, Wausan-ro 94, Mapo-gu, Seoul 04066, Korea; hyolol@naver.com (H.K.); sjaewoo10@daum.net (J.S.)

[2] Centre for Water Resource Cycle Research, Korea Institute of Science and Technology, Hwarangno 14-gil 5, Seongbuk-gu, Seoul 02792, Korea; seocklee@kist.re.kr

[3] KWR Watercycle Research Institute, Groningenhaven 7, Nieuwegein 3433 PE, and Copernicus Institute for Sustainable Development and Innovation, Utrecht University, Heidelberglaan 2, 3584 CS Utrecht, The Netherlands; Stef.Koop@kwrwater.nl (S.K.); Kees.van.Leeuwen@kwrwater.nl (K.v.L.)

[4] Water Recycle Research Division, Seoul Water Institute, Cheonho-daero 716-10, Gwangjin-gu, Seoul 04981, Korea; youngjune@seoul.go.kr

* Correspondence: jeryang@hongik.ac.kr; Tel.: +82-2-320-1639

Received: 17 April 2018; Accepted: 22 May 2018; Published: 24 May 2018

Abstract: Many cities are facing various water-related challenges caused by rapid urbanization and climate change. Moreover, a megacity may pose a greater risk due to its scale and complexity for coping with impending challenges. Infrastructure and governance also differ by the level of development of a city which indicates that the analysis of Integrated Water Resources Management (IWRM) and water governance are site-specific. We examined the status of IWRM of Seoul by using the City Blueprint®Approach which consists of three different frameworks: (1) Trends and Pressures Framework (TPF), (2) City Blueprint Framework (CBF) and (3) the water Governance Capacity Framework (GCF). The TPF summarizes the main social, environmental and financial pressures that may impede water management. The CBF assesses IWRM of the urban water cycle. Finally, the GCF identifies key barriers and opportunities to develop governance capacity. The results indicate that nutrient recovery from wastewater, stormwater separation, and operation cost recovery of water and sanitation services are priority areas for Seoul. Furthermore, the local sense of urgency, behavioral internalization, consumer willingness to pay, and financial continuation are identified as barriers limiting Seoul's governance capacity. We also examined and compared the results with other mega-cities, to learn from their experiences and plans to cope with the challenges in large cities.

Keywords: Integrated Water Resources Management; water management sustainability; urban resilience; urban water cycle; water governance

---

1. Introduction

Globally, more than half of the world's population resides in urban areas, and this figure is projected to increase to 66% by 2050 [1]. Cities are important engines of innovation and wealth creation, as well as sources of improved efficiencies for the use of materials and energy [2]. On the other hand, primarily due to the concentration of people in a relatively small area, cities also act as centres of intense resource consumption and pollution [3,4]. Rapid urbanization along with the effects of climate change creates multiple challenges regarding water quality, water scarcity, and flooding resulting in high vulnerability and, sometimes, unforeseen consequences [5]. Actually, these risks are amplified in cities that lack the necessary infrastructure and/or institutional arrangements with the adaptive capacity to cope with these challenges [6,7]. A sustainable city thus requires appropriate and efficient management and control of a large variety of issues, notably the availability of sufficient clean freshwater and the protection against flooding as a prerequisite for health, economic development and social well-being of their inhabitants [7].

Water management and water governance challenges are often more prominent in larger cities [6]. Twenty-five million people—50% of the population of the Republic of Korea—reside in the metropolitan area of Seoul, which is amongst the largest urban regions in the world [8]. The city of Seoul has undergone extensive growth over the past half-century and has grown into a prosperous metropolis. The city's growth has been accompanied by the development and adoption of advanced water technologies and water policies. However, continuous efforts are necessary to improve Seoul's water management to cope with pressures that constantly change and may aggravate due to climate change, aging infrastructures, and evolving social demands. Moreover, due to its complex geomorphology [9] and a high spatiotemporal variability in hydro-climatic conditions, water management in Korea has always been challenging [10].

The City Blueprint®Approach (CBA) has been developed to assess the sustainability of Integrated Water Resources Management (IWRM) in a municipality [11,12]. The CBA consists of three assessment frameworks: (1) the Trends and Pressures Framework (TPF), which summarizes the principal social, environmental and financial pressures that impedes water management, (2) the City Blueprint Framework (CBF), which provides an overview of the performances of IWRM, and, (3) the water Governance Capacity Framework (GCF), which identifies key barriers and opportunities in urban water governance (Figure 1). The CBF has been used extensively since its development for rapid baseline assessments in about 70 cities around the globe. This allows for a comparison with other cities and facilitates city-to-city learning on strategic planning, exchange of knowledge, experiences, and best practices [13]. Results for 45 municipalities and regions in 27 different countries have been published [14], and, a recent update with references to publications and presentations for 70 municipalities and regions in 37 countries is available as an E-Brochure (European Commission: Brussels, Belgium) on the EIP (European Innovation Partnerships) Water website [15].

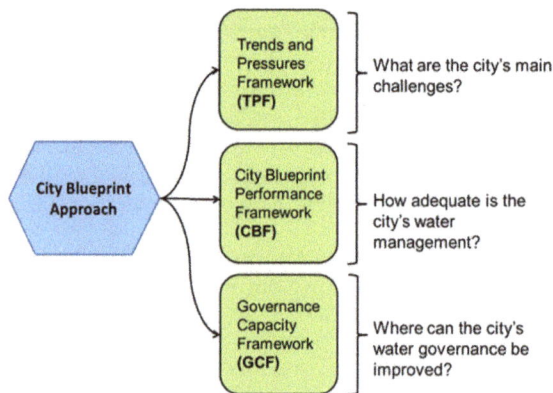

Figure 1. Overview of the City Blueprint Approach which consists of three complementary diagnostic frameworks to assess the urban water cycle management and governance [12–14].

The aim of this study is to identify barriers, enablers, and city-to-city learning opportunities to improve Seoul's water management and resilience. Also, regarding the scale of Seoul, and given that scale matters for tackling water management challenges, we compare CBF results with other megacities that were examined in earlier studies [14,15]. This comparative study will allow Seoul to learn from other well-managed cities and improve on weaknesses that were identified through this assessment.

2. Materials and Methods

2.1. Study Area

Korea is located in the northeastern part of the Asian continent, and Seoul, the capital of Korea, is in the northwestern part of the country. Seoul has been the capital of the country for more than 600 years since the foundation of the Joseon Dynasty in 1394. The geographical area of Seoul has expanded throughout history with the increasing population, and has shown explosive growth since the end of the Korean War in the early 1950s [16]. The total area of Seoul is 605.2 km^2 with a population of 10,112,070 as of 2017 [17]. The population of Seoul increased from 1.7 million in 1950 to 10 million in 1992, with an average growth rate of 278,583/year. Since then, the population has stabilized at around 10 million due to elevated housing prices and the government policy of controlling urban sprawl by constructing satellite cities and towns around Seoul [16]. Currently, the population density is around 17,200/km^2, which has been sustained for a decade. However, this number is still 70% greater than the average population density of the 34 other megacities worldwide (10,100/km^2). Although the population density within the administrative boundary of Seoul has stabilized since the early 1990s, the population of the Seoul metropolitan area, which includes several large satellite cities, keeps increasing and is expected to grow further in coming decades.

Seoul has four distinct seasons with the average temperature varying from −1 °C during the winter season (from December to February) to 25 °C during the summer season (from June to August). About 65% of the annual rainfall is concentrated in summer due to the monsoon. While the highly variable hydro-climatic conditions have already posed many water-related challenges in Seoul, climate change effects are also apparent in precipitation and air temperature records [18]. While the mean annual rainfall before the 1950s was around 1230 mm, it has now increased to around 1400 mm. The frequency and intensity of torrential rainfall in summer also increased, resulting in a greater intra-annual rainfall variability [18]. The mean air temperature increased from 10.4 °C in 1909 to 13.4 °C in 2014 by an overall rate of 0.0238 °C/year, which is higher than global trends of 0.0066 ~0.0189 °C/year [16,19]. This trend has become more significant since the 1950s due to rapid urbanization and, correspondingly, the urban heat island (UHI) effect [16,20]. Since the UHI effect increases energy consumption, health problems (e.g., heat strokes), and surface water quality deterioration, a rapid increase in air temperature poses a serious IWRM challenge [21].

2.2. City Blueprint Approach

In order to assess the trends and pressures, IWRM and the governance capacities of Seoul, we applied the CBA (Figure 1). Detailed information about the data sources, the calculations and examples are provided in three questionnaires available on the EIP Water website [15].

2.2.1. Trends and Pressure Framework (TPF)

Each city has its own unique social, financial and environmental background. As such, cities' performance regarding urban water management should be carefully assessed based on the context that has shaped the current state of infrastructure and governance for urban water management. The TPF aims to provide a concise understanding of these contextual trends and pressures that affect water management of a city [13]. It is evaluated with 12 indicators, which are divided over social, environmental, and financial categories (Table 1). Each indicator is scaled from 0 to 4 points, where a higher score indicates stronger pressure or concern. Note that many of these indicators are evaluated based on the ranking of the city among all countries, thus a specific score does not necessarily imply its absolute pressure state [13,15]. After each indicator is scored, these scores are classified into five categories: no concern (0–0.5), little concern (0.5–1.5), medium concern (1.5–2.5), concern (2.5–3.5), and great concern (3.5–4). More detailed descriptions of the indicators, data requirements and sample calculations as well as a critical discussion on its limitations, can be found elsewhere [13,14].

Table 1. Basic method and features of the Trends and Pressures Framework and City Blueprint Framework [13,15,22].

Trends and Pressures Framework (TPF)		
Goal	Baseline assessment of social, environmental, and financial pressures	Indicators
Framework	Social pressures	1. Urbanization rate 2. Burden of disease 3. Education rate 4. Political instability
	Environmental pressures	5. Flooding 6. Water scarcity 7. Water quality 8. Heat risk
	Financial pressures	9. Economic pressure 10. Unemployment rate 11. Poverty rate 12. Inflation rate
Data	Public data or data provided by Seoul Metropolitan Government	
Scores	0: no concern, 1: little concern, 2: medium concern, 3: concern and 4: great concern	
Overall score	Trends and Pressure Index (TPI), the arithmetic mean of 12 indicators. Indicators scoring a concern or great concern (3 or 4 points) are communicated as a priority	
City Blueprint Framework (CBF)		
Goal	Baseline performance assessment of the state of IWRM	
Framework	Twenty-five indicators divided over seven broad categories: 1. Water quality 2. Solid waste 3. Basic water services 4. Wastewater treatment 5. Infrastructure 6. Climate robustness 7. Governance	
Data	Public data or data provided by the water and wastewater utilities and cities based on a questionnaire [15]	
Scores	0 (low performance) to 10 (high performance)	
Overall score	Blue City Index (BCI), the geometric mean of 25 indicators	
Governance Capacity Framework (GCF)		
Goal	Baseline assessment of the governance capacity of a city	
Framework	Five challenges: (1) water scarcity, (2) flood risk, (3) wastewater treatment, (4) solid waste treatment, and (5) UHI. In each water challenge, 27 indicators are divided over nine broad categories: 1. Awareness 2. Useful knowledge 3. Continuous learning 4. Stakeholder engagement process 5. Management ambition 6. Agents of change 7. Multi-level network potential 8. Financial viability 9. Implementing capacity	
Data	Policy documents, scientific literature, and interviews Total interviewees: 10 (academia: 5, practitioners or civil servants: 5) 2~3 interviewees for each challenge	
Scores	'very encouraging (++)' to 'very limiting (−−)'	

2.2.2. City Blueprint Framework (CBF)

The CBF comprises 25 indicators divided over seven broad categories: (I) water quality, (II) solid waste, (III) basic water services, (IV) wastewater treatment, (V) infrastructure, (VI) climate robustness and (VII) governance (Table 1) [5,6]. The indicators are scored on a scale between 0 (very poor) to 10 (excellent). The geometric average of these 25 indicators is the Blue City Index (BCI) [6,7]. As the CBF was developed as the first framework of the CBA [11], the applications of this methodology in many municipalities and regions have been published [14,15]. Details regarding data, calculation of each index, and scaling methods are described in Koop and Van Leeuwen [13] and on the EIP Water website [15]. For the assessment of Seoul, most of the data were collected from public sources. For indicators that require self-assessment, the relevant materials and data were collected by interviewing experts in the Seoul Water Institute and the Seoul Metropolitan Government.

2.2.3. Governance Capacity Framework (GCF)

The sustainability of any resource management regime depends on the institutional capacity that enables adaptive management that can cope with external shocks and pressures [23,24]. The GCF was developed as the third framework of the CBA to assess the governance capacity of a city that allows or limits its sustainable management of water [12,23]. The GCF aims to identify the key enabling or limiting of governance conditions regarding five main urban water challenges that are relevant to urbanization and climate change. These challenges include (1) water scarcity, (2) flood risk, (3) wastewater treatment, (4) solid waste treatment, and (5) UHI [23]. For each challenge, the GCF assesses nine governance conditions, each of which includes three indicators. Each indicator is evaluated by a Likert-scale scoring method which ranges from 'very encouraging' (++) to 'very limiting' (−−) (Table 1). Since its development, the GCF has been successfully operationalized in several cities including Amsterdam, Quito, Ahmedabad, and New York City [22,25–27]. More details on the methodology are reported in Koop et al. [23].

The GCF indicator scoring was done through two steps: (1) preliminary scoring based on an analysis of policy documents and scientific literature, and (2) confirmatory scoring based on qualitative semi-structured interviews and surveys with experts to obtain additional details on the governance for each water challenge. The respondents were categorized as government personnel and academic scholars. Ten respondents were carefully selected based on their relevance to each of the five water challenges. Several respondents from the government sector had professional experience in multiple categories (e.g., flood risk and wastewater treatment). In those cases the interviewees were allowed to respond to multiple water challenges, resulting in at least two to three responses for each water challenge (Table 1).

3. Results

3.1. Trends and Pressures of Seoul

All TPF indicators of Seoul ranged from no concern (0–0.5) to medium concern (1.5–2.5), except heat risk, for which the indicator score was 2.72 (concern). The indicators categorized as medium concern included education rate, political instability, water scarcity, and economic pressure, with respective scores of 1.70, 1.92, 1.67 and 2.12. The arithmetic mean of all indicators, i.e., the Trends and Pressures Index (TPI) was 0.90, which is rather low and comparable to cities in the Netherlands and Sweden [15]. Among the 11 Asian cities analyzed with CBF, Singapore, with a TPI of 1.0, and Taipei, with a TPI of 1.4, were most comparable to Seoul. However, the other eight Asian cities (with TPIs of 1.9~2.6) face greater concerns, generally due to social pressure from high urbanization rates, environmental pressure from water scarcity, flooding, and heat risk, and financial pressure from low GDPs [28]. A full overview of TPI scores for 70 municipalities and regions, including 11 Asian cities, is provided in the most recent version of the E-Brochure [15].

3.2. City Blueprint of Seoul

The CBF presents a snapshot, i.e. the current performance of a city regarding IWRM. The geometric mean of all 25 CBF indicators, i.e. the Blue City Index, for Seoul is 7.3 (Figure 2). Based on a hierarchical clustering analysis of CBF indicator scores of 45 municipalities, Koop and Van Leeuwen [14] identified five different levels of sustainability of IWRM in cities worldwide: (1) cities lacking basic water services (BCI 0–2), (2) wasteful cities (BCI 2–4), water efficient cities (BCI 4–6), resource efficient and adaptive cities (BCI 6–8), and (5) water-wise cities (BCI 8–10). According to this categorization, Seoul is classified as a 'resource efficient and adaptive city.' Moreover, among the 70 cities assessed so far, Seoul has one of the highest BCI scores. However, our analysis reveals that there are also opportunities for improvement. The specific areas where improvement can be made are represented by relatively low indicator scores. Since many of the indicators obtained a full score of 10, we arbitrarily regarded any score less than six as the criterion for selecting areas for further improvement. Indicators that scored lower than six included nutrient recovery, average age of the sewer network, operation cost recovery, and stormwater separation (Figure 2a).

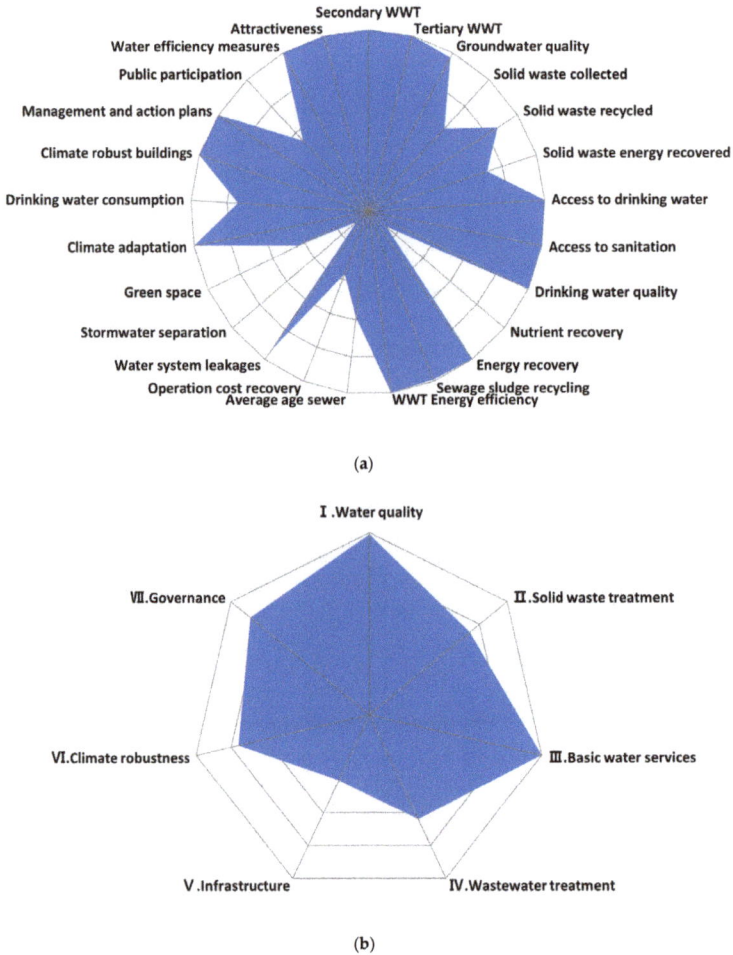

(a)

(b)

Figure 2. The City Blueprint of Seoul (**a**) based on 25 indicator scores and (**b**) the average scores of the seven categories. The Blue City Index, the geometric mean of all 25 indicators, is 7.3.

Among the seven broad categories, wastewater treatment (IV) and infrastructure (V) have average scores of 6.4 and 4.0, respectively (Figure 2b). In particular, infrastructure includes three indicators, i.e. stormwater separation, average age of sewer, and operating cost recovery, where improvements can be made. In other words, infrastructure improvement is thought to be an effective measure to enhance IWRM in Seoul.

3.3. The Water Governance Capacity of Seoul

Table 2 shows the results of GCF analysis for five urban water challenges in Seoul, whereas Figure 3 summarizes the average of each indicator score for all five challenges. According to our analysis, four indicators, i.e., indicator 1.2 local sense of urgency, indicator 1.3 behavioral internalization, indicator 8.2 consumer willingness to pay, and indicator 8.3 financial continuation, were found to be limiting (Table 3). Furthermore, the governance capacity for water scarcity and UHI was relatively low with a few indicators that limited the governance capacity. Specifically, the indicators 1.2 and 1.3 were found to limit the capacity to govern the challenges of water scarcity, wastewater treatment, and UHI. In addition, the indicators 8.2 and 8.3 limited the capacity to govern flood risk and solid waste treatment. Water scarcity was the only challenge with five limiting governance indicator, i.e., indicators 1.2, 1.3, 6.3, 7.1, and 7.3.

Table 2. Results of the water governance capacity analysis of Seoul.

Dimension	Conditions	Indicators	Water Scarcity	Flood Risk	Waste Water Treatment	Solid Waste Treatment	Urban Heat Islands
Knowing	1. Awareness	1.1 Community Knowledge	0	+	0	++	0
		1.2 Local sense of urgency	−	+	−	0	−
		1.3 Behavioral internalization	−	+	−	+	−
	2. Useful knowledge	2.1 Information availability	+	+	0	0	0
		2.2 Information transparency	+	+	0	+	+
		2.3 Knowledge cohesion	+	+	+	+	+
	3. Continuous learning	3.1 Smart monitoring	+	+	+	+	0
		3.2 Evaluation	+	0	0	+	0
		3.3 Cross-stakeholder learning	0	0	+	+	0
Wanting	4. Stakeholder engagement process	4.1 Stakeholder inclusiveness	+	+	+	+	+
		4.2 Protection of core values	0	+	+	+	+
		4.3 Progress and variety of options	+	0	0	+	+
	5. Management ambition	5.1 Ambitious realistic management	0	0	0	+	0
		5.2 Discourse embedding	0	+	0	+	+
		5.3 Management cohesion	0	+	0	+	0
	6. Agents of change	6.1 Entrepreneurial agents	0	+	+	+	0
		6.2 Collaborative agents	0	+	+	0	0
		6.3 Visionary agents	−	++	+	+	0
Enabling	7. Multi-level network potential	7.1 Room to maneuver	−	+	+	0	0
		7.2 Clear division of responsibilities	0	0	+	+	0
		7.3 Authority	−	+	+	+	0
	8. Financial viability	8.1 Affordability	++	0	++	+	+
		8.2 Consumer willingness to pay	0	−	0	0	−
		8.3 Financial continuation	0	−	0	0	−
	9. Implementing capacity	9.1 Policy instruments	0	+	+	+	+
		9.2 Statutory compliance	0	+	+	+	+
		9.3 Preparedness	+	+	+	+	+

Yellow: limiting (−); light green: encouraging (+); dark green: very encouraging (++).

Table 3. Overview of the four most limiting governance indicators.

1.2 Local sense of urgency.

To what extent do actors have a sense of urgency, resulting in widely supported awareness, actions, and policies that address the water challenge?
The perception regarding this indicator varied considerably between the stakeholders. Few experts and NGOs have recognized the uncertain threats from climate change and urbanization, and express their increasing concerns for the future. However, most of the general public does not feel this urgency about these water-related challenges.

1.3 Behavioral internalization

To what extent do local communities and stakeholders try to understand, react, anticipate and change their behavior in order to contribute to solutions regarding the water challenge?
Although actions to improve urban water-related resilience (e.g. separate collection, green roofs, green space) exist, measures are only taken under external pressure, including restraints and economic incentives.

8.2 Consumer willingness to pay

How is expenditure regarding the water challenge perceived by all relevant stakeholders (i.e., is there trust that the money is well-spent)?
Differences in awareness of the urgency of water challenges in communities determine the willingness to pay for measures.
In general, rates of cost recovery in each neighborhood of the city are lower than the actual costs, even when funds are provided by the national or local governments, leading the neighborhood to maintain the status quo.

8.3 Financial continuation

To what extent do financial arrangements secure long-term, robust policy implementation, continuation, and risk reduction?
To deal with future water challenges, long-term strategies have been planned in a ten-year cycle. However, since financial resource allocation support and maintain the status quo, there is a lack of resources to deal with prevention of unpredictable future risks. Furthermore, some water challenges that seem relatively minor issues for the communities do not receive sufficient financial resources for research and improvement.

3.4. Comparison with Other Cities

A full overview of BCI scores for 70 municipalities and regions, including 11 Asian cities, is provided in Figure 4. Cities with BCIs higher than Seoul are Singapore, and some cities in the Netherlands (e.g., Amsterdam and Groningen) and Sweden (e.g., Helsingborg, Malmo, Kristianstad, and Stockholm). As the major purpose of the CBA is city-to-city learning, i.e., improving implementation capacities of cities and regions by sharing best practices [15], these cities can be prime candidates for benchmarking. However, except for Singapore (with a population of 5.7 million in 2018), the scales of the other cities are much smaller than Seoul. The city with the largest population among these cities is Amsterdam with a population of 850,000, which is less than 10% of that of Seoul (or than 4% of the metropolitan area of Seoul). Also, all cities with BCIs lower than Seoul but higher than 6.0 are still not comparable to Seoul by scale. As many urban water management policies and plans are constrained by the scale of a city, e.g. large-scale replacement of sewer networks, we chose to limit the comparative analysis to megacities of a comparable size, i.e., Istanbul, London, and New York City (NYC). These are megacities with approximately 8–15 million inhabitants.

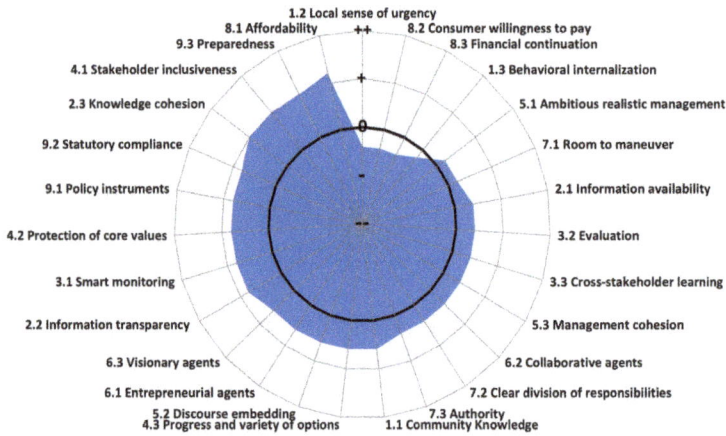

Figure 3. Result of the Governance Capacity Framework (GCF) of Seoul. The 27 indicators are organized clockwise around the spider web by most limiting (−−) to most encouraging (++) for the overall governance capacity.

The comparison of the 4 megacities is shown in Figure 5. The BCIs are highest for Seoul (7.3), then London (5.3), New York City (NYC) (4.8), and Istanbul (3.5). The common category with a high score is basic water services (Figure 5; indicators 7–9). On the contrary, wastewater treatment (indicators 10–13) and infrastructure (indicators 14–17) showed high variability among these cities. More specifically, in the wastewater treatment category, London and NYC showed better performance for the nutrient recovery (indicator 10) compared to Seoul.

In the category of infrastructure, NYC showed a higher indicator score than Seoul for stormwater separation. The type of sewer system depends upon the history of infrastructure installment of a city, and typically younger drainage systems are better separated in stormwater and sewage systems. Thus, Istanbul shows high indicator scores for both the average age of the sewer and stormwater separation. However, in NYC, the score for stormwater separation is relatively higher than the average age of the sewer system, which is exactly opposite to Seoul. This implies that there are opportunities to learn from NYC if Seoul is to improve its sewer system by expanding the portion of separate stormwater systems. Operation cost recovery is an indicator for which Seoul scores lower than London.

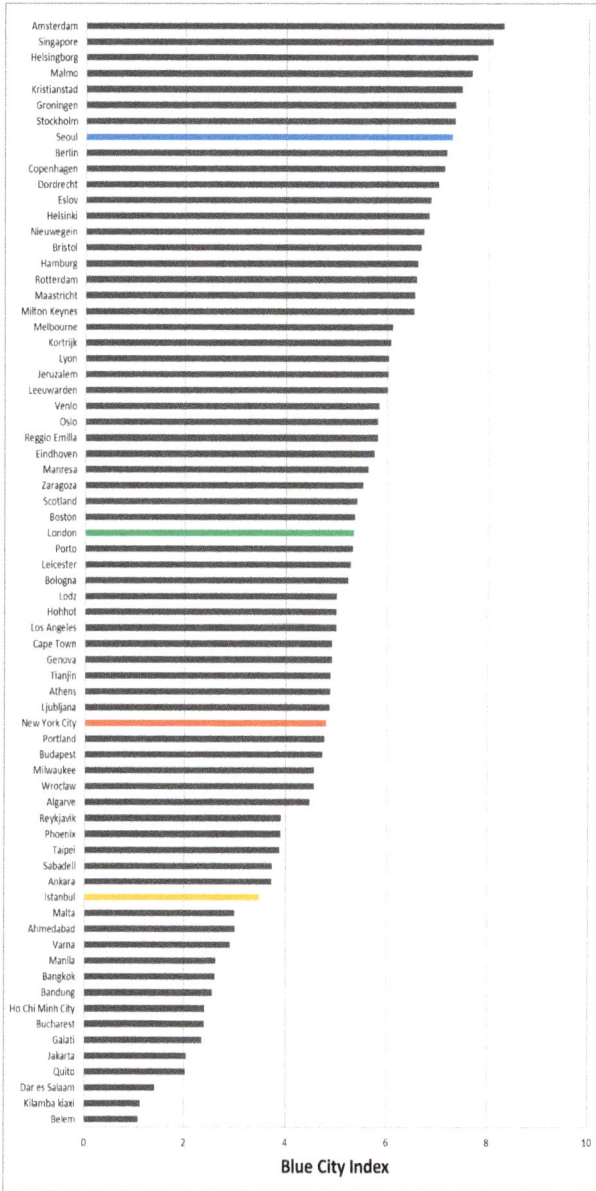

Figure 4. Results of the City Blueprint analysis of 70 municipalities and regions in 37 different countries. The Blue City Index, the geometric mean of 25 indicators of the City Blueprint, has been calculated according to Koop and Van Leeuwen [13–15]. The BCIs of Seoul, New York City, London, and Istanbul are highlighted.

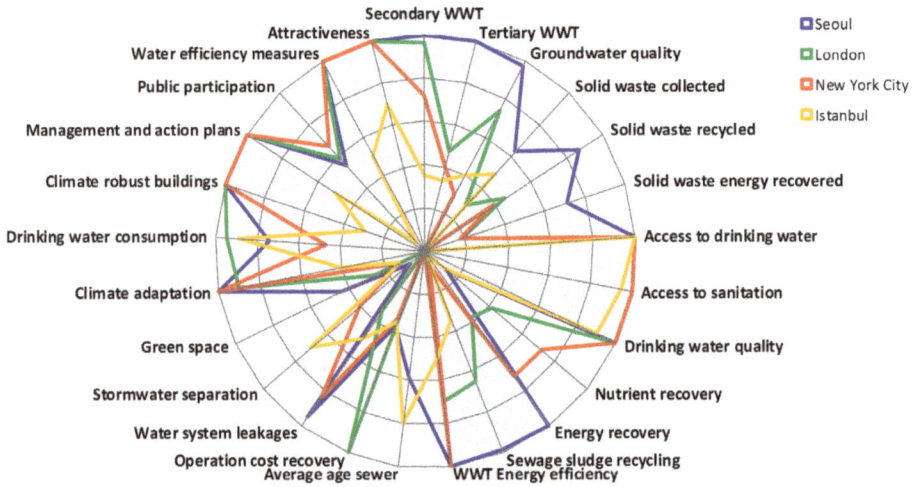

Figure 5. Comparison of 25 indicator scores between Seoul, New York City, London, and Istanbul.

4. Discussion

4.1. The Challenges for Seoul

Analyzing water management and water governance of a megacity, especially when its infrastructure has been developed over several decades, provides unique insights that may not be identified in smaller and younger cities. Due to its large scale and past propensity to build centralized infrastructures driven by economic efficiency, a full-scale replacement or renovation of these infrastructures to cope with changing conditions may not be economically nor technically feasible in the short-term. Thus, finding an innovative way to increase resilience in cities may be required and offers learning opportunities to other megacities, especially those on rapid development trajectories.

South Korea is recognized for its fast, intense economic development and industrialization. This has been accompanied by rapid urbanization along with an extensive installation of urban infrastructure in Seoul [29,30]. Due to the urgent need to provide essential water services for the rapidly growing population in the city, however, past water policies have focused on the expansion of water infrastructure in a quantitative manner without deliberation for long-term sustainable water management in the urban environment. This urban development process has been successful as reflected by the CBA analysis, which categorized Seoul as a 'resource efficient and adaptive city' with a BCI score of 7.3. However, climate change and aging water infrastructure act as drivers for further change and new emerging challenges, which call for further and continuous adaptation and improvement in infrastructure, policies, and practices of urban water management and governance [7,12]. Improving resilience is the big challenge. Based on our analysis, we provide some suggestions for priorities to improve IWRM in Seoul, which could potentially transform Seoul from a 'resource efficient and adaptive city' to a 'water-wise city' in the near future.

4.2. Nutrient Recovery

Nutrient recovery is one of the indicators which offers clear opportunities for improving Seoul's IWRM as it was shown as the only weakness in the category of wastewater treatment (Figure 1). Nutrient recovery is necessary in Korea for several reasons: (1) phosphorus is nonrenewable and a limited resource [31]; (2) Korea entirely depends on the imports of phosphorus; (3) phosphorus removal from wastewater will significantly contribute to the reduction of eutrophication of surface waters. Thus, the introduction of technology for recovering nutrients from wastewater is an effective option for coping with diminishing resources, while simultaneously reducing eutrophication and improving surface water quality. However, nutrient recovery from wastewater treatment is a rather recent technology which still needs further improvement to become economically feasible and to meet regulations for the quality of recovered materials in many countries. London and NYC recover phosphorus by producing biosolids. In the UK, 3–4 million tons of biosolids, which is about 75% of sewage sludge production, are applied annually to agricultural land [32]. Also, NYC produces approximately 1200 tons of biosolids every day. In 1988, US Federal government banned the ocean disposal of biosolids, and NYC needed to find alternative uses for this material. The NYC Department of Environmental Protection implemented a program to beneficially use most of the biosolids to fertilize crops and improve soil conditions for plant growth [33].

However, biosolids also contain chemical contaminants such as heavy metals and persistent organic chemicals, which limit the use of biosolids is many countries. This is one of the reasons that biosolids are not yet actively used in Seoul. Similar problems were observed in Amsterdam, and the produced struvite ($MgNH_4PO_4 \cdot 6H_2O$) can now be applied as fertilizer in parks and sports fields, preventing contaminants from entering the food chain [34,35]. Currently, in Korea, nutrient recovery from wastewater treatment plants (WWTPs) is done only by recycling through earthworm rearing and composting. The national legislation allows the use of these composts for landscaping of gardens, parks, etc., and not for edible and feed production purposes, as is the case in Amsterdam [34,35]. In 2014, the Seoul Metropolitan Waterworks Research Institute succeeded in developing a device for recovering phosphorus from sewage, but it has not yet been applied on a commercial scale mainly due to the economic feasibility.

The recovery of nutrients is still not common for WWTPs, even though recovery technologies are available. Countries like the Netherlands, Denmark, and Germany upgraded their plants recently. Although there are many economic, technical, and legislative issues to overcome for recovering nutrients from WWTPs, the limited availability of phosphorous, which is essential for food production, imposes a potential future geo-political risk. Currently, the economic viability and safety issues for using the recovered materials are the major barriers that hinder the active introduction of nutrient recovery facilities in Seoul. However, given that several cities, including Amsterdam, already installed the technology successfully [34,35], a stronger willingness of the government to achieve a long-term IWRM will be the most critical decision factor for enabling the new technology to become economically viable.

4.3. Operation Cost Recovery

Water infrastructure is the most expensive infrastructure in cities [12], which means securing an adequate long-term financial condition is necessary for effective maintenance and improvement thereof [36]. However, as the indicator of the operation cost recovery in Seoul shows, it may be a major obstacle hindering large-scale improvement of the sewer systems [37].

Among the other megacities, London is the only city with a higher score than Seoul. The primary reason for London's higher operation cost recovery is privatization in water sectors since 1989. Although London's water and sewerage charges are set by the regulator 'Ofwat', a non-ministerial government department that protects the interests of consumers, the charges should ensure the profitability of water companies. Total annual charges for drinking water and sewerage in London were 3.98 USD/m^3 in 2013 [38].

On the contrary, the total annual charge of water services in Seoul is extremely low with 0.53 USD/m^3 in 2013 [39,40] which makes the realization rate of drinking water and sewerage 89% and 67%, respectively. According to an OECD (Organisation for Economic Co-operation and Development) survey, the average water price of 114 cities in OECD countries was 3.84 USD/m^3 and the water price of Korea is the lowest among the OECD countries [38]. The low water pricing in Korea hinders secure reinvestment of resources for introducing new water infrastructure and improving old water facilities, such as those seen in other indicators. This may pose a serious threat to the long-term water security and supply stability in Seoul, which may result in a decline in service levels. Given that climate change tends to increase the vulnerability of water infrastructure in various ways, securing enough recovery from operation costs will be an essential option to provide water services sustainably to Seoul citizens. A political discussion and decision on a sufficient water price is one of the key components for improving urban resilience to unexpected future risks [37].

4.4. Sewer Systems

During the 1970s and 80s, the majority of the sewer system in Seoul was installed with combined sewers. Since 2000, the installment of separate sewer systems has been given priority. However, as the sewer maintenance project is limited to only small redeveloping areas, it is unlikely to increase the portion of the separated sewer system in the short term unless there is a substantial change in local water policy. Also, adverse effects, such as land subsidence, from the aging of pipe systems are escalating [16]. Thus, along with expanding separate sewer systems, giving higher priority to the replacement of old pipes may reduce the water-related risks significantly. The relevance of infrastructure maintenance is high as observed by both the OECD [7] and UN-Habitat [6].

Combined sewer systems are common in many cities, such as Seoul, London, and NYC. However, due to a continued increase in the impermeable surface area that increases stormwater runoff and higher peak precipitation due to climate change, there is a high likelihood of combined sewer overflow (CSO) [41]. NYC is also concerned about CSOs as 60~70% are combined sewer systems. Since it is a daunting task to change the existing infrastructure in such large cities, London and NYC have tried to implement various alternative ways to deal with CSOs. These include the construction of sewer tunnels or CSO retention tanks, upgrades in key WWTPs, and the development of green infrastructure [42].

Seoul and NYC are trying to increase the portion of separate sewer systems, but it will take a long time due to the scale of construction. Also, only a partial retrofitting of the whole system may result in misconnections between the different drainage types, which can cause undesired effects on sewer management and water quality. In this situation, implementing an alternative way to deal with combined sewers may serve as a solution for large cities.

4.5. Implications from the Governance Capacity Analysis

Even if weaknesses of specific sectors for the water management of a city can be identified (e.g., by CBF), if there is a barrier in the governance for disseminating core information and promoting future actions for fixing existing weaknesses, a city may not be able to cope with the challenges that threaten the sustainable provision of water services. Our analysis of the governance capacity indicated that awareness and financial viability are the weakest governance conditions in Seoul. Specifically, there is a low local sense of urgency and behavioral internalization for the challenges of water scarcity, wastewater treatment, and urban heat islands (UHI). Furthermore, there is a low consumer willingness to pay and financial continuation for the challenges of flood risk and UHI.

There are various efforts in education, promoting the engagement of local stakeholders, and using media for dispersing the information on water challenges that the citizens may face in the near future. In spite of these efforts, however, raising the local sense of urgency and behavioral internalization is a difficult task. Ironically, the most effective boost in awareness can be achieved if the local communities are frequently exposed to water threats and have experiences encountering inconveniencies within water services. As one of the cities with well-equipped water infrastructures, Seoul has overcome various urban water problems that have been the norm in the past. As unprecedented changes are expected, as indicated by our CBF analysis, it is important to disseminate existing and newly obtained information to the public as a governance condition for adequately managing urban resilience. Resilience is not a static property of a system but requires constant adaptation and transformation [43]. Keeping the status quo will not ensure the sustainability of IWRM of Seoul in the coming decades, as pressures are likely to aggravate. This will require strong governance for raising consumer willingness to pay and financial continuation. Without such a sense of urgency as water infrastructures currently perform well, efforts for enhancing financial capacity and continuity may raise political objection. Therefore, local stakeholders must perceive that resilience of the current water infrastructure in Seoul may be adequate to respond to pressures of the past, but not to pressures of the future. Therefore, in addition to mitigation, it is essential to develop adequate adaptive responses as a means of moderating damages or realizing opportunities associated with climate change [12,44].

5. Conclusions

We examined the current status of IWRM of Seoul by using the City Blueprint Approach in order to explore the options for improvement of water management and water governance in Seoul. This study also sets a good example to the challenges faced in IWRM in megacities, especially those with well-established water infrastructure. Our analysis revealed that there are several options to achieve better IWRM in Seoul. These include nutrient recovery, stormwater separation, and operation cost recovery, for which vigorous investments with high priority may provide strong opportunities for improvement.

When compared to other megacities, stormwater separation is a common weakness. Because of the large scale of these cities, their urban environment is complicated by the intertwined structure of multiple infrastructures, reducing the ability to manage infrastructure adaptively. While it is imperative to increase the portion of separated drainage systems in a city, especially to reduce the risk from combined sewer overflows, the experiences in other cities show that there are also other sustainable options such as expanding green infrastructure. When implemented with a long-term view on the sustainability of IWRM in cities, these options can also enhance other sectors, such as increasing the green area ratio and reducing impervious surface areas, thereby improving e.g. recreation, attractiveness and the livability of the city.

Resource recovery is another indicator that can improve Seoul's IWRM, especially regarding the depletion of phosphorous reserves in the near future [14]. Due to the early stage of establishment, nutrient recovery technologies are still not common in use. Low recovery rates, economic feasibility, and limited applicability of end-products by regulations are barriers that making cities reluctant to deploy the technology on a commercial scale. Nonetheless, megacities, where large-scale WWTPs are typically in operation, are the right places since the large flow rate of wastewater can improve the phosphorous recovery rate with high economic efficiency and profitability and, at the same time, improve surface water quality. Regarding its necessity in the coming decades, the benefits that a city gains include securing the depleting resources and relative advantages of a mega-scale for improving the efficiency, Seoul can benefit from embarking on a path to promote the development and installation of nutrient recovery technologies.

Several potential barriers—the local sense of urgency, behavioral internalization, consumer willingness to pay, and financial continuation—that may retard the efforts for improving urban water management sustainability were also identified through the water GCF. This finding is especially important for cities that rely heavily upon the current system with a false sense of security. A resilient city requires an adequate preparedness for the occurrence of future threats, as well as an adaptive capacity to cope with continuously changing pressures [14,44,45].

Many cities are facing similar water-related challenges. Although Seoul gained a high overall BCI score, it does not necessarily ensure the sustainability of IWRM in the future. Water-related risks constantly evolve due to non-stationary conditions resulting from urban dynamics and climate change, and it requires timely adaptation as well as the transformation of a city to respond to the changing environment [14,44,45]. The challenges of megacities may be proportional to their scale, but large cities are also known as centers of innovation. Finding sustainable and prompt solutions can be stimulated by sharing the experiences and knowledge of multiple cities that are trying to cope with these challenges [15].

Author Contributions: H.K. and J.S. acquired and analyzed data; J.P., K.v.L., and H.K. wrote the paper; K.v.L. and S.K. conceived and designed the analysis; S.L. and Y.J.C. and contributed to conception of the paper and interpretation of data.

Acknowledgments: This research was supported by Basic Science Research Program through the National Research Foundation of Korea (NRF) funded by the Ministry of Science and ICT (NRF-2016R1C1B1011770). The City Blueprint Action Group is part of Watershare (https://www.watershare.eu/) and the European Innovation Partnership on Water of the European Commission (http://www.eip-water.eu/City_Blueprints). Part of this research was carried out in the context of the POWER project. The European Commission is acknowledged for funding POWER in H2020-Water Under Grant Agreement No. 687809.

Conflicts of Interest: The authors declare no conflict of interest.

References

1. United Nations (UN). *World Urbanization Prospects: The 2014 Revision*; Department of Economic and Social Affairs (DESA), Population Division, Population Estimates and Projections Section, UN: New York, NY, USA, 2014.

2. Bettencourt, L.M.A.; Lobo, J.; Helbing, D.; Kühnert, C.; West, G.B. Growth, innovation, scaling, and the pace of life in cities. *Proc. Natl. Acad. Sci. USA* **2007**, *104*, 7301–7306. [CrossRef] [PubMed]

3. McDonald, R.I.; Green, P.; Balk, D.; Fekete, B.M.; Revenga, C.; Todd, M.; Montgomery, M. Urban growth, climate change, and freshwater availability. *Proc. Natl. Acad. Sci. USA* **2011**, *108*, 6312–6317. [CrossRef] [PubMed]

4. Yeh, C.T.; Huang, S.L. Global urbanization and demand for natural resources. In *Carbon Sequestration in Urban Ecosystems*; Rattan, L., Bruce, A., Eds.; Springer: Berlin, Germany, 2012; pp. 355–371. [CrossRef]

5. Hunt, A.; Watkiss, P. Climate change impacts and adaptation in cities: A review of the literature. *Clim. Chang.* **2011**, *104*, 13–49. [CrossRef]

6. UN-Habitat. *Urbanization and Development: Emerging Futures*; UN Habitat: Nairobi, Kenya, 2016; ISBN 9789211333954.

7. Organisation for Economic Co-operation and Development (OECD). *Water and Cities: Ensuring Sustainable Futures*; Organization for Economic Co-operation and Development: Paris, France, 2015.

8. Korean Statistical Information Service (KOSIS). Available online: http://kosis.kr/index/index.jsp (accessed on 23 May 2018).

9. Kim, D.; Park, T.S.; Park, J.; Lee, S.O. A river environment index for Korean national rivers: Rationale, methods and application. *Water Policy* **2014**, *16*. [CrossRef]

10. Jung, I.; Bae, D.; Kim, G. Recent trends of mean and extreme precipitation in Korea. *Int. J. Climatol.* **2011**, *31*, 359–370. [CrossRef]

11. Van Leeuwen, C.J.; Frijns, J.; van Wezel, A.; van de Ven, F.H.M. City Blueprints: 24 Indicators to Assess the Sustainability of the Urban Water Cycle. *Water Resour. Manag.* **2012**, *26*, 2177–2197. [CrossRef]

12. Koop, S.H.A.; van Leeuwen, C.J. The challenges of water, waste and climate change in cities. *Environ. Dev. Sustain.* **2017**, *19*, 385–418. [CrossRef]

13. Koop, S.H.A.; vanP Leeuwen, C.J. Assessment of the Sustainability of Water Resources Management: A Critical Review of the City Blueprint Approach. *Water Resour. Manag.* **2015**, *29*, 5649–5670. [CrossRef]

14. Koop, S.H.A.; van Leeuwen, C.J. Application of the Improved City Blueprint Framework in 45 Municipalities and Regions. *Water Resour. Manag.* **2015**, *29*, 4629–4647. [CrossRef]

15. European Innovation Partnerships (EIP) Water. *City Blueprints: Improving Implementation Capacities of Cities and Regions*; E-Brochure; European Commission: Brussels, Belgium; Available online: http://www.eip-water. eu/City_Blueprints (accessed on 4 April 2018).

16. Lee, J.Y.; Kwon, K.D.; Raza, M. Current water uses, related risks, and management options for Seoul megacity, Korea. *Environ. Earth Sci.* **2018**, *77*. [CrossRef]

17. Seoul Metropolitan Government Seoul Statistics. Available online: http://stat.seoul.go.kr/jsp3/index.jsp (accessed on 12 August 2018).

18. Kim, H.S.; Chung, Y.S.; Tans, P.P.; Yoon, M.B. Climatological variability of air temperature and precipitation observed in South Korea for the last 50 years. *Air Qual. Atmos. Health* **2016**, *9*, 645–651. [CrossRef]

19. Rebetez, M.; Reinhard, M. Monthly air temperature trends in Switzerland 1901–2000 and 1975–2004. *Theor. Appl. Climatol.* **2008**, *91*, 27–34. [CrossRef]

20. Cheon, J.Y.; Ham, B.S.; Lee, J.Y.; Park, Y.; Lee, K.K. Soil temperatures in four metropolitan cities of Korea from 1960 to 2010: Implications for climate change and urban heat. *Environ. Earth Sci.* **2014**, *71*, 5215–5230. [CrossRef]

21. Son, J.Y.; Lane, K.J.; Lee, J.T.; Bell, M.L. Urban vegetation and heat-related mortality in Seoul, Korea. *Environ. Res.* **2016**, *151*, 728–733. [CrossRef] [PubMed]

22. Feingold, D.; Koop, S.; van Leeuwen, K. The City Blueprint Approach: Urban Water Management and Governance in Cities in the US. *Environ. Manag.* **2018**, *61*, 9–23. [CrossRef] [PubMed]

23. Koop, S.H.A.; Koetsier, L.; Doornhof, A.; Reinstra, O.; van Leeuwen, C.J.; Brouwer, S.; Dieperink, C.; Driessen, P.P.J. Assessing the Governance Capacity of Cities to Address Challenges of Water, Waste, and Climate Change. *Water Resour. Manag.* **2017**, *31*, 3427–3443. [CrossRef]

24. Folke, C.; Carpenter, S.; Elmqvist, T.; Gunderson, L.; Holling, C.S.; Walker, B. Resilience and Sustainable Development: Building Adaptive Capacity in a World of Transformations. *Ambio.* **2002**, *31*, 437–440. [CrossRef] [PubMed]

25. Gawlik, B.M.; Easton, P.; Koop, S.; van Leeuwen, K.; Elelman, R. *Urban Water Atlas for Europe*; European Commission, Publications Office of the European Union: Luxembourg, Luxembourg, 2017.

26. Schreurs, E.; Koop, S.; van Leeuwen, K. Application of the City Blueprint Approach to assess the challenges of water management and governance in Quito (Ecuador). *Environ. Dev. Sustain.* **2017**, *20*, 509–525. [CrossRef]

27. Koop, S.H.A.; Aartsen, M.; Goswami, B.; Oost, J.; Schmidt, G.; van Leeuwen, C.J. *City Blueprint Approach: Key opportunities for India's urban water challenges*; Water Digest: New Delhi, India, 22 March 2017; pp. 66–73. Available online: http://thewaterdigest.com/Emagazine_2017/magazine.html (accessed on 20 May 2018).

28. Rahmasary, A.N.; Robert, S.; Chang, I.S.; Jing, W.; Park, J.; Bluemling, B.; Koop, S.; van Leeuwen, K. Overcoming the challenges of water, waste and climate change in Asian cities. *Environ. Manag.* **2018**, in review.

29. Choi, I.C.; Shin, H.J.; Nguyen, T.; Tenhunen, J. Water Policy Reforms in South Korea: A Historical Review and Ongoing Challenges for Sustainable Water Governance and Management. *Water* **2017**, *9*, 717. [CrossRef]

30. Yoo, S.H.; Yang, C.Y. Role of Water Utility in the Korean National Economy. *Int. J. Water Resour. Dev.* **1999**, *15*, 527–541. [CrossRef]

31. Cordell, D.; Drangert, J.O.; White, S. The story of phosphorus: Global food security and food for thought. *Glob. Environ. Chang.* **2009**, *19*, 292–305. [CrossRef]

32. UKWIR Biosolids: Good Practice Guidance. Available online: http://www.water.org.uk/publications/%0Areports/biosolids-good-practice-guidance (accessed on 23 February 2017).

33. Swanson, R.L.; Bortman, M.L.; O'Connor, T.P.; Stanford, H.M. Science, policy and the management of sewage materials. The New York City experience. *Mar. Pollut. Bull.* **2004**, *49*, 679–687. [CrossRef] [PubMed]

34. Van der Hoek, J.P.; Struker, A.; De Danschutter, J.E.M. Amsterdam as a sustainable European metropolis: integration of water, energy and material flows. *Urban Water J.* **2017**, *14*, 61–68. [CrossRef]

35. Van Leeuwen, C.J.; Sjerps, R.M.A. The City Blueprint of Amsterdam: an assessment of integrated water resources management in the capital of the Netherlands. *Water Sci. Technol. Water Supply* **2015**, *15*, 404. [CrossRef]

36. IWA. *International Statistics for Water Services*; International Water Association: London, UK, 2014.

37. Tyler, S.; Moench, M. A framework for urban climate resilience. *Clim. Dev.* **2012**, *4*, 311–326. [CrossRef]

38. OECD. *Environment at a Glance 2015: OECD Indicators*; Organization for Economic Co-operation and Development: London, UK; Paris, France, 2015.

39. Ministry of Environment (MOE). *2015 Statistics of Sewerage*; Ministry of Environment: Sejong City, Korea, 2015.

40. Ministry of Environment (MOE). *2014 Statistics of waterworks*; Ministry of Environment: Sejong City, Korea, 2015.

41. Nilsen, V.; Lier, J.A.; Bjerkholt, J.T.; Lindholm, O.G. Analysing urban floods and combined sewer overflows in a changing climate. *J. Water Clim. Chang.* **2011**, *2*, 260–271. [CrossRef]

42. One New York: The Plan for a Strong and Just City. 2015. Available online: http://www.adaptationclearinghouse.org/resources/one-new-york-the-plan-for-a-strong-and-just-city-one-nyc.html (accessed on 24 May 2018).

43. Park, J.; Seager, T.P.; Rao, P.S.C.; Convertino, M.; Linkov, I. Integrating Risk and Resilience Approaches to Catastrophe Management in Engineering Systems. *Risk Anal.* **2013**, *33*, 356–367. [CrossRef] [PubMed]

44. Watkiss, P.; Bosello, F.; Buchner, B.; Catenacci, M.; Goria, A.; Kuik, O.; Karakaya, E. *Climate Change: The Cost of Inaction and the Cost of Adaptation*; EEA Technical Report No 13/2007; European Environment Agency: Copenhagen, Denmark, 2007. [CrossRef]

45. Hoekstra, A.Y.; Buurman, J.; van Ginkel, K.C.H. Urban water security: A review. *Environ. Res. Lett.* **2018**, *13*, 53002. [CrossRef]

water ![MDPI logo]

Article

Evaluation of Water Governance Processes Required to Transition towards Water Sensitive Urban Design—An Indicator Assessment Approach for the City of Cape Town

Boipelo Madonsela [1], Stef Koop [2,3], Kees van Leeuwen [2,3] and Kirsty Carden [1,*

[1]   Department of Civil Engineering and Future Water Research Institute, University of Cape Town,
     Private Bag X3, Rondebosch 7701, South Africa; mdnboi002@myuct.ac.za

[2]   KWR Watercycle Research Institute, Groningenhaven 7, 3430 BB Nieuwegein, The Netherlands;
     stef.koop@kwrwater.nl (S.K.); kees.van.leeuwen@kwrwater.nl (K.v.L.)

[3]   Copernicus Institute of Sustainable Development Utrecht University, Princetonlaan 8a,
     3508TC Utrecht, The Netherlands

*   Correspondence: kirsty.carden@uct.ac.za; Tel.: +27-21-650-5317

Received: 21 December 2018; Accepted: 31 January 2019; Published: 7 February 2019

Abstract: In the face of water related risks resulting from climate change and rapid urbanization, water resources in South African cities have increasingly come under pressure. Following the most recent drought period (2015–2018), local authorities such as the City of Cape Town are being tasked with restructuring policy to include climate change adaptation strategies to adapt more adequately and proactively to these new challenges. This paper describes an evaluation of the water governance processes required to implement Water Sensitive Urban Design (WSUD) in Cape Town—with a specific focus on the barriers to, and opportunities for, those processes related to wastewater treatment, flood risk and the pressing issue of water scarcity. The City Blueprint Approach (CBA) was selected as the indicator assessment approach for this task. The CBA is a set of diagnostic tools comprising the Trends and Pressures Framework, the City Blueprint Framework and the Governance Capacity Framework. This was applied to Cape Town based on in-depth interviews and publicly available information. The analysis revealed that smart monitoring, community knowledge and experimentation with alternative water management technologies are important when considering uncertainties and complexities in the governance of urban water challenges. We conclude that there is potential for Cape Town to transition to a water sensitive city through learning from this experimentation and by implementing WSUD strategies that address water scarcity following the shifts in governance caused by the 2015–2018 drought.

Keywords: Cape Town; City Blueprint Approach; water governance; water scarcity; water sensitive cities; climate change adaptation

---

1. Introduction

Cities globally are more and more becoming hotspots for risk and disaster [1], mainly as a result of rapid urbanization, population growth and the impacts of climate change. South Africa is a semi-arid country, with rainfall being seasonal and distributed unevenly [2]. It experiences a rainfall average of less than 500 mm/year (compared to a global average of 869 mm), making it the 30th driest country in the world [2,3]. Increasing water demand is also putting pressure on the allocation and management of water resources in South African cities [3]. Recently (2015–2018), a country-wide drought resulted in severe water shortages in many parts of South Africa, most notably affecting the Western Cape province and the City of Cape Town (CoCT). In early 2018, with the main storage dams predicted to decline to critically low levels, the city announced plans for "Day Zero", that is, the stage at which water storage levels reached 13.5%, when the municipal water supply would largely be shut off.

Local authorities are increasingly being tasked with restructuring policy to include climate change adaptation strategies to deal more adequately and proactively with these new challenges. Conventional, top-down and fragmented water management paradigms are no longer able to adequately address water challenges in the current context of uncertainty and complexity [4]. A shift towards adaptive approaches to urban water management has been proposed in order to address these complexities whilst ensuring the satisfactory delivery of water services to citizens [4]. One such approach is termed Water Sensitive Urban Design (WSUD), which encompasses all aspects of the urban water cycle including stormwater management, wastewater treatment and water supply, and *"represents a significant shift in the way water and related environmental resources and water infrastructure are considered in the planning and design of cities . . . "* [5]. The principles of WSUD have gained importance in terms of guiding cities around the world in the socio-technical transformations of conventional approaches needed to aid transitions towards becoming Water Sensitive Cities [6,7].

Applying and implementing WSUD principles in South African cities is challenging owing to factors such as fragmented institutional structures within municipalities (e.g., different local government departments working in "silos"), social constraints, as well as financial and human resource limitations [8]. Water challenges often transcend administrative boundaries and involve many different departments and/or organizations each with different responsibilities and interests; therefore, a problem-oriented diagnostic analysis is required instead of focusing on individual water management departments only [9]. In this paper, we analyse the overall management and governance (at a local authority level using the CoCT as a case study) of some of the major water challenges that characterise urban South Africa. We aim to contribute to a better understanding of the barriers to, and opportunities for, improving the governance capacity to address the pressing issues of water scarcity, wastewater treatment and flood risk in South African cities. These particular challenges were selected based on their links to integrated urban water cycle management as the main principle of WSUD [6]. To achieve this aim, the City Blueprint Approach (CBA) was selected as an appropriate means of evaluating the required governance processes for a water sensitive Cape Town. The CBA is an indicator assessment tool comprising the Trends and Pressures Framework (TPF), the City Blueprint Framework (CBF) and the Governance Capacity Framework (GCF) [9,10]. The current (2015–2018) water crisis and history of frequent flood events (particularly in low-lying informal areas) in Cape Town exemplify the relevance of this analysis and may also provide valuable insights for other cities in South Africa dealing with similar water challenges. Hence, the overall objective of this paper is to identify where the CoCT can improve its water governance processes in its transition to a Water Sensitive City.

This paper first provides a detailed explanation of the methods undertaken in applying the CBA to Cape Town. Secondly, the paper presents the results of the CBF and the GCF assessments of water scarcity, flood risk and wastewater treatment respectively. The discussion provides a critical refection on the results and presents the implications for Cape Town's transition towards water sensitivity. We conclude with the most significant points in the water governance analysis.

2. Materials and Methods

The City Blueprint Approach was selected as an appropriate means to fulfill the research aim of evaluating the water governance processes required to implement Water Sensitive Urban Design (WSUD) in Cape Town. The CBA comprises the TPF, CBF and GCF (see Figure 1). It was developed by the KWR Watercycle Research Institute in cooperation with Utrecht University, The Netherlands [11] and acknowledges that every city has its own social, financial and environmental setting in which water managers have to operate.

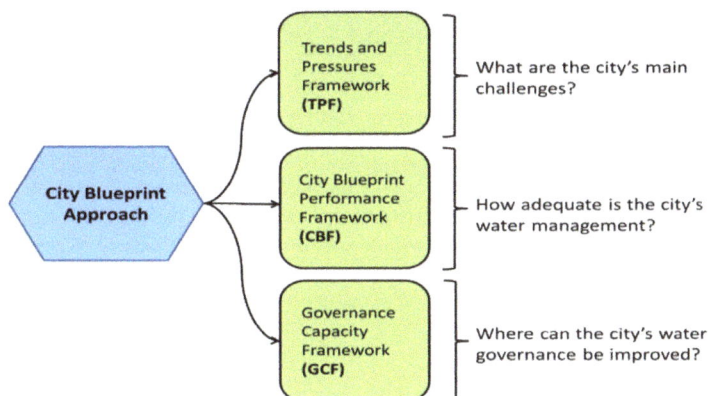

Figure 1. City Blueprint Approach (Koop & Van Leeuwen, 2017 [11]).

Other examples of assessment frameworks that aim to enhance cities' transitions towards being water sensitive include, inter alia, the Water Sensitive Cities (WSC) Index by the Cooperative Research Centre for Water Sensitive Cities and the Principles for Water Sensitive Cities by the International Water Association (IWA). The development of the WSC Index involved multiple development phases aimed at improving its functionality, including the prototyping of the Index as applied to two local authorities in Melbourne, Australia. The feedback from the two pilot studies was used to improve its functionality, usability, benefits and reliability [12]. Unlike the WSC Index and the CBA, the IWA Principles do not provide a sustainable water management assessment for cities; instead the principles provide a framework which is intended to guide city officials to implement and develop their urban water visions and strategies for water sensitive transitions [13].

The benefit of using the CBA in this assessment relates to the fact that it was first applied to 45 cities in 27 countries before undergoing a critical revision based on the learning experiences obtained during this process [14]. This revision included: (1) the updating of existing indicators; (2) ensuring that individual indicators make an equal contribution to the final score (sustainability measurement); (3) ensuring that indicator results are easy to understand by the end-user; and (4) developing a separate supplementary framework which supports the undertaking of the main framework. The improved tool which emanated from the revision was applied to the CoCT and forms part of a larger study to further assess the feasibility of this approach.

The TPF comprises 12 descriptive indicators and eight additional sub-indicators divided into social, environmental and financial categories. Each indicator is scaled from 0 to 4 points, where a higher score represents a higher urban pressure or concern. For seven indicators and sub-indicators a scoring method is applied based on international quantitative standards. These include the WHO [15] scoring for burden of disease as well as the World Bank's scoring for primary education, political instability, unemployment, poverty, and inflation [16–19]. Finally, the scoring system for groundwater scarcity and surface water scarcity from the FAO [20] was adopted. These international standardized indicators are available for most countries. The TPF indicator score is based on the city's score amongst all available country scores. As such, all available country scores were ranked and linearly standardized on a scale from 0 to 4 points. An equation that best fitted this ranking (lowest correlation coefficient: r = 0.97) was used to calculate the 0–4 point score for the CoCT. These scores are not normative and only provide an indication of urban pressures with respect to global trends.

The CBF consists of twenty-five indicators, which are divided into seven comprehensive categories: (1) water quality; (2) solid waste; (3) basic water services; (4) wastewater treatment; (5) infrastructure; (6) climate robustness; and (7) governance. Each indicator is scored on a scale from 0 (low performance) to 10 (high performance). The calculation method for each indicator is publicly available [10,21].

Data for the Cape Town assessment was sourced online, predominantly from publicly available reports, local policy documents sourced from the city's website, as well as through interviews with city officials. The geometric mean of all 25 indicators, the Blue City Index (BCI) was also calculated. Detailed information on indicator selection and scoring methods for both the TPF and the CBF are provided by Koop & Van Leeuwen [10,14], whereas data requirements, data sources and examples are provided on the EIP Water website [21].

The GCF consists of nine categories each with three indicators, which together were used to determine the governance capacity required to address three selected water challenges: water scarcity, flood risk and wastewater treatment (as will be outlined in the discussion section). Each of the 27 indicators was scored according to a Likert scale to gauge the subjective opinions and values of respondents and the analysis of the publicly available documents [9]. The scale ranges from very encouraging (++) to very limiting (−). A detailed description of each indicator's pre-defined question is provided in Table 1. For the indicator-specific Likert scale and link to the literature, we refer to [21]. The scoring for each indicator was based on three steps:

1. A preliminary score was given and substantiated by argumentation based on publicly available reports, local policy documents, local legislation and online articles.
2. Based on a stakeholder analysis, the main actors involved in each of the water challenges were selected and interviewees were selected accordingly. In-depth interviews were recorded and used to improve the written substantiation in order to refine each indicator score. A total of nine separate in-depth interviews were conducted, five of which were with city officials, two with academics and another two with local water experts. The interview questions focused on three themes relating to knowledge, management practices and implementing capacity.
3. Interviewees were asked to provide constructive feedback, additional arguments and information sources to further improve the accuracy of the indicator scores.

Although the methods employed helped to fulfil the research aim, there were still unavoidable limitations to the research. Firstly, the data for this research was gathered during the time when the CoCT was experiencing a severe water crisis (2017). For this reason, organizing and scheduling interviews with city officials proved to be a challenging task. Scheduling interviews depended on the availability of respondents and their willingness to participate in the study. During the data collection period, city officials were especially busy and therefore only a limited number were able to participate. For this reason, secondary data sources, discussions with other interviewees and follow up emails were also relied on for information. In addition, it is acknowledged that the interview responses may have been influenced in some way by the ongoing water crisis. Despite this being the case, all responses from the respondents offered an extremely valuable contribution to this research.

The scoring of the qualitative indicators of the GCF by the researcher was to some degree subjective, thus increasing the potential danger of the indicators not being scored accurately. With the intention of ensuring that this process was carried out as effectively as possible, the justification for each indicator score as well as the sources used to score the indicator was recorded by the researcher. This information was reviewed by the supervisor of this research as well as an academic from the KWR Watercycle Research Institute.

Table 1. Overview of pre-defined questions to be answered by the researcher based on a triangular approach consisting of three steps: (1) literature review; (2) in-depth interviews with selected experts; and (3) feedback procedure. The full details of the Likert scoring are provided at [21].

Indicator	Pre-Defined Question
1.1 Community knowledge	To what extent is knowledge regarding the current and future risks, impacts, and uncertainties of the water challenge dispersed throughout the community and local stakeholders which may results in their involvement in decision-making and implementation?
1.2 Local sense of urgency	To what extent do actors have a sense of urgency, resulting in widely supported awareness, actions, and policies that address the water challenge?
1.3 Behavioural internalization	To what extent do local communities and stakeholders try to understand, react, anticipate and change their behaviour in order to contribute to solutions regarding the water challenge?
2.1 Information availability	To what extent is information on the water challenge available, reliable, and based on multiple sources and methods, in order to meet current and future demands so as to reveal information gaps and enhance well-informed decision-making?
2.2 Information transparency	To what extent is information on the water challenge accessible and understandable for experts and non-experts, including decision-makers?
2.3 Knowledge cohesion	To what extent is information cohesive in terms of using, producing and sharing different kinds of information, usage of different methods and integration of short-term targets and long-term goals amongst different policy fields and stakeholders in order to deal with the water challenge?
3.1 Smart monitoring	To what extent is the monitoring of process, progress, and policies able to improve the level of learning (i.e., to enable rapid recognition of alarming situations, identification or clarification of underlying trends)? Or can it even have predictive value?
3.2 Evaluation	To what extent are current policy and implementation continuously assessed and improved, based on the quality of evaluation methods, the frequency of their application, and the level of learning?
3.3 Cross-stakeholder learning	To what extent are stakeholders open to and have the opportunity to interact with other stakeholders and deliberately choose to learn from each other?
4.1 Stakeholder inclusiveness	To what extent are stakeholders interact in the decision-making process interaction (i.e., are merely informed, are consulted or are actively involved)? Are their engagement processes clear and transparent? Are stakeholders able to speak on behalf of a group and decide on that group's behalf?
4.2 Protection of core values	To what extent: (1) is commitment focused on the process instead of on early end-results? (2) do stakeholders have the opportunity to be actively involved? (3) are the exit procedures clear and transparent? (All three ensure that stakeholders feel confident that their core values will not be harmed.)
4.3 Progress and variety of options	To what extent are procedures clear and realistic, are a variety of alternatives co-created and thereafter selected from, and are decisions made at the end of the process in order to secure continued prospect of gain and thereby cooperative behaviour and progress in the engagement process?
5.1 Ambitious and realistic management	To what extent are goals ambitious (i.e., identification of challenges, period of action considered, and comprehensiveness of strategy) and yet realistic (i.e., cohesion of long-term goals and supporting flexible intermittent targets, and the inclusion of uncertainty in policy)?
5.2 Discourse embedding	To what extent is sustainable policy interwoven in historical, cultural, normative and political context?
5.3 Management cohesion	To what extent is policy relevant for the water challenge, and coherent regarding: (1) geographic and administrative boundaries; and (2) alignment across sectors, government levels, and technical and financial possibilities?
6.1 Entrepreneurial agents	To what extent are the entrepreneurial agents of change enabled to gain access to resources, seek and seize opportunities, and have influence on decision-making?
6.2 Collaborative agents	To what extent are actors enabled to engage, build trust and collaboration, and connect business, government, and other sectors, in order to address the water challenge in an unconventional and comprehensive way?
6.3 Visionary agents	To what extent are actors in the network able to manage and effectively push forward long-term and integrated strategies which are adequately supported by interim targets?

Table 1. *Cont.*

Indicator	Pre-Defined Question
7.1 Room to manoeuvre	To what extent do actors have the freedom and opportunity to develop a variety of alternatives and approaches (this includes the possibility of forming ad hoc, fit-for-purpose partnerships that can adequately address existing or emerging issues regarding the water challenge)?
7.2 Clear division of responsibilities	To what extent are responsibilities clearly formulated and allocated, in order to effectively address the water challenge?
7.3 Authority	To what extent are legitimate forms of power and authority present that enable long-term, integrated and sustainable solutions for the water challenge?
8.1 Affordability	To what extent are water services and climate adaptation measures available and affordable for all citizens, including the poorest?
8.2 Consumer willingness to pay	How is expenditure regarding the water challenge perceived by all relevant stakeholders (i.e., is there trust that the money is well-spent)?
8.3 Financial continuation	To what extent do financial arrangements secure long-term, robust policy implementation, continuation, and risk reduction?
9.1 Policy instruments	To what extent are policy instruments effectively used (and evaluated), in order to stimulate desired behaviour and discourage undesired activities and choices?
9.2 Statutory compliance	To what extent is legislation and compliance, well-coordinated, clear and transparent and do stakeholders respect agreements, objectives, and legislation?
9.3 Preparedness	To what extent is the city prepared (i.e. there is clear allocation of responsibilities, and clear policies and action plans) for both gradual and sudden uncertain changes and events?

3. Results

3.1. Trends and Pressures of Cape Town

Table 2 shows the scores of each of the twelve indicators of the TPF, ranging from 0 to 4 (0 indicating the lowest degree of concern and 4 the highest degree of concern). The TPF indicators for Cape Town that scored between 2.5 and 3.5, that is, representing areas of concern, were burden of disease, water scarcity, sea water intrusion and salinization, river peak discharges, and inflation.

Table 2. Trends and Pressures Framework analysis for Cape Town. Indicator scores range from 0 to 4 (0 indicating the lowest degree of concern and 4 the highest degree of concern).

Category	Indicators	Sub-Indicators	Indicator Scores
1. Social pressures	1. Urbanization rate		1.725
	2. Burden of disease		3
	3. Education rate		2.45
	4. Political instability		2.104
2. Environmental pressures	5. Water scarcity	5.1 Fresh water scarcity	3
		5.2 Ground water scarcity	1
		5.3 Salinization and/or seawater intrusion	3
	6. Flood risk	6.1 Urban drainage flood	1
		6.2 Sea-level rise	0
		6.3 River peak discharges	3
	7. Water quality	7.1 Surface water quality	0.632
		7.2 Biodiversity	1.28
	8. Heat risk		0.4
3. Financial pressures	9. Economic pressure		3.54
	10. Unemployment rate		4.1
	11. Poverty rate		1.81
	12. Inflation rate		3.17

Fresh water scarcity is an important factor, as Cape Town relies primarily on surface water sources and water scarcity can negatively impact the socio-economic aspects of a city [8]. It has the potential to affect human health by increasing the burden of disease. A score of 3 was reported for the salinization and/or seawater intrusion indicator, highlighting the fact that Cape Town's groundwater sources are vulnerable to salinization. Together with seawater intrusion, this can influence the salinity of groundwater and thus the water quality of freshwater aquifers. This is especially important in the CoCT as the City's future water supply augmentation plans include groundwater abstraction. The indicator score for river peak discharges indicates that flood risk is also an area of concern for water management in Cape Town. Floods have social, economic and environmental consequences; this includes loss of human life; increase in water-borne diseases as well as damage to infrastructure. This may result in certain economic activities coming to a halt as well as disruption of service delivery such as electricity, wastewater treatment, health care, education and the supply of clean water. Indicators that received scores between 3.5 and 4, representing increasing levels concern, were economic pressure and unemployment. The City's unemployment rate scored as a significant area of concern for Cape Town and has an impact on the ability of low-income citizens to afford and pay for water and sanitation services, which is an important revenue stream that enables the CoCT to implement projects and programs such as water-infrastructure maintenance. The TPF assessment provided insights into the environmental, social and economic aspects of Cape Town, over which the city has limited influence, although they do provide the context within which the city water managers must operate.

3.2. City Blueprint of Cape Town

The CBF scores are presented in Figure 2, which gives an indication of the management of Cape Town's water cycle. Figure 2 shows the scores for each of the twenty-five indicators, ranging from 0 at the center of the circle increasing outwards to 10. The overall city score (Blue City Index) of 4.9 reflects the fact that Cape Town is currently categorized as a water efficient city (according to Koop & Van Leeuwen [14]).

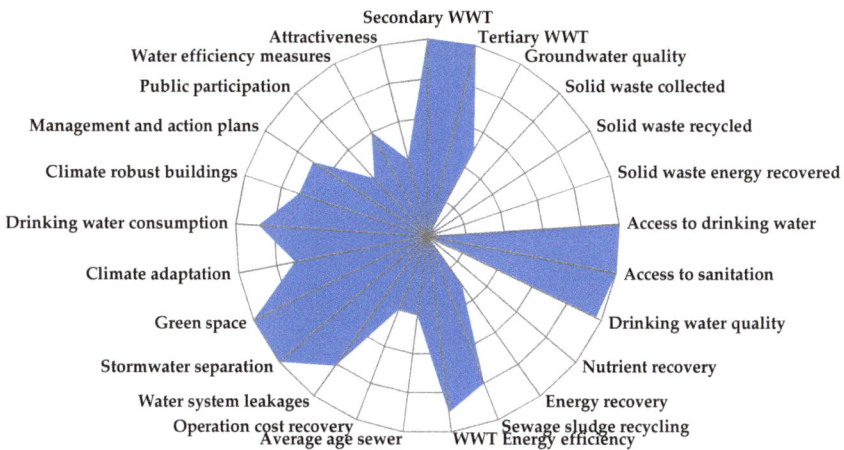

Figure 2. City Blueprint Framework analysis for Cape Town. Various components of urban water management are integrated into a framework of 25 indicators that are scored from 0 (low performance: inner circle) to 10 (high performance: outer circle). The overall score, the Blue City Index is 4.9.

The CBF assessment presents a snapshot of the performance of Cape Town's water system to illustrate the strengths and weaknesses of Cape Town's water management. The City scored relatively well on leakage control as only 9% of water is lost through system leakages, compared to the national South African average of 25% [22], as well as international cities such as Quito, Ecuador which are losing around 30% [23]. Similarly, Cape Town performs well on the delivery of services, scoring 100% for access to sanitation and drinking water, as well as on the percentage of wastewater/sewage that is treated. All sewage generated in the city is treated to some level at one of 23 wastewater treatment works—delivered either through formal sewage networks, or through alternative collection systems for informal settlements, such as chemical, portable and container toilets. Notwithstanding these high levels of wastewater treatment, the fact that Cape Town has low energy and nutrient recovery levels from these treatment processes highlights a major area for improvement. At present there is energy recovery at only one of the City's wastewater treatment plants, with methane biogas from the anaerobic digesters used on site. Currently there is no nutrient recovery as a separate item from wastewater treatment processes; nutrients are contained in the sludge which is used for agricultural processes or taken to landfill. There is considerable room to improve the city's solid waste treatment, as only ~10% of the city's waste is recycled [24]). Another potential area of concern is the fact that sewer networks in Cape Town are 40 years old on average [25]. This increases the probability for blockages and leakages in sewers and substantially increases the costs to refurbish and replace the extensive underground network over the next decade.

3.3. GCF Analysis of Cape Town

Table 3 shows the scores for the GCF assessment for Cape Town for each of the twenty-seven indicators, based on the responses to interview questions and publicly-available information. The analysis of the CBF for Cape Town provided a basis for the selection of the three water challenges that were analyzed in depth using the GCF. Although Cape Town scores well on access to drinking water and drinking water consumption (Figure 2), the current water crisis draws attention to the specific need to analyze and understand the governance of water scarcity in Cape Town. Similarly, in spite of adequate access to water and sanitation services in informal settlements (Figure 2), drainage and flood risk remain serious issues of concern; we therefore deemed it important to further analyze the governance of flood risk in Cape Town.

Furthermore, the CBF results show that there is room for improvement in energy recovery from wastewater treatment. Given these points, an in-depth analysis of the governance of water scarcity, flood risk and wastewater treatment is important as these challenges have an effect on the varying needs of society such as flood protection, human and environmental health, and water resources.

Table 3. Governance Capacity Framework assessment for Cape Town. Each indicator is scored according to a Likert scale: – very limiting; - limiting; 0 indifferent; + encouraging; ++ very encouraging.

Category	Indicators		Water Scarcity	Flood Risk	Wastewater Treatment
1. Awareness	1.1	Community knowledge	0	++	0
	1.2	Local sense of urgency	+	+	++
	1.3	Behavioral internalization	0	+	-
2. Useful knowledge	2.1	Information availability	+	+	++
	2.2	Information transparency	0	+	+
	2.3	Knowledge cohesion	0	+	+
3. Continuous learning	3.1	Smart monitoring	+	0	+
	3.2	Evaluation	-	0	0
	3.3	Cross-stakeholder learning	+	+	+

Table 3. *Cont.*

Category		Indicators	Water Scarcity	Flood Risk	Wastewater Treatment
4. Stakeholder engagement processes	4.1	Stakeholder inclusiveness	-	0	0
	4.2	Protection of core values	0	0	0
	4.3	Progress and variety of options	+	0	+
5. Management ambition	5.1	Ambitious and realistic management	0	+	++
	5.2	Discourse embedding	-	+	+
	5.3	Management cohesion	+	+	+
6. Agents of change	6.1	Entrepreneurial agents	0	+	+
	6.2	Collaborative agents	++	++	++
	6.3	Visionary agents	0	+	+
7. Multi-level network potential	7.1	Room to manoeuver	+	+	+
	7.2	Clear division of responsibilities	0	0	0
	7.3	Authority	+	+	+
8. Financial viability	8.1	Affordability	0	-	-
	8.2	Consumer willingness to pay	+	++	++
	8.3	Financial continuation	0	-	+
9. Implementing capacity	9.1	Policy instruments	0	0	+
	9.2	Statutory compliance	0	-	+
	9.3	Preparedness	0	++	+

4. Discussion

The focus of this paper is on the governance capacity of water scarcity, flood risk and wastewater treatment in Cape Town, hence the findings of the GCF assessment for Cape Town for the three water management challenges are provided in the sections that follow.

4.1. Water Scarcity

Cape Town is a coastal city in South Africa, with a Mediterranean-type climate, causing it to experience hot, dry summers and mild, wet winters. The city relies heavily on winter rainfall as 98% of Cape Town's water supply is supplied by surface water sources from 6 major dams on the outskirts of the city [26]. The current water crisis has resulted in most governance processes related to water scarcity in the city being in a constant state of change during the course of 2017/18. For this reason, the results of the governance assessment of water scarcity in Cape Town reveal the current situation and may not be representative of a typical year—even though it provides a useful "worst case" scenario assessment.

A respondent who is a former employee of the CoCT expressed that water scarcity was not sufficiently high on the city's agenda before 2017. This is despite the fact that climate change research has consistently placed emphasis on the possibilities of changing rainfall patterns in the Western Cape/Cape Town region, with likely adverse impacts on water resource availability for the region [27,28]. Consequently, the City's management ambitions before the drought were largely focused on service delivery objectives such as providing water service points (taps) in informal settlements and maintenance of infrastructure. This is reflected in the CoCT's annual Water Services Development Plan [22]. In addition, a respondent who is a city official stated that planning for water supply management for the year 2017 was based on a best-case scenario of receiving optimal rainfall. This indicates that unchanging situations were assumed and therefore planning for severe drought conditions was limited. Cape Town therefore scored as 0 (indifferent) for indicator 5.1 ambitious and realistic management of water scarcity. This is also due to the fact that long-term goals to augment the city's water supply by using groundwater, methods of desalination and water reclamation for potable use have been part of water resources planning processes since June 2007 as part of the Western Cape Water Reconciliation Strategy [29], however there are no signs that these long-term plans are being supported by intermittent targets. This resulted in the city being forced to implement plans for augmentation in a short time span of six to eighteen months in the face of the current drought. This has proven to be a learning opportunity, as reflected in the fact that Cape Town has recently adopted a new water management scenario termed the "New Normal" in which the city has been classified as a permanent drought region. Consequently, the city will no longer exclusively rely on surface water sources, and resilience to climatic uncertainty is being pursued in its future planning.

Cape Town's low score on indicator 5.1 ambitious and realistic management further reveals that although the city scored as + (encouraging) on both indicator 3.1 smart monitoring and 3.3 cross-stakeholder learning these two governance aspects were not used to enhance planning and decision making to reduce the city's long-term vulnerability to drought events. Similarly, the city scored as + (encouraging) on indicator 3.3 Cross-stakeholder learning as respondents stated that cross-stakeholder learning is valued, and results are incorporated to enhance optimal decision making. For instance, various University of Cape Town-related research initiatives with the City, such as Mistra Urban Futures, FRACTAL and Climate Change Think Tank, which facilitate better understanding of issues of climate change and sustainability at the city scale between city officials and academics, have been undertaken. Such cross-stakeholder learning initiatives involve a two-way learning system between academics and city officials through innovative knowledge sharing practices. Despite the fact that such programs do not continue indefinitely due to issues such as lack of funding (indicator 8.3), nevertheless, the knowledge from such research-based initiatives is still useful to enhance decision making on urban sustainability issues such as water scarcity in Cape Town and could be applied to other cities in the country. The GCF results also help to illustrate that although Cape Town scored well on access to drinking water, water scarcity in the City is in fact still a serious challenge.

4.2. Flood Risk

Flooding is a common phenomenon during Cape Town's rainy season, particularly in informal settlements and expansive low-lying areas such as the Cape Flats, which are prone to extreme flooding events. For this reason, there is a great sense of urgency to address flood risk in Cape Town. Enhancing community knowledge and including local communities in addressing flood risk is high on the agenda for local authorities. An annual multi-departmental 'Winter Readiness Program' led by the City's Disaster Risk Management Centre (DRMC) is run before the onset of each rainy season. The program aims to implement various measures to mitigate flood risk during the season whilst raising awareness and increasing community involvement. Practical tips such as how to raise flooring in homes and diverting flood water away from shacks are distributed to residents of informal settlements. While planning for the 2016 rainy season, 34 high flood risk areas including informal settlements were identified for running the program. A component of the program focuses on clearing stormwater infrastructure of solid waste to ensure its functionality. Community members are employed to litter-pick and remove sand from drainage systems and the banks of channels. In 2017, 1805 temporary jobs were created and R35 million (approximately $2.4 million) was spent on these cleaning programs. Information regarding flood risk is also distributed on the city's website. The city's DRMC compiled a series of educational pamphlets named the "Flood-wise pamphlets" which are also made available on the website. These address issues such as understanding the causes of flooding, practical solutions to prevent flooding and health issues related to flooding. Hence Cape Town scored as + (encouraging) on the indicators belonging to the category "awareness" (Table 3).

In addition to the CoCT making strides in addressing flood risk in informal settlements, the city's stormwater department has also devised two policies, the Management of Urban Stormwater Impacts

Policy (MUSIP) [30] and the Flood Plain and River Corridor Management Policy (FPRCMP) [31], which aim to address the challenge of flood risk in formal developments and quality of stormwater runoff from developments. For this reason, Cape Town scores as + (encouraging) on indicator *5.2 Discourse embedding* as the city uses different methods to address flood risk in different contexts. The MUSIP aims to *"minimize the undesirable impacts of stormwater runoff from developed areas by introducing WSUD principles to urban planning..."* [30]. The objective of the policy is for all Greenfield development sites, Brownfield development sites > 50,000 m² and Brownfield development sites < 50,000m² with a total impervious surface > 15% of site to include a Sustainable Drainage Systems (SuDS) component which achieves the objectives set out by the policy. The FPRCMP aims to *"manage development adjacent to watercourse and wetlands taking cognizance of the flood regime . . . "* [31]. The policy objective is to set back developments beyond floodplain zones, geomorphological buffers and ecological zones as per the conditions and requirements of the policy. The development of the MUSIP and the FPRCMP illustrates that there is a growing understanding of the complexity and uncertainty related to flood risk and awareness that the development of innovative approaches is crucial. Hence Cape Town also scores as + (encouraging) on indicators 5.1 Ambitious and Realistic Management and 5.3 Management Cohesion.

The MUSIP provides a degree of freedom to agents of change to explore new alternatives and to seize more high-risk opportunities. This is revealed by the City's score for condition 6 Agents of change as + (encouraging). Entrepreneurial agents, that is, consultants who design and implement SuDS technologies in new developments, are essentially given the freedom to experiment with alternative technologies when implementing these, as the policy does not prescribe what technologies are to be installed. The policy only requires effective technologies which adhere to the policy's SuDS objectives, thereby creating an enabling environment for implementation. This will aid in driving change as respondents emphasize that experimentation is crucial in legitimizing alternative technologies which may be otherwise doubted. Despite the fact that the MUSP and the FPRCMP have been developed to stimulate desired behavior and discourage undesired behavior, the implementation of these policies (indicator 9.2 statutory compliance) remains difficult for various reasons. For instance, local government lacks the human resources to check compliance to policy on the ground. Also, in developments where SuDS are successfully installed maintenance of technologies proves to be difficult resulting in ineffective performance.

4.3. Wastewater Treatment

Sixty-six percent of the water consumed by Cape Town ends up at twenty-three wastewater treatment works from where the final treated effluent is discharged back into the environment (CoCT, 2017). The wastewater undergoes treatment processes to ensure that the effluent released into rivers, the ocean and other water bodies meets prescribed standards. Ensuring that the quality of the effluent is of acceptable quality to be discharged into the environment requires rigorous monitoring of the process and functioning of the wastewater treatment systems. The city scored as + (encouraging) on indicator 3.1 smart monitoring with regard to the governance of wastewater treatment. The GCF reveals that smart monitoring is essential in ensuring that other governance aspects such as statutory compliance (indicator 9.2), preparedness for risk and adequate service delivery (indicator 9.3) are carried out successfully.

The quality of the effluent being discharged from the wastewater treatment works in Cape Town is monitored by the City on a continuous basis and the results are provided in the annual Water Services Development Plan. Effluent quality is also reported in accordance with the National Department of Water and Sanitation's (DWS) license requirements by way of the Green Drop certification program for wastewater treatment works nationwide. The Green Drop system is an incentive-based method which grants Green Drop Status to Water Service Authorities (in this case the CoCT) based on their level of compliance with wastewater legislation and other best practices as required by DWS. The most recent Green Drop report was published in 2014, in which Cape Town scored a Green Drop status of 89.7% (Good) based on compliance to Green Drop criteria at all of the City's wastewater treatment facilities [32]. Not only is smart monitoring applied to wastewater effluent quality, but a register of non-compliance incidents at wastewater treatment facilities is also included in the annual Water Services Development Plan [22]. The register provides a clear definition of the problem, the cause of the problem and remedial actions taken. The precautionary principle is adopted for dealing with risks such as non-compliance incidents, as a departmental Risk Management Register is compiled for the water and sanitation department where action plans are provided for potential risks. This has resulted in Cape Town scoring as + (encouraging) on indicator 9.3 Preparedness. Continuous monitoring of wastewater effluent quality and monitoring of non-compliance incidents enhances the city's preparedness in dealing with both sudden and gradual deviations in wastewater treatment processes.

Continuous monitoring of effluent is also important, as effluent is only being discharged into the environment. Approximately 8% of the total volumes of treated wastewater are currently re-used by more than 160 industrial and commercial customers [26]. The CoCT has been promoting the re-use of treated effluent by using an incentive-based method of selling treated water at a price lower than that of potable water. The drought crisis has prompted the city to plan to increase the percentage of treated effluent being used. Not only this, another important area of focus for the city is reclamation of potable water from treated effluent in order to augment drinking water supplies. Wastewater treatment processes are therefore also being used to promote conservation of the City's limited potable water supply. Hence Cape Town scores as encouraging for indicator 7.1 Room to maneuver, as the city has the opportunity to develop alternatives to address water scarcity in the city.

5. Conclusions

The City Blueprint assessment of Cape Town illustrates the importance of considering uncertainties and complexities in the governance processes related to urban water challenges. The governance of wastewater treatment and that of flood risk in Cape Town already embraces uncertainties, and as a result the level of preparedness to deal with unexpected disaster and risk is deemed adequate. On the other hand, consideration of the uncertainty and complexity in the governance of water scarcity has been lacking over the years. This is revealed by the City's attempts to implement augmentation schemes in a short time span of six to eighteen months during 2017/18 to address the water crisis. Our study has also revealed that information transparency and access to information by the public, through social media, posters in public spaces and on the City's website, plays an important role in educating the public about water challenges and can be used as a tool to encourage behavioral change regarding water scarcity. Although the CoCT makes a concerted effort to ensure that information and knowledge is disseminated to the public, most information is only available online, for example through the 'Water Dashboard' feature provided on the City's website, which may limit some citizen's access to information. With this being said, it is recommended that information be provided in places which are easily and frequently accessed by the public such as schools and shops. In the same light, this point suggests reflection of the GCF methodology which mainly considers the type of knowledge which is available to the public without considering the fact that communication platforms can limit access to information.

Furthermore, the GCF illustrates the potential for Cape Town to adopt the principles of WSUD. The 'non-conventional' nature of WSUD options, such as the use of nature-based solutions and green infrastructure for water supply, stormwater management and wastewater treatment, means that local authorities may deem it more risky than conventional (grey) water infrastructure. Therefore, successful implementation of policies such as the Management of Urban Stormwater Impacts Policy and the Flood Plain and River Corridor Management Policy, which are already underpinned by principles of WSUD, can be used to give more credibility to the approach. It is crucial for a South African city like Cape Town *"to determine what water sensitivity means in SA taking into account poverty, inequality, lack of services and context specific challenges"* [33]. This assessment has shown that great effort has already been made in the CoCT to embed local context into addressing water challenges, thus illustrating the potential for a transition towards water sensitivity. Lastly, the drought is causing a shift in governance processes related to water scarcity and has resulted in the adoption of a new management scenario, the 'New normal'. There is therefore potential for the principles of WSUD to be implemented in new strategies relating to water scarcity in Cape Town.

Author Contributions: Conceptualization, B.M., S.K. and K.v.L.; methodology, B.M., S.K., K.C.; software, S.K.; formal analysis, B.M.; investigation, B.M.; resources, K.C.; data curation, B.M.; writing—original draft preparation, B.M.; writing—review and editing, S.K., K.v.L., K.C.; supervision, K.C.; project administration, K.C.; funding acquisition, K.v.L., K.C.

Funding: This project was partially funded by the South African Water Research Commission (WRC) as part of Project K5/2413—Development and management of a Water Sensitive Design Community of Practice program. Part of this research was also funded by the POWER project. The European Commission is acknowledged for Funding POWER in H2020-Water under Grant Agreement No. 687809.

Acknowledgments: This paper would not have been published without the support provided by the programme committee of the International Young Water Professionals Conferences (www.iwaywpconference.org) organized by IWA, WISA and YWP-ZA.

Conflicts of Interest: The authors declare no conflict of interest. The funders had no role in the design of the study; in the collection, analyses, or interpretation of data; in the writing of the manuscript, or in the decision to publish the results.

References

1. Wamsler, C.; Brink, E.; Rivera, C. Planning for climate change in urban areas: From theory to practice. *J. Clean. Prod.* **2013**, *50*, 68–81. [CrossRef]
2. Friedrich, E.; Pillay, S.; Buckley, C.A. Carbon footprint analysis for increasing water supply and sanitation in South Africa: A case study. *J. Clean. Prod.* **2009**, *17*, 1–2. [CrossRef]
3. *National Water Resource Strategy: Water for an Equitable and Sustainable Future*; Department of Water Affairs: Pretoria, South Africa, 2013; Volume 53, pp. 1689–1699.
4. Ferguson, B.C.; Frantzeskaki, N.; Brown, R.R. A strategic program for transitioning to a Water Sensitive City. *Landsc. Urban Plan.* **2013**, *117*, 32–45. [CrossRef]
5. Fletcher, T.D.; Shuster, W.; Hunt, W.F.; Ashley, R.; Butler, D.; Arthur, S.; Trowsdale, S.; Barraud, S.; Semadeni-Davies, A.; Bertrand-Krajewski, J.L.; et al. SUDS, LID, BMPs, WSUD and more—The evolution and application of terminology surrounding urban drainage. *Urban Water J.* **2015**, *12*, 525–542. [CrossRef]
6. Wong, T.H.F.; Brown, R.R. The water sensitive city: Principles for practice. *Water Sci. Technol.* **2009**, *60*, 673–682. [CrossRef] [PubMed]
7. Rijke, J.; Farrelly, M.; Brown, R.; Zevenbergen, C. Configuring transformative governance to enhance resilient urban water systems. *Environ. Sci. Policy* **2013**, *25*, 62–72. [CrossRef]
8. Armitage, N.; Fisher-Jeffes, L.; Carden, K.; Winter, K.; Naidoo, V.; Spiegel, A.; Mauck, B.; Coulson, D. *Water Sensitive Urban Design (WSUD) for South Africa: Framework and Guidelines*; WRC Report TT 558/14; Water Research Commission: Pretoria, South Africa, 2014; pp. 1–58.
9. Koop, S.H.A.; Koetsier, L.; Van Doornhof, A.; Reinstra, O.; Van Leeuwen, C.J.; Brouwer, S.; Dieperink, C.; Driessen, P. Assessing the governance capacity of cities to address challenges of water, waste and climate change. *Water Resour. Manag.* **2017**, *31*, 3427–3443. [CrossRef]
10. Koop, S.H.A.; Van Leeuwen, C.J. Assessment of the sustainability of water resources management: A critical review of the City Blueprint approach. *Water Resour. Manag.* **2015**, *29*, 5649–5670. [CrossRef]
11. Koop, S.H.A.; Van Leeuwen, C.J. The challenges of water, waste and climate change in cities. *Environ. Dev. Sustain.* **2017**, *19*, 385–418. [CrossRef]
12. Beck, L.; Brown, R.; Chesterfield, C.; Dunn, G.; de Haan, F.; Lloyd, S.; Rogers, B.; Urich, C.; Wong, T. Beyond benchmarking: A water sensitive cities index. *OzWater* **2016**, *16*, 10–12.
13. Internation Water Association. *Principles for Water Wise Cities*; Internation Water Association: London, UK, 2016.
14. Koop, S.H.A.; Van Leeuwen, C.J. Application of the improved city blueprint framework in 45 municipalities and regions. *Water Resour. Manag.* **2015**, *29*, 4629–4647. [CrossRef]
15. WHO Metrics: Disability-Adjusted Life Year (DALY). 2018. Available online: https://www.who.int/healthinfo/global_burden_disease/metrics_daly/en/ (accessed on 20 December 2018).

16. World Bank Primary Completion Rate, Total (% of Relevant Age Group). 2012. Available online: http://data.worldbank.org/indicator/SE.PRM.CMPT.ZS/countries/1W-AO?display=default (accessed on 20 December 2018).

17. World Bank Worldwide Governance Indicators. 2013a. Available online: http://info.worldbank.org/governance/wgi/indexaspx#faq (accessed on 20 December 2018).

18. World Bank Inflation, Costumers Price (Annual %). 2013b. Available online: http://data.worldbank.org/indicator/FP.CPI.TOTL.ZG (accessed on 20 December 2018).

19. World Bank Unemployment Total (% of Total Labor Force) (Modelled ILO Estimate). 2014. Available online: http://data.worldbank.org/indicator/SL.UEM.TOTL.ZS (accessed on 20 December 2018).

20. FAO Food and Agriculture Organization of the United Nations: AQUASTAT Database. 2015. Available online: http://www.fao.org/nr/water/aquastat/data/query/index.html;jsessionid=B022D1C2732DF571D2A384B57E0128D6 (accessed on 20 December 2018).

21. European Commission European innovation Partnership on Water. 2017. Available online: http://www.eip-water.eu/City_Blueprints (accessed on 20 February 2018).

22. *City of Cape Town: Water Services Development Plan—IDP Water Sector Input Report*; City of Cape Town: Cape Town, South Africa, 2016.

23. Schreurs, E.; Koop, S.; van Leeuwen, K. Application of the City Blueprint Approach to assess the challenges of water management and governance in Quito (Ecuador). *Environ. Dev. Sustain.* **2017**, *20*, 509–525. [CrossRef]

24. *National Waste Information Baseline Report*; Department of Environmental Affairs, Republic of South Africa: Pretoria, South Africa, 2012.

25. Madonsela, B.; University of Cape Town, Cape Town, South Africa. Respondent 5. Personal communication, 2018.

26. *Water Services and the Cape Town Urban Water Cycle*; City of Cape Town: Cape Town, South Africa, 2017.

27. Ziervogel, G.; Shale, M.; Du, M. Climate change adaptation in a developing country context: The case of urban water supply in Cape Town. *Clim. Dev.* **2010**, *2*, 94–110. [CrossRef]

28. Mukheibir, P.; Ziervogel, G. Developing a Municipal Adaptation Plan (MAP) for climate change: The city of Cape Town. *Environ. Urban.* **2007**, *19*, 143–158. [CrossRef]

29. *Western Cape Water Supply System: Reconciliation Strategy Study*; Department of Water and Sanitation: Pretoria, South Africa, 2007.

30. *Management of Urban Stormwater Impacts Policy*; City of Cape Town: Cape Town, South Africa, 2009.

31. *Floodplain and River Corridor Management Policy*; City of Cape Town: Cape Town, South Africa, 2009.

32. *Green Drop Progress Report 2014*; Department of Water & Sanitation: Pretoria, South Africa, 2014.

33. Fisher-Jeffes, L.; Carden, K.; Armitage, N.P.; Spiegel, A.; Winter, K.; Ashley, R. Challenges Facing Implementation of Water Sensitive Urban Design in South Africa. In Proceedings of the 7th Conference on Water Sensitive Urban Design, Mebourne, Australia, 20–23 February 2012; pp. 1–8.

Article

Governing Non-Potable Water-Reuse to Alleviate Water Stress: The Case of Sabadell, Spain

Marketa Šteflová [1], Steven Koop [2,3], Richard Elelman [4], Jordi Vinyoles [5] and Kees Van Leeuwen [2,3,*]

[1] University College of Utrecht, Utrecht University, Campusplein 1, 3584 ED Utrecht, The Netherlands; m.steflova@students.uu.nl
[2] KWR Watercycle Research Institute, Groningenhaven 7, 3433 PE Nieuwegein, The Netherlands; stef.koop@kwrwater.nl
[3] Copernicus Institute of Sustainable Development and Innovation, Utrecht University, Heidelberglaan 2, 3584 CS Utrecht, The Netherlands
[4] Fundació CTM Centre Tecnològic, Plaça de la Ciencia, 2, 08243 Manresa, Spain; richard.elelman@ctm.com.es
[5] CASSA, Companyia d'Aigües de Sabadell, S.A., Concepció, 20-08202 Sabadell, Spain; jvinyoles@cassa.es
* Correspondence: kees.van.leeuwen@kwrwater.nl

Received: 2 May 2018; Accepted: 5 June 2018; Published: 6 June 2018

Abstract: The world will experience an estimated 40% freshwater supply shortage by 2030, converting water scarcity into one of the principal global challenges that modern society faces. Urban water reuse is recognized as a promising and necessary measure to alleviate the growing water stress in many regions. The transformation to widespread application of water-reuse systems requires major changes in the way water is governed, and countries such as Spain already find themselves involved in this process. Through the systematic assessment of the city of Sabadell (Spain), we aim to identify the main barriers, opportunities and transferable lessons that can enhance governance capacity to implement systems for non-potable reuse of treated wastewater in cities. It was found that continuous learning, the availability and quality of information, the level of knowledge, and strong agents of change are the main capacity-building priorities. On the other hand, awareness, multilevel network potential and implementing capacity are already well-established. It is concluded that in order to undertake a widespread application of water-reuse practices, criteria examining water quality according to its use need to be developed independently of the water's origin. The development and implementation of such a legislative frame should be based on the experience of local water-reuse practices and continuous evaluation. Finally, the need for public engagement and adequate pricing mechanisms are emphasized.

Keywords: water-reuse; governance capacity; water management; water scarcity

---

1. Introduction

Approximately four billion people experience severe water scarcity during at least one month per year, while over 500 million people face severe water scarcity, with water consumption exceeding the renewable resource by a factor of two or more [1]. The agricultural sector is responsible for 70% of the world's water consumption whereas industry and municipalities withdraw 19% and 11%, respectively. However, Hoekstra et al. estimates that agriculture accounts for up to 92% of the global urban water footprint [2]. Large increases in freshwater demand can be expected in the next decade, particularly for industrial production in urban areas [3]. These developments are estimated to lead to a 50% increase in water demand in developing countries by 2025 [4] and a 40% freshwater supply shortage worldwide by 2030 [5]. A wide plethora of factors including population growth, diets shifting towards water-intensive foods such as meat, groundwater depletion, salinization caused by unsustainable irrigation, as well as saltwater intrusion and sea level rise, which further increase the pressure on freshwater resources [1–7]. The impact of water scarcity is expected to lead to substantial unemployment. In particular, water-dependent labor in arid and semi-arid areas—95% of which are agricultural jobs—will be affected. This type of unemployment may lead to food insecurity and social instability that could trigger large scale involuntary migration towards cities and across continents [8]. These risks urge for water-use efficiency and water-reuse.

Wastewater is increasingly being considered, a largely untapped resource for freshwater and raw materials that can alleviate water stress [7]. At present, high-income countries treat 70% of their wastewater, upper-middle-income countries 38% and lower-middle-income countries 28%. Only 8% of the wastewater in low-income countries undergoes any kind of treatment [9]. Altogether, this means that an estimated 80% of the wastewater is released into the environment untreated [6] which leads to eutrophication, biodiversity loss and can threaten drinking water, fisheries, aquaculture and tourism [10]. Reusing treated wastewater therefore has a large potential to alleviate water stress. Moreover, since 80% of all wastewater is not treated, many Waste Water Treatment (WWT) systems have yet to be built. Thus, WWT systems and reuse systems may be a promising solution package to improve public health, reduce water pollution and alleviate water scarcity in particular in water-scarce urban regions.

The European Commission has recognized the untapped potential of treated wastewater, and has called for "closing the loop" through a circular economy approach [11]. It recognizes and enables the reuse of wastewater as a safe solution to reduce water demand. From the total reused tertiary treated water only 2.3% is for potable purposes whereas most water is reused for irrigation (52%), industry (19.3%) and non-potable urban applications (8.3%) [12].

In order to reduce water stress by a wider application of wastewater-reuse practices in cities, major changes are required in the way the water cycle is governed at the local, regional and national level. Spain is in the middle of such a transformation process where various multi-level governance barriers and challenges emerge [13]. Despite the fact that a number of water reuse applications have already been developed and established in many countries, the widespread adoption or 'mainstreaming' of water reuse practices appears to be slow and various technical and non-technical barriers have been published [14–17]. In environmental governance literature, a plethora of social factors and conditions have been identified that impede or enhance climate adaptation such as water reuse schemes [18–20]. However, most identified conditions are based on conceptual and theoretical considerations with a lack of emphasis on empirical validation [18,21,22]. In addition, concepts and definitions are often inconsistent, non-specific and in part, overlap each other [19,20]. If findings are not organized in a common framework, isolated knowledge will not cumulate [23]. Hence, a diagnostic framework is required that facilitates the accumulation of coherent knowledge that could improve the understanding of the barriers, opportunities and lessons beyond the case study itself [19]. Water scarcity issues transcend administrative boundaries and involve many stakeholders. The capacity to collaborate, to collectively overcome different barriers, is therefore essential. A promising way to consistently analyze the main barriers and opportunities that might emerge in the adoption of water reuse schemes is through the concept of governance capacity. There are multiple definitions of governance capacity. However, a few common traits can be defined [24]. First, capacity refers to the ability of actors to jointly act in the face of collective challenges. Second, capacity is the product of actors' interaction that is influenced by the socio-institutional setting. Third, actors' values, culture and interests shape their interactions and influence collective problem-solving. Accordingly, we apply the definition of Koop et al. who defined governance capacity as 'a set of key governance conditions that should be developed to enable change that will be effective in finding dynamic solutions for water challenges in cities' [24].

In this paper, we focus on the governance capacity of cities to alleviate water stress by applying wastewater-reuse for non-potable purposes. Through a case study in the city of Sabadell (Spain), this paper aims to identify the main barriers, opportunities and transferable lessons that can enhance the governance capacity to implement systems for non-potable reuse of treated wastewater in cities.

This paper has the following structure. Section 2 describes the applied methodology and Section 3 provides the results of the case study. Section 4 provides a discussion of the results and reflects on possible transferable lessons from Sabadell for other cities in Spain, the Mediterranean and other water-stressed regions. The main conclusions are provided in Section 5.

2. Methodology

Based on an extensive literature review, Koop et al. [24] developed a diagnostic framework in order to assess the most important conditions that together determine the capacity to govern water challenges. The Water Governance Capacity Framework (GCF) consists of three dimensions, nine conditions and 27 indicators (Table 1). The "knowing" dimension relates to the need to be aware, understand, and learn about the risks and impacts of policy and strategic choices. The "wanting" dimension refers to the need for actors to commit, cooperate, act upon ambitions and use their skills to find solutions. The "enabling" dimension refers to network, resources, and instruments that actors require to realize their ambitions.

Table 1. Water Governance Capacity Framework (GCF) [24].

Dimensions	Conditions	Indicators
Knowing	1 Awareness	1.1 Community knowledge 1.2 Local sense of urgency 1.3 Behavioural internalization
	2 Useful knowledge	2.1 Information availability 2.2 Information transparency 2.3 Knowledge cohesion
	3 Continuous learning	3.1 Smart monitoring 3.2 Evaluation 3.3 Cross-stakeholder learning
Wanting	4 Stakeholder engagement process	4.1 Stakeholder inclusiveness 4.2 Protection of core values 4.3 Progress and variety of options
	5 Management ambition	5.1 Ambitious and realistic management 5.2 Discourse embedding 5.3 Management cohesion
	6 Agents of change	6.1 Entrepreneurial agents 6.2 Collaborative agents 6.3 Visionary agents
Enabling	7 Multi-level network potential	7.1 Room to manoeuver 7.2 Clear division of responsibilities 7.3 Authority
	8 Financial viability	8.1 Affordability 8.2 Consumer willingness-to-pay 8.3 Financial continuation
	9 Implementing capacity	9.1 Policy instruments 9.2 Statutory compliance 9.3 Preparedness

Each indicator has its own pre-defined question and indicator-specific 5-point Likert scale, ranging from very encouraging (++) to very limiting (−−) of the overall governance capacity to address a water challenge. A detailed description of each indicator's pre-defined question is provided in Table 2. For the indicator-specific Likert scale and link to the literature, we refer to reference [25]. By substantiating the scores of each indicator according to a triangular approach, the findings are validated in a standardized and reproducible way. This triangular approach consists of three steps:

1. A desk study of scientific literature, official government sources, policy documents and grey literature resulting in a report of the substantiated preliminary Likert score of each indicator.
2. The construction of a standardized importance/influence matrix to identify stakeholders, categorize them, and specify their roles and responsibilities [26]. In this matrix, importance refers to the priority given to satisfy the needs and interests of a stakeholder. Influence refers to the power of stakeholders to enhance or impede a policy, plan or objective. The importance/influence matrix consists of four classes: (1) crowd (low importance and low influence); (2) context (low importance and high influence); (3) subjects (high importance and low influence); and (4) key players (high importance and high influence). For each class, at least one stakeholder representing the government, the market and civil society were selected as suggested by Lange et al. [27]. A coding system is applied in this paper to refer to maintain anonymity, where [SR001], [SR002], [SR003] and so on refer to the conducted interviews. The interviews were conducted face-to-face, lasted approximately 1 hour each and were recorded to increase the accuracy of the information gathered.
3. All interviewees were asked for their reactions to the indicator scores and their respective explanations. Their feedback took the form of additional information and they were asked to support their statements with reports, policy references, arguments etc. Based on the incorporation of the aforementioned further input, the final indicator scores were determined.

The selected stakeholders represented the organizations Simbiosy, the General Water Society of Barcelona (SGAB), the Consortium of Integrated Water Management of Catalunya (CONGIAC), Sabadell's wastewater treatment plant, Riusec (EDAR RIUSEC), the University of Barcelona, the Polytechnic University of Catalonia, the local water service utility, Aigues Sabadell (CASSA), the Institute of Environmental Assessment and Water Research (IDÆA), the Technical Service Consortium of the Costa Brava, Figueres City Council, Sabadell City council, Barcelona Provincial Government and the Catalan Water Agency. Experts from the identified key stakeholders were selected for semi-structured interviews in order to gather the information to score the indicators and also to receive follow-up questions for clarification or to better understand the content. People with different roles, expertise and responsibilities were selected to reduce the risk of bias and in order to unravel socially desirable responses. A coding system is applied in this paper to consistently refer to these anonymized interviews. The overall indicator scores were determined based on the separate interview scores and the collection of additional information that may ratify or provide nuance to the interview findings. Altogether, 16 interviews were conducted during the period, 6 June–6 July, 2017.

Table 2. Overview of pre-defined questions to be answered by the researcher based on a triangular approach consisting of three steps: (1) literature review; (2) in-depth interviews with selected experts; (3) feedback procedure. The full details of the Likert scoring are provided on the EIP Water website [25].

Indicator	Pre-Defined Question
1.1 community knowledge	To what extent is knowledge regarding the current and future risks, impacts, and uncertainties of the water challenge dispersed throughout the community and local stakeholders who may result in their involvement in decision-making and implementation?
1.2 Local sense of urgency	To what extent do actors have a sense of urgency, resulting in widely supported awareness, actions, and policies that address the water challenge?
1.3 Behavioural internalization	To what extent do local communities and stakeholders try to understand, react, anticipate and change their behaviour in order to contribute to solutions regarding the water challenge?
2.1 Information availability	To what extent is information on the water challenge available, reliable, and based on multiple sources and methods, in order to meet current and future demands so as to reveal information gaps and enhance well-informed decision-making?
2.2 Information transparency	To what extent is information on the water challenge accessible and understandable for experts and non-experts, including decision-makers?

Table 2. *Cont.*

Indicator	Pre-Defined Question
2.3 Knowledge cohesion	To what extent is information cohesive in terms of using, producing and sharing different kinds of information, usage of different methods and integration of short-term targets and long-term goals amongst different policy fields and stakeholders in order to deal with the water challenge?
3.1 Smart monitoring	To what extent is the monitoring of process, progress, and policies able to improve the level of learning (i.e., to enable rapid recognition of alarming situations, identification or clarification of underlying trends)? Or can it even have predictive value?
3.2 Evaluation	To what extent are current policy and implementation continuously assessed and improved, based on the quality of evaluation methods, the frequency of their application, and the level of learning?
3.3 Cross-stakeholder learning	To what extent are stakeholders open to and have the opportunity to interact with other stakeholders and deliberately choose to learn from each other?
4.1 Stakeholder inclusiveness	To what extent are stakeholders interacting in the decision-making process (i.e., are they merely informed, are they consulted or are they actively involved)? Are their engagement processes clear and transparent? Are stakeholders able to speak on behalf of a group and decide on that group's behalf?
4.2 Protection of core values	To what extent (1) is commitment focused on the process instead of on early end-results? (2) do stakeholders have the opportunity to be actively involved? (3) are the exit procedures clear and transparent? (All three ensure that stakeholders feel confident that their core values will not be harmed.)
4.3 Progress and variety of options	To what extent are procedures clear and realistic, are a variety of alternatives co-created and thereafter selected from, and are decisions made at the end of the process in order to secure continued prospect of gain and thereby cooperative behaviour and progress in the engagement process?
5.1 Ambitious and realistic management	To what extent are goals ambitious (i.e., identification of challenges, period of action considered, and comprehensiveness of strategy) and yet realistic (i.e., cohesion of long-term goals and supporting flexible intermittent targets, and the inclusion of uncertainty in policy)?
5.2 Discourse embedding	To what extent is sustainable policy interwoven in historical, cultural, normative and political context?
5.3 Management cohesion	To what extent is policy relevant for the water challenge, and coherent regarding (1) geographic and administrative boundaries; and (2) alignment across sectors, government levels, and technical and financial possibilities?
6.1 Entrepreneurial agents	To what extent are the entrepreneurial agents of change enabled to gain access to resources, seek and seize opportunities, and have an influence on decision-making?
6.2 Collaborative agents	To what extent are actors enabled to engage, build trust and collaborate, and connect business, government, and sectors, in order to address the water challenge in an unconventional and comprehensive way?
6.3 Visionary agents	To what extent are actors in the network able to manage and effectively push forward long-term and integrated strategies which are adequately supported by interim targets?
7.1 Room to manoeuvre	To what extent do actors have the freedom and opportunity to develop a variety of alternatives and approaches (this includes the possibility of forming ad hoc, fit-for-purpose partnerships that can adequately address existing or emerging issues regarding the water challenge)?
7.2 Clear division of responsibilities	To what extent are responsibilities clearly formulated and allocated, in order to effectively address the water challenge?
7.3 Authority	To what extent are legitimate forms of power and authority present that enable long-term, integrated and sustainable solutions for the water challenge?
8.1 Affordability	To what extent are water services and climate adaptation measures available and affordable for all citizens, including the poorest?
8.2 Consumer willingness to pay	How is expenditure regarding the water challenge perceived by all relevant stakeholders (i.e., is there trust that the money is well spent)?
8.3 Financial continuation	To what extent do financial arrangements secure long-term, robust policy implementation, continuation, and risk reduction?
9.1 Policy instruments	To what extent are policy instruments effectively used (and evaluated), in order to stimulate desired behaviour and discourage undesired activities and choices?
9.2 Statutory compliance	To what extent is legislation and compliance, well-coordinated, clear and transparent and do stakeholders respect agreements, objectives, and legislation?
9.3 Preparedness	To what extent is the city prepared (i.e., there is clear allocation of responsibilities, and clear policies and action plans) for both gradual and sudden uncertain changes and events?

3. Case Study Description

At present, around 11% of total treated wastewater is reused in Spain [11]. Spain has experienced several episodes of water stress during the 1990s and the early 21st century. In particular, several acute droughts led to domestic water cuts and at times required the use of sea-going water tankers from different locations in the Mediterranean coast [13]. In order to alleviate water stress, the central government and regional governments have promoted desalination plants and have devised a National Plan for Water-reuse [13]. In particular, before 2011, approximately 50 municipalities in Catalunya had approved local regulations to promote decentralized reuse systems. The actions for the use of reclaimed water in Spain, mainly consist in transporting it for specific uses, such as the watering of golf courses and public gardens, the cleaning of streets, or for agriculture and industry. The implementation of a distribution network for reclaimed water, coexisting with the drinking water network, has, to date, been applied in only a few cities, such as Madrid and Sabadell. Only in Sabadell is water supplied for the use of flushing toilets.

With a population of over 208,000 people, Sabadell is the co-capital and second largest city of the County of Valles Occidental in Catalonia, Spain [28]. It is situated 22 km north of Barcelona, in the basins of the rivers Ripoll and Riusec, both integrated within the Besos River Basin. It is a highly commercial and industrial city that acts as a driving force for economic and urban development. A dual network is already applied in a large part of the city that separately distributes drinking water and treated non-potable water from the EDAR Riusec treatment plant and from groundwater sources. The second WWT plant of the city, Riu Ripoll, returns treated wastewater upstream of the Ripoll River, aiming to restore the ecological flow. Together these plants treated 22,544 m^3 day^{-1} and 14,170 m^3 day^{-1} respectively in 2017 [29]. Nonetheless, the total amount of treated non-potable water supplied through the dual network is only around 274 m^3 day^{-1}.

The governance of the water sector in Sabadell is composed of both private and public stakeholders. In Spain, the national and regional governments mandate the normative and legislative contexts. Nonetheless, each municipality is responsible for the management of the water in its jurisdiction. Thus this role falls in the hands of the City council of Sabadell. This municipality, among others, has subcontracted the private company CASSA to do this. In addition, Water of Sabadell (CASSA) has recently become part of AGBAR (Aguas Barcelona), which in turn is predominantly owned by Suez Environment. The stakeholders with high influence and the most interest were identified as the Catalan Water Agency (state), the Provincial Government of Barcelona (state), CASSA (market) and the City Council of Sabadell (state). The stakeholders with a high interest but low influence are EDAR Water Treatment plant (state/market), Network of Cities & Towns for Sustainability (Civil Society), Consortium of Besos Tordera and the Catalan Association of Friends of Water (Civil Society).

A desk study of Sabadell's Integrated Water Resources Management (IWRM)—called a City Blueprint—was performed within the European POWER project (https://www.power-h2020.eu/) and indicated that the city is vulnerable to heat risk and water scarcity (Figure 1 [30]). In addition, financial pressures such as high unemployment (18.4%) and a moderate average GDP per capita (25,684 USD/year) could affect urban water management investments. Sabadell has a high drinking water quality, with 187/187 samples that meet the quality standards [29]. Furthermore, Sabadell's drinking water consumption of 96 L per person per day is one of the lowest rates in Europe of domestic water consumption. The average age of the pipes of the drinking water distribution network is 38 years, so some areas require refurbishment. Non-revenue water accounts for 19.4%.

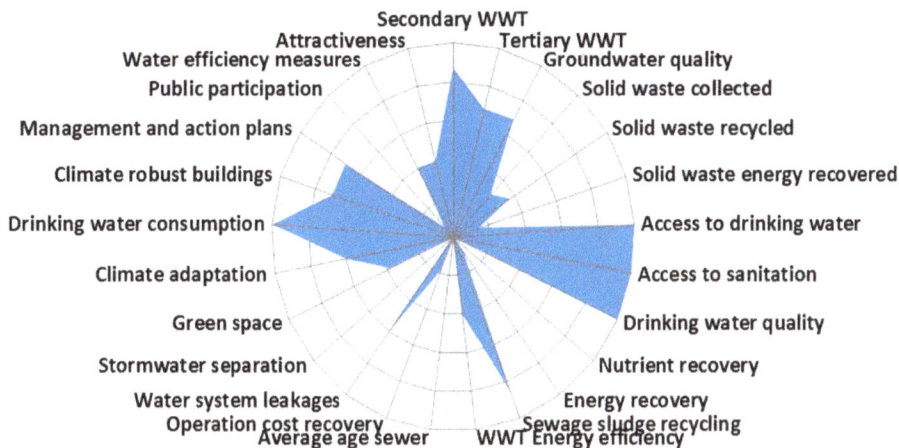

Figure 1. Spiderweb of the City Blueprint analysis of the city of Sabadell, Spain. The scores range from 0 (poor performance; centre of the circle) to 10 (high performance; periphery of the circle). The results are reported in detail in the study by Šteflová [30].

4. Results

The results of the Governance Capacity Framework on site research show that the multi-level governance system of Sabadell is complex and that the reuse of treated wastewater for non-potable purposes is progressive, but not yet widely adopted in the centralized water governance system in the area of Catalonia. Figure 2 summarizes the results of Sabadell's multi-level governance capacity to alleviate water stress by wastewater-reuse schemes. The indicators are ranked from most limiting to most encouraging concerning the capacity to govern practices of water reuse. The limiting and encouraging conditions are presented systematically in accordance with Table 1.

Condition 1: Awareness

The level of knowledge about the region's water scarcity and the amplifying impact of climate change are found to be relatively high [SR015]. However, there is little understanding with regards to how the water is used and distributed within the region, the linkages and interdependencies in relation to weather patterns, land use or environmental processes (indicator 1.1). Accordingly, the impacts on the water quality of rivers, groundwater, and the risks and uncertainties associated with the increasing water scarcity are largely underestimated [SR002-SR011-SR013-SR015-SR016]. The general sense of urgency of water stress is moderate amongst the citizens [SR001-SR011; indicator 1.2]. Nevertheless, water conservation strategies (indicator 1.3) such as grey water-reuse on a household level is widely applied [SR001-SR002-SR004], which is reflected in the city's low per capita water consumption of around 96 L person^{-1} day^{-1} ([SR006-SR011] [29]). The latter results from the fact that historically the region has experienced many droughts, and conservation strategies are engraved into the collective memory of the region [SR005-SR008-SR010-SR011-SR012-SR014-SR016].

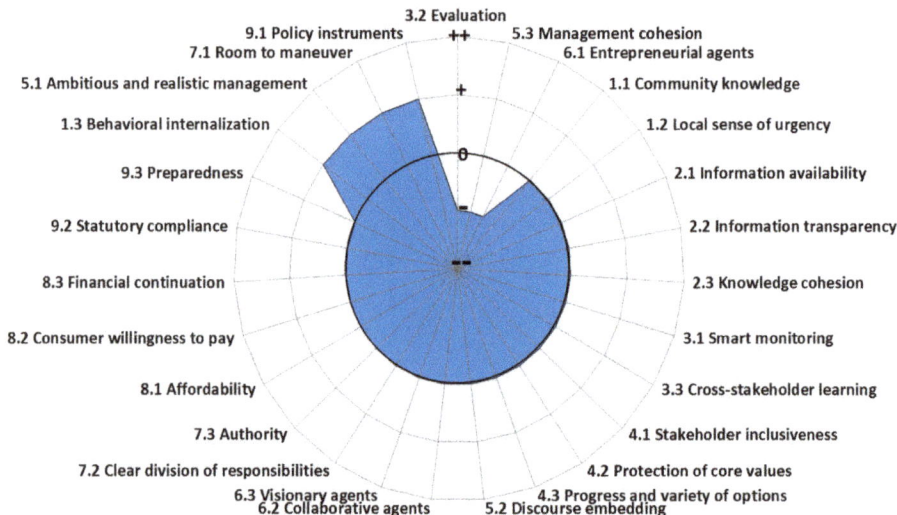

Figure 2. Results of the governance capacity to alleviate water stress by wastewater-reuse schemes in the city of Sabadell, Spain. The 27 indicators are organized clockwise around the spider web circle by most limiting (− −) to most encouraging (++).

Condition 2: Useful Knowledge

The perception regarding the availability, transparency and cohesion of the information varied considerably between the stakeholders. It was revealed that citizens have limited access to information (indicator 2.1) and that available information is difficult to locate (indicator 2.2). The regional meteorological information and the water reservoir data published by the Catalan Water Agency (ACA) and the regional administration are available but not presented in a way that is intelligible for the general public [SR011-SR014-SR009]. Furthermore, much of the accessible information is not fully up-to-date or has a technical nature [SR004-SR007-SR010-SR012-SR014-SR016]. At the city level there have been strong attempts to improve transparency and information services such as water saving tips on the back of water bills, education programs and conferences for all who are interested ([SR001-SR002] [30]). In general, the publicly available information is somewhat limited, which may be a by-product of limited incentives for stakeholders to communicate with citizens [SR007-SR010-SR011-SR012-SR014-SR016]. Consequently, the available knowledge about water scarcity and reuse practices is not cohesive and demonstrates the existence of different perceptions amongst stakeholders (indicator 2.3).

Condition 3: Continuous Learning

The local water supplier CASSA has an advanced monitoring system (indicator 3.1) that can rapidly recognize alarming situations such as potable water contamination or leakages, and to some extent is also able to recognize long-term patterns of consumption, water flow and water quality [SR001-SR012-SR013]. Nonetheless, a more regional, national or cross-sectorial monitoring and evaluation of the water sector is largely lacking, leading to fragmented knowledge [SR011-SR014]. According to one interviewee, this is an important reason for the lack of transparency and the sharing of information between stakeholders [SR007]. Evaluation of policy (indicator 3.2) occurs on an infrequent basis [SR008], it can be non-directional and susceptible to political shifts [SR013]. The evaluation procedure is rigid, in particular with respect to the environmental laws, water distribution regulations and administrative aspects. In consequence, many norms are rather outdated and limit the application of water-reuse schemes [SR005-SR011-SR012]. For example, despite the higher quality standards of treated secondary wastewater compared to other water sources, it is still prohibited to use treated secondary wastewater as a resource for drinking water [SR001]. Criteria are largely based on the origin of the water source (e.g., freshwater or recycled wastewater) instead of formulating quality standards for different use categories. Finally, it was found that learning between stakeholders (indicator 3.3) occurs on the technical level, and often with respect to narrowly defined topics. It is not very common that cross-stakeholder learning takes place on a strategic, administrative or financial level [SR005-SR012]. Many stakeholders are reluctant to share information due to the sector's competitiveness. Subsequently, cross-stakeholder learning is limited to a small alignment of stakeholders with similar interests [SR003-SR008-SR009-SR011].

Condition 4: Stakeholder Engagement Process

On the regional scale it is found that only a few and mostly conventional stakeholders are included in the decision-making process (indicator 4.1) which ultimately is bilateral and dominated by the national and regional government and ACA [SR002-SR004-SR005-SR007-SR010-SR011-SR012]. Most stakeholders, including academia for instance, are often only informed or consulted instead of engaged in decision making [SR005-SR012-SR013-SR015]. The decision-making process can be described as top-down with little opinion forthcoming from the local level. For example, local suggestions for more practical water quality norms for water-reuse schemes are not yet widely included in national guidelines. The water consumers or citizens have little active involvement or participation in the decision making process, which poses risks that their interests and core values can be harmed (indicator 4.2). However, citizen engagement is improving substantially (indicator 4.3) and new bottom-up initiatives as well as collaborations with grassroot organizations are appearing, such as ecological/green activist groups or basin associations (e.g., the Tordera River Basin Association and the Amics de l'Aigua civic organization).

Condition 5: Management Ambition

Sabadell aims to be one of the leading cities in water-reuse practices in Europe (indicator 5.1-[SR001-SR013] [28]). However, the city has to deal with fragmented, sometimes contradicting policies that affect water-reuse practices. In particular, different guidelines exist originating from different government levels resulting in insufficient management cohesion (indicator 5.3). The national and regional policy ambitions regarding water conservation are moderate and most interviewees indicated that the statutory compliance to these policies is suboptimal [SR003-SR010-SR015]. Water quality and distribution norms are still determined by unilateral decisions, and a lack of sectorial, geographic and administrative alignment can be seen. Urban space construction permits have been released in the past even though they go against environmental efforts and restrictions for potential flood damage prevention [SR008]. Low water fees for agricultural purposes contradict with the efforts to lower the agricultural sector's water consumption. Such discrepancies between national laws and the municipal administrative and geographical context can be observed and reflect a somewhat limited discourse embedding [indicator 5.2-SR007-SR009-SR014-SR016]. Ultimately, this leads to major inefficiencies and barriers in implementation of reuse schemes [condition 9-SR013].

Condition 6: Agents of Change

Agents of change are found to have a limited impact on the overall governance capacity. There is little room for local entrepreneurial agents in the region's water sector (indicator 6.1). Water is ultimately publically administrated and even when subcontracted by a private company such as CASSA, the services are provided by monopolistic clusters and thus the sector is difficult to enter for entrepreneurs [SR007-SR008-SR009-SR012]. There is some sector-specific entrepreneurial space, particularly with respect to the technology development [SR011-SR013-SR016]. However, in most cases entrepreneurs who enter the market would have to work with a larger, already established stakeholder [SR001]. Most expertise is accumulated in research centres [SR002] and many entrepreneurs export their goods and services abroad to areas such as Latin America [SR009]. Most collaborative agents (indicator 6.2) are active in close collaboration between a limited number of traditional stakeholders [SR003-SR014]. More recently, new and more inclusive collaborations are starting to take place. For example, public communication strategies with respect to household wastewater disposal have been established between CASSA, Sabadell City Council and ACA. This type of collaboration is often brief and established for very specific issues or events [SR001-SR016]. A frequent barrier for closer collaboration is the divergent interests of the involved stakeholders [SR016]. Finally, it is found that there does not exist a unifying long-term vision or strategy, which transcends different levels of decision making or enables continuity beyond the political mandate of 4 years [indicator 6.3-SR013-SR014]. The leading body, ACA, is only recently recovering from a chaotic organizational period and thus their role as a visionary body has yet to be realized. At present, there is no entity that assumes the responsibility for leading the country's or region's water sector towards the implementation of water-reuse schemes or other water-conserving strategies. Rather, there is a great diversity in local initiatives that aim to implement water-reuse practices.

Condition 7: Multi-Level Network Potential

It was found that stakeholders have only limited room to manoeuvre (indicator 7.1) due to inter-organizational difficulties and the strict regulations and procedural demands of the ACA. Particularly with respect to the use or distribution of water permits, implementation is difficult [SR007-SR011]. A distinction between the private and public entities can be observed. Public management has more rigid structures and procedures whereas private management typically has more room and independence to determine strategies and to experiment [SR001-SR010-SR013]. Private companies can use this internally to remain flexible and to improve continuously [SR009-SR014-SR016]. In Sabadell, room to manoeuvre is also ensured by a long-term contract between the water supplier CASSA and the city council. The division of responsibilities (indicator 7.2) in Sabadell is mostly clear but also somewhat inflexible [SR005-SR007-SR008-SR011-SR012-SR014-SR015]. Water management is primarily the responsibility of each individual city council, with ACA and the Catalan government merely inspecting that municipalities comply with existing legislation. This leads to divergent, contradictory and overlapping approaches between different levels of decision-making [SR009-SR010-SR013-SR016]. There are also gaps identifiable, in particular with respect to financial responsibilities. It is unclear which organizations will finance the necessary infrastructural refurbishments [SR014]. ACA's authority is strong [indicator 7.3-SR008-SR011-SR015-SR016]. However, some argue that ACA's procedural demands may hinder the progress of the water sector, and procedures sometimes resonate with political shifts [SR015-SR001-SR007] due to organizational and financial uncertainty [SR014]. Partly as a result of this, the region of Catalonia lacks a clear centralized visionary leadership to enhance water-reuse practices [SR010-SR011]. On the other hand, the role of ACA is primarily to act as an arbiter that balances and arbitrates between the interests of different stakeholders [SR008]. However, with respect to water-reuse practices, strict control of existing regulations impedes further progression. ACA is in the position to unify the many different municipal approaches to implementing water-reuse schemes. Given the many experiences and promising results it would appear to be the opportune moment to develop a coherent legislative framework for water-reuse practices in cities.

Condition 8: Financial Viability

Basic water services are accessible for everyone. These services are either affordable or there are funds that support the most marginalized communities (indicator 8.1). Affordability for climate change adaptation is low because the possibility that citizens can apply to reduce the current and future impacts of climate change-induced water stress are rather limited. Although non-potable water reuse is feasible, legislation limits its distribution and only those living in a few specific areas of the city that possess dual networks can make use of it [SR001]. The water consumer's willingness to pay (indicator 8.2) is however also restricted to basic services, and willingness to pay extra for an extension of the dual networks is relatively low, with citizens indicating that these extra investments should be financially supported by the government in consultation with the service provider CASSA [SR001-SR002-SR012-SR015]. The structure that can ensure financial continuity (indicator 8.3) of water-reuse practices (indicator 8.3) was observed to be very diverse among stakeholders and governance levels. At the national, Catalan, provincial and municipal level the continuity of funding for any climate change adaptation measures—including water-reuse practices—is largely susceptible to temporary, ad hoc, short-term policies [SR005-SR012-SR014]. At present, infrastructure refurbishment requires an estimated €3 million in Sabadell [SR013]. In order to build new infrastructure to promote the use of non-potable water, additional financial support will be required.

Condition 9: Implementing Capacity

Implementing capacity has been found to have both encouraging and discouraging factors. Some policy instruments (indicator 9.1), such as a progressive tax on water consumption or a connection tax for new water distributions, are being used effectively to incentivize low water consumption. Statutory compliance (indicator 9.2) is adequate, the legislation is generally respected and all stakeholders comply because noncompliance is an unnecessary risk [SR005-SR007]. There is good understanding of the dynamics of reservoirs and a lot of experience from past drought events that can be employed to mitigate future drought events in the region. Water-reuse practice is increasingly being recognized as a promising and necessary climate adaptation measure. In municipalities such as Sabadell, important progress has been made to implement water-reuse schemes. Sabadell has a long-term mitigation strategy and the city has demonstrated a capacity to respond efficiently to water stress in the past. However, given that it has rained in the past months and that water reservoirs are full, ensuring water supply for the region for the next 2 years, the issue of increased water stress has temporarily been put aside [SR005-SR014-SR004]. Hence, structural preparation (indicator 9.3) and infrastructural investments to mitigate water stress within a 30 or 50 year time frame appears to be an inevitable necessity. Some interviewees expressed their concern that the region is ill prepared for the impacts of climate change in the long run and current efforts seem to be insufficient [SR002-SR003-SR015]. Hence, the ability of Catalonia to prepare for future water stress within the next decades will be largely dependent on the willingness to learn from and enable local water-reuse practices.

5. Discussion

In order to reduce water stress by a wider application of wastewater-reuse practices, major changes are required in the way the water cycle is governed at the local, regional, national and international level. Spain is currently experiencing this transformation, with different multi-level governance barriers and challenges emerging. Governance capacity is essential to collaborate effectively and to overcome these barriers. Our results demonstrate that Sabadell has gained considerable experience and has demonstrated a willingness to extend the reuse of municipal wastewater. The study has identified barriers and opportunities, revealing common factors and trends that are likely to manifest themselves in other municipalities not only in Spain, but throughout the Mediterranean and other water-scarce regions.

5.1. Coherent Legislative Frame Required to Support Water-Reuse

Through our case study of Sabadell, we found that legislative restrictions and inconsistencies imposed by regional and national regulations impeded the city's water reuse practices. In Europe, the lack of common water-reuse criteria has been identified as an important factor restricting the expansion of water-reuse practices [31]. France, Cyprus, Greece, Italy, Portugal and Spain do have national policies on water reuse that focus mainly on agricultural irrigation. However, more focus on criteria for other applications are necessary to exploit the full potential of water reuse. Several regions in Spain already widely apply water-reuse [31]. Royal Decree 1620/2007 provides the legal clearance and the Wastewater-reuse National Plan gives orientation and proposes procedures and criteria for the following applications: (a) urban uses such as garden irrigation, street cleaning, and fire fighting, as well as household applications such as toilet flushing; (b) irrigation of agricultural crops and use in aquaculture; (c) industrial uses such as in cooling towers and evaporative condensers; (d) leisure; and (e) environmental applications including aquifer recharge, forest irrigation, and wetlands maintenance [32,33]. However, overly strict legislative water quality standards together with a demanding licensing process slow down Spanish municipal efforts to reuse treated wastewater [34]. A reoccurring issue is that norms discriminate water, based on its source rather than on its properties, which is unfounded given the advanced treatment methods that ensure adequate water quality. In order to realize a widespread application of water-reuse practices, it is necessary to develop reclaimed water quality criteria per use category, independently of the water's origin [31]. The legislative framework tends to be overly precautious, which does not reflect, nor support, the ambitions and efforts of local stakeholders aiming to implement this innovation. At present, municipal experiences with water reuse are not fully taken into account in the evaluation of regional and national policy. Moreover, periodical shifts in political parties further impede the development of a coherent policy that supports water-reuse schemes [SR013-SR014]. Hence, guidelines and learning experiences are divided between a heterogeneous range of stakeholders, indicating that improved coordination through coherent regulation could boost the mainstreaming of municipal reuse of treated wastewater.

5.2. Realising Public Engagement for Non-Potable Water-Reuse

The public's awareness and understanding of the safety and applicability of reclaimed water is essential to the success of any water-reuse program [15] and requires public engagement [35]. The vast majority of Spanish, European and Western societies have organized urban water systems, applying a centralized approach. Such systems allow for the large-scale collection and treatment of wastewater making it cost efficient due to their economy of scale [13]. On the other hand, alternative, decentralized systems typically give citizens the responsibility to collect, treat or dispose their water. Decentralized reuse systems are in most cases a more expensive option [13–36], in particular in urban areas where such investments in centralized water infrastructure have already been made [36]. However, there remains a persistent lack of public engagement in centralized water systems because the user, who is disconnected from water treatment, takes these services for granted, and is often not aware of the challenges and risks involved. The level of public acceptance is a key factor in the implementation of water-reuse schemes [15]. Numerous reasons why the public has a tendency to be reluctant in supporting general water-reuse, potable as well as non-potable, have been studied across the globe [14]. Common results are that potential consequences are unknown due to their (perceived) limited experience, implementation is seen as being irreversible, and citizens feel that they have little control over the process [35]. The source of reused water is viewed as being unnatural and even toxic, which leads to the irrational, well recognized psychological "Yuck factor" [36,37], "naturalness" and "contagion" phenomena [38]. Ultimately, public acceptance and support for water-reuse schemes is greatly a matter of the "interplay between trust, risk perception and emotional reaction" [38]. The more the citizens are informed and the more that they trust the authorities the lower risk they perceive, and are thus more likely to support the schemes [39].

Building public support requires the construction of holistic narratives that appeal to the affective reactions as well as to the cognitive ones, changing the citizen's perception of wastewater as "dirty waste" to "resource" [38]. It is therefore a task that requires multi-level collaboration, i.e., interdepartmental efforts, stakeholder engagement and citizen participation. It might be crucial to clearly distinguish between reuse for potable and non-potable purposes. Winning trust and earning support for water-reuse for potable purposes involves overcoming many hurdles. However, the potential to reduce domestic and industrial water consumption is greater when contemplating a situation that does not require drinking water quality standards, as is the case in Sabadell. Hence, the implementation of non-potable water reuse is often the first choice; for example, in Europe 97.7%

of water reuse is for non-potable purposes. In Sabadell the public attitude towards non-potable water-reuse was found to be quite positive. This confirms earlier research in Sabadell [40,41]. Due to previous experience regarding water stress, water conservation and the use of alternative sources for purposes such as flushing the toilet, washing the car or watering the garden, are already widely applied. It might explain the general positive public attitude towards water-reuse schemes at a municipal level. Hence, centralized water-reuse schemes may be more accepted by citizens if it is combined with household systems of rainwater harvesting, water conservation and grey water recycling.

5.3. Pricing Water

Despite a modest awareness of water safety and a general acceptance of water reuse practices, Sabadell's citizens and stakeholders were not willing to pay more in order to extend the duel distribution network in the city. Pricing water services is always a trade-off between the basic human right to affordable water and cost recovery. In fact, definitions of cost recovery can be misleading as cost can be deferred to the future or transferred to the environment, leading to environmental degradation. Hence, the question whether the price includes all operations, maintenance, and capital costs that can ensure freshwater availability in the long-term is decisive for the continuity of water services, in particular in water-stressed regions. Examples of external factors that can be excluded are water scarcity, social and financial issues, and environmental burdens such as wastewater or wastewater effluent disposal into the environment [15]. At the same time, affordability for everyone has to be ensured [13] which emphasize the need for steering mechanisms. Reclaimed water is often priced just below the drinking water price in order to make it more attractive to consumers but this also poses issues concerning the recovery of costs [38–41]. The expansion of Sabadell's water-reuse practices is hindered because the costs would result in a higher price of reclaimed water that could not compete with drinking water. Moreover, treated wastewater is still perceived as inferior by the consumer and thus few citizens are willing to pay the same or a higher price for it compared to water from primary sources. Major policy instruments such as a progressive tax on water consumption or the inclusion of the costs of reclaiming water in the consumer price for drinking water are used effectively in Sabadell to incentivize low water consumption and ensure affordability. At the same time, the existing water infrastructure requires substantial investments that may have consequences on the water price. Given the economic situation, it is a challenge to refurbish the existing water infrastructure, enhance water-reuse practices and maintain affordability of water services for everyone. In order to take up this challenge, co-financing from different stakeholders including citizens, local businesses, as well as funding from regional, national or European authorities might be required to combat water stress and continue water services in the long run.

5.4. Research Limitations and Key Priorities

The governance capacity analysis applied in Sabadell reveals a number of highly interconnected and interrelated governance processes. As a consequence, some of the framework's indicators interrelate as well. Some hypothetical "ideal" situations will not result only in very encouraging (++) indicator scores. For example, the role of entrepreneurial, collaborative, and visionary agents of change (indicators 6.1, 6.2 and 6.3) are context-dependent. Visionary agents may be more useful in times of crisis, whereas collaborative agents are more valuable in initiating new collaborations, and entrepreneurial agents operate best in flexible and open governance networks [21,22,24,27]. Hence, situations may occur in which entrepreneurial and collaborative agents of change are very encouraging (++) while, as a consequence, visionary agents are less prominently active and score lower. Another important interrelation is between indicators 5.1 ambitious and realistic goals and 9.2 statutory compliance, because it is easier to comply with goals that are not ambitious [24]. The governance capacity analysis applies a triangular approach of knowledge co-production consisting of three steps: (1) a desk study; (2) interviews and (3) feedback from interviewees. The detailed reporting and transparent research steps ensure reproducibility. However, because the number of interviewees is moderate and interviewees with different backgrounds sometimes contradict each other, the scoring justification for indicators 1.1 community knowledge, 1.2 local sense of urgency and 1.3 behavioural internalization, were limited due to the lack of available open source reports, documents and scientific literature. For these three indicators, a survey would provide more substantiated and accurate scores.

6. Conclusions

In line with the United Nations, we conclude that wastewater can be considered as a reliable and largely untapped resource that has much potential to alleviate water stress. In order to enable the wider application of wastewater-reuse practices, major changes are required in the way the water cycle is governed at the local, regional, national and international level. Spain is in the middle of a process of transformation wherein different multi-level governance barriers and challenges emerge. Through a case study in the city of Sabadell (Spain), this paper has aimed to identify the principal barriers, opportunities and transferable lessons that can enhance governance capacity in order to implement systems for the non-potable reuse of treated wastewater in cities. Overall, it was found that citizens in Sabadell do not fully understand the possible risks and impacts of water stress, but have a positive attitude towards the reuse of treated wastewater for non-potable purposes. In order to overcome the different operational barriers identified in the implementation of water-reuse practices, a coherent legislative framework is required that applies quality criteria per use category independently of the water's origin. Such a framework should be developed and implemented based on the local experiences of municipalities, such as Sabadell, through a regular evaluation process that transcends political cycles and overcomes the current fragmentation of interests, responsibilities and tasks between stakeholders. This requires the further development of governance capacity across multi-level governance layers. The results from Sabadell indicate that in particular with respect to knowledge provision, policy evaluation and learning, the development of a more coherent policy strategy for the wider application of water-reuse practices is required. These barriers and opportunities may provide learning lessons for other municipalities in Spain and throughout the Mediterranean and other water-scarce regions in the world.

Author Contributions: M.Š. conducted the interviews, extensively studied case-related documents and policy documents, and was first contributor and coordinator of the paper's writing process. S.K. is developer of the diagnostic governance capacity framework applied in this paper. He contributed to the interview preparation, data analysis and the writing of the paper. R.E. and J.V. facilitated the interview process and contributed to the data analysis and critically reviewed the draft papers. K.V.L. contributed to the data interpretation, and contributed to the conception and final review of the paper.

Acknowledgments: Our gratitude goes to those interviewed in Spain, whose participation in interviews was crucial to this investigation; Adriana Sanz, Albert Tauler, Albert Testart, Anja Berling, Dolors Vinyoles, Elisabeth Santacruz, Enric Coll, Enric Vazquez, Joan de Pablo, Jordi Agustí, Jordi Vinyoles, Lluís Sala, Oriol Ferrer, Roberto Ordás, Xavier Ludovid and Xavier Turro. The Governance Capacity Framework is part of City Blueprint Approach developed at KWR Watercycle Research Institute in the context of Watershare (http://www.watershare.eu). The City Blueprint Action Group is part of the European Innovation Partnership on Water of the European Commission (http://www.eip-water.eu/City_Blueprints) The European Commission is acknowledged for funding POWER in H2020-Water Under Grant Agreement No. 687809.

Conflicts of Interest: The authors declare no conflict of interest.

References

1. Mekonnen, M.M.; Hoekstra, A.Y. Four billion people facing severe water scarcity. *Sci. Adv.* **2016**, *2*, e1500323. [CrossRef] [PubMed]
2. Hoekstra, A.Y.; Mekonnen, M.M.; Chapagain, A.K.; Mathews, R.E.; Richter, B.D. Global monthly water scarcity: Blue water footprints versus blue water availability. *PLoS ONE* **2012**, *7*, e32688. [CrossRef] [PubMed]
3. United Nations Educational, Scientific and Cultural Organization (UNESCO). *World Water Assessment Programme (WWAP). The United Nations World Water Development Report 2015: Water for a Sustainable World*; UNESCO: Paris, France, 2015; p. 11. Available online: http://unesdoc.unesco.org/images/0023/002318/231823E.pdf (accessed on 9 February 2018).

4. United Nations Educational, Scientific and Cultural Organization (UNESCO). *The United Nations World Water Development Report 4: Managing Water under Uncertainty and Risk*; UNESCO: Paris, France, 2012; p. 265. Available online: http://unesdoc.unesco.org/images/0021/002156/215644e.pdf (accessed on 9 February 2018).

5. 2030 Water Resources Group. *Charting Our Water Future. Economic Frameworks to Inform Decision-Making*; 2030 Water Resources Group: West Perth, NY, USA, 2009; p. 5. Available online: http://www.2030wrg.org/wp-content/uploads/2014/07/Charting-Our-Water-Future-Final.pdf (accessed on 9 February 2018).

6. Koop, S.H.; Van Leeuwen, C.J. The challenges of water, waste and climate change in cities. *Environ. Dev. Sustain.* **2017**, *19*, 385–418. [CrossRef]

7. United Nations Educational, Scientific and Cultural Organization (UNESCO). *The United Nations World Water Development Report 2017. Wastewater the Untapped Resource*; UNESCO: Paris, France, 2017; Available online: http://unesdoc.unesco.org/images/0024/002471/247153e.pdf (accessed on 9 February 2018).

8. United Nations Educational, Scientific and Cultural Organization (UNESCO). *Migration and Its Interdependencies with Water Scarcity, Gender and Youth Employment*; UNESCO: Paris, France, 2017; Available online: http://unesdoc.unesco.org/images/0025/002589/258968E.pdf (accessed on 9 February 2018).

9. Satoa, T.; Qadir, M.; Yamamotoe, S.; Endoe, T.; Zahoor, A. Global, regional, and country level need for data on wastewater generation, treatment, and use. *Agric. Water Manag.* **2013**, *130*, 1–13. [CrossRef]

10. Ligtvoet, W.; Hilderink, H.; Bouwman, A.; Puijenbroek, P.; Lucas, P.; Witmer, M. *Towards a World of Cities in 2050. An Outlook on Water-Related Challenges*; Background Report to the UN-Habitat Global Report; Environmental Assessment Agency (PBL): The Hague, The Netherlands, 2014.

11. EU Water Directors. *Guidelines on Integrating Water-Reuse into Water Planning and Management in the Context of the WFD*; Common Implementation Strategy for the Water Framework Directive and the Foods Directive; European Commission: Brussels, Belgium, 2016.

12. International Water Management Institute (IWMI). *CGIAR Research Program on Water, Land and Ecosystems. Global Experiences in Water-Reuse*; Resource Recovery & Reuse Series 4; IWMI: Colombo, Sri Lanka, 2014; p. 5. Available online: http://www.indiaenvironmentportal.org.in/files/file/resource_recovery_and_reuse_0.pdf (accessed on 9 February 2018).

13. Domenech, L.; March, H.; Saurí, D. Degrowth initiatives in the urban water sector? A social multi-criteria evaluation of non-conventional water alternatives in Metropolitan Barcelona. *J. Clean. Prod.* **2013**, *38*, 44–55. [CrossRef]

14. Smith, H.M.; Brouwer, S.; Jeffrey, P.; Frijns, J. Public Responses to water-reuse—Understanding the Evidence. *J. Environ. Manag.* **2018**, *207*, 43–50. [CrossRef] [PubMed]

15. Sanz, L.A.; Gawlik, B.M. *Water-Reuse in Europe—Relevant Guidelines, Needs for and Barriers to Innovation. JRC Science and Policy Reports*; European Commission: Brussels, Belgium, 2014; Available online: file:///C:/Users/Marketa/Downloads/lb-na-26947-en-n.pdf (accessed on 15 March 2018).

16. Van Rensburg, P. Overcoming global water reuse barriers: The Windhoek experience. *Int. J. Water Resour. Dev.* **2015**, *4*, 622–636. [CrossRef]

17. Organization for Economic Cooperation and Development. *OECD Principles on Water Governance*; OECD: Paris, France, 2015.

18. Biesbroek, G.R.; Klostermann, J.E.M.; Termeer, C.J.A.M.; Kabat, P. On the nature of barriers to climate change adaptation. *Reg. Environ. Chang.* **2013**, *13*, 1119–1129. [CrossRef]

19. Plummer, R.; Crona, B.; Armitage, D.R.; Olsson, P.; Tengo, M.; Yudina, O. Adaptive comanagement: A systematic review and analysis. *Ecol. Soc.* **2012**, *17*, 11. [CrossRef]

20. Eisenack, K.; Moser, S.C.; Hoffmann, E.; Klein, R.J.T.; Oberlack, C.; Pechan, A.; Rotter, M.; Termeer, C.J.A.M. Explaining and overcoming barriers to climate change adaptation. *Nat. Clim. Chang.* **2014**, *4*, 867–872. [CrossRef]

21. Pahl-Wostl, C. A conceptual framework for analysing adaptive capacity and multi-level learning processes in resource governance regimes. *Glob. Environ. Chang.* **2009**, *19*, 354–365. [CrossRef]

22. Van Rijswick, M.; Edelenbos, J.; Hellegers, P.; Kok, M.; Kuks, S. Ten building blocks for sustainable water governance: An integrated method to assess the governance of water. *Water Int.* **2014**, *39*, 725–742. [CrossRef]

23. Ostrom, E. A general framework for analyzing sustainability of social-ecological systems. *Science* **2009**, *24*, 419–422. [CrossRef] [PubMed]

24. Koop, S.H.A.; Koetsier, L.; Van Doornhof, A.; Van Leeuwen, C.J.; Brouwer, S.; Dieperink, C.; Driessen, P.J. Assessing the governance capacity of cities to address challenges of water, waste, and climate change. *Water Resour. Manag.* **2017**, *31*, 3427–3443. [CrossRef]

25. European Commission. *European Innovation Partnership on Water. Indicators of the Governance Capacity Framework*; European Commission: Brussels, Belgium, 2018; Available online: http://www.eip-water.eu/City_Blueprints (accessed on 22 May 2018).

26. Department for International Development (DFID). *Tools for Development. A Handbook for Those Engaged in Development Activities*; DFID: London, UK, 2003.

27. Lange, P.; Driessen, P.P.J.; Sauer, A.; Bornemann, B.; Burger, P. Governing towards sustainability-conceptualizing modes of governance. *J. Environ. Policy Plan.* **2013**, *15*, 403–425. [CrossRef]

28. IEEE. IEEE Affiliated Smart City Profile—Sabadell, Spain. 2017. Available online: http://smartcities.ieee.org/affiliated-cities/sabadell-spain.html (accessed on 9 February 2018).

29. Aigues Sabadell. *Memoria de Desenvolupament Sostenible*; Aigues Sabadell: Barcelona, Spain, 2016.

30. Šteflová, M. *Barriers, Opportunities and Transferable Lessons that Can Be Identified from Sabadell's Wastewater Recycling Network in Efforts of Alleviating Water Stress in Spain*; Utrecht University Repository: Utrecht, The Netherlands, 2017.

31. Paranychianakis, N.V.; Salgot, M.; Snyder, S.A.; Angelakis, A.N. Water-reuse in EU States: Necessity for Uniform Criteria to Mitigate Human and Environmental Risks. *Crit. Rev. Environ. Sci. Technol.* **2015**, *45*, 1409–1468. [CrossRef]

32. Government of Spain. *Real Dicrete No. 1620/2007 of the 7th of December, with Which the Legislation of Reutilizing Treated Waters Is Established*; BOE num. 294; Government of Spain: Madrid, Spain, 2007.

33. Water-Reuse National Plan. *Plan Nacional de Reutilizaci on de Aguas*; Ministry of the Environment, and Rural and Marine Environments: Madrid, Spain, 2010.

34. Paranychianakis, N.V.; Kotselidou, O.; Vardakou, E.; Angelakis, A.N. *Greek Regulations on Wastewater Reclamation and Reuse*; Hellenic Union of Municipal Enterprises for Water Supply and Sewage: Larissa, Greece, 2009.

35. Hartley, T.W. Public perception and participation in water-reuse. *Desalination* **2005**, *187*, 115–126. [CrossRef]

36. Bichai, F.; Grindle, A.K.; Murthy, S.L. Addressing barriers in the water recycling innovation system to reach water security in arid countries. *J. Clean. Prod.* **2018**, *171*, S97–S109. [CrossRef]

37. Ching, L. A lived-experience investigation of narratives: Recycled drinking water. *Int. J. Water Resour. Dev.* **2016**, *32*, 637–649. [CrossRef]

38. Miller, G. Integrated concepts in water-reuse: Managing global water needs. *Desalination* **2005**, *187*, 65–75. [CrossRef]

39. Beierle, T. The quality of stakeholder-based decisions. *Risk Anal.* **2002**, *22*, 739–749. [CrossRef] [PubMed]

40. Frijns, J.A.G.; Smith, H.M.; Brouwer, S.; Garnett, K.; Elelman, R.; Jeffrey, P. How governance regimes shape the implementation of water-reuse schemes. *Water* **2016**, *8*, 605. [CrossRef]

41. Jimenez, B.; Asano, T. *Water-Reuse: An International Survey of Current Practice, Issues and Needs*; IWA Publishing: London, UK, 2008.

water

MDPI

Article

Assessing Sustainability of Wastewater Management Systems in a Multi-Scalar, Transdisciplinary Manner in Latin America

Lucía Benavides [1,*], Tamara Avellán [1,*], Serena Caucci [1], Angela Hahn [1], Sabrina Kirschke [1] and Andrea Müller [1,2]

[1] Institute for Integrated Management of Material Fluxes and of Resources (UNU-FLORES), United Nations University, 01067 Dresden, Germany; caucci@unu.edu (S.C.); hahn@unu.edu (A.H.); kirschke@unu.edu (S.K.); mueller@unu.edu (A.M.)

[2] Chair of Environmental Development and Risk Management, Technische Universität Dresden, 01217 Dresden, Germany

[*] Correspondence: benavides@unu.edu (L.B.); avellan@unu.edu (T.A.); Tel.: +49-157-323-88794 (T.A.)

Received: 30 December 2018; Accepted: 24 January 2019; Published: 31 January 2019

Abstract: Wastewater management in Latin America faces great challenges to reach a sustainable state. Although enough infrastructure has been built to treat around 40% of wastewater, only between 15–20% is effectively treated, and abandoned or defective infrastructure is a common sight. Data about current conditions at specific sites is quite fragmented, when existing. This leads to challenges in management, decision making and planning for sustainable options. We argue that a main obstacle is the lack of a regionally relevant sustainability assessment framework that allows for a holistic understanding of wastewater management as a nexus problem. We therefore developed a comprehensive framework to (1) understand current conditions (2) involve stakeholders and (3) point to pathways to improve wastewater management in the Americas. Building on literature review and stakeholder involvement, we constructed a multi-scalar extended dataset framework that is adaptable to different study sites using specific criteria. Sustainability was assessed through a "distance-to-target" approach. Social and economic variables were the lowest ranking in both cases, with technical variables generally performing better. Although some dimensions of sustainability are performing acceptably, others, such as social and economic, are general low to very low performing. This means, when looked at in an integrated manner, neither of the wastewater management systems analysed can be considered sustainable. Here we present the approach itself, the results of its application in two pilot sites in Latin America, and our recommendation to shift waste water management into sustainability.

Keywords: assessment framework; sustainability assessment; baseline assessment; co-design; stakeholder involvement; wastewater management

---

1. Introduction

Wastewater and Its Management in Latin America

Wastewater management systems (WWMS) serve multiple functions within their cities. They channel and treat the wastewater produced by their customers, reduce the pollution load to the environment and the catchment they are embedded in and thus safeguard it and its inhabitants from detrimental health effects. Usually citizens only notice them when they do not provide those services. Wastewater treatment systems can, in addition, provide resources, such as bioenergy from biogas produced during the decomposition of organic matter, irrigation water or stabilized sludge to be used as fertilizer. Understanding the risks and benefits that a wastewater treatment system can offer to its community is not limited to the technical understanding of its components. It demands understanding the multiple dimensions of sustainability, understood as 'the maintenance of economic well-being, protection of the environment and prudent use of natural resources, and equitable social progress which recognizes the just needs of all individuals, communities, and the environment' [1].

In Latin America, 80% of the population lives in urban areas, with small cities (up to half a million inhabitants) growing the most rapidly [2]. Exact data on sanitation and treatment coverage are not readily available [3], but it is known that wastewater treatment is in general poor, with infrastructure to treat around 40% of municipal wastewater having been built, but less than 20% of that wastewater effectively being treated [4,5]. Commonly built solutions have been centralised wastewater treatment plants (WWTP), which may satisfy the demand of highly populated areas, but do not necessarily comply with the new expectations about water recycling and reuse, and of nutrient recovery [6], as requested by Sustainable Development Goal (SDG) 6.3 or the New Urban Agenda adopted at the latest Habitat III Conference [7].

Tackling the deficit of safely treated wastewater is an urgent matter: Clean water and access to safe sanitation for all is one of the targets decided by the global community within the Sustainable Development Goals (SDG 6.2) [8]. In Latin America large cities concentrate the largest shares of population, but when it comes to issues in the water management services, rural areas and small- and medium-sized cities are the most affected zones, especially regarding sanitation and wastewater treatment [4]. Small- and medium-sized cities are defined according to population, varying in proportion to each country's size, with a maximum of 1 million inhabitants for Latin American cities [3]. These types of cities show high urbanization rates, being the fastest growing urban areas [9]. This means that the established urban management systems have to consider the growth projections and adapt to keep up with the growing water demand and wastewater generation. Therefore, sustainable options for wastewater management for small- to medium-sized cities are urgently needed.

The SludgeTec project, a multinational partnership (the United Nations University's Institute for Integrated Management of Material Fluxes and of Resources—UNU FLORES, the Universidad de San Carlos de Guatemala—USAC, the Mexican Trust Fideicomiso de Infraestructura Ambiental de los Valles de Hidalgo in Tepeji, Mexico—FIAVHI, and the Technische Universität Dresden-TUD, aimed for international experts and local stakeholders to co-design a sustainable wastewater treatment and management options for two pilot areas in the Americas: Los Cebollales WWTP in Panajachel, Lake Atitlan, Guatemala and Tlaxinacalapan WWTP in Tepeji, State of Hidalgo, Mexico. Research was carried out between November 2017 and February 2019 by a multi-disciplinary and international team of researchers and practitioners.

To achieve the project's objective (co-designing sustainable options), it was first necessary to accurately assess current sustainability, that is, to describe baseline conditions. Establishing baselines is crucial for scientifically sound sustainability interventions [10], and is a key practice in many environmental fields, as it allows to evaluate the change in time of given parameters and therefore to track project success, for example. Without a baseline, it is impossible to carry out "before and after" comparisons [11]. Furthermore, a baseline assessment can be very useful in informing and engaging stakeholders [9], and a powerful way to gather and centralize otherwise dispersed data, assess data availability for a given topic, and eventually, socialize knowledge. This is particularly relevant in a region where data scarcity is known to be an issue.

The importance of baseline setting being clear, we were confronted with the non-existence of a comprehensive guideline to describe baseline and assess the sustainability of WWMS. Guidelines exist on the broad and very general steps to be followed in establishing a baseline [12], and on the data items to be considered in the assessment of specific components of a WWMS, such as finance, technical issues, etc. [13,14]. There has also been some research to systematise the indicators and data items needed for technology options evaluation [15–17]. However, the guidelines analysed during our literature review focus mostly on single dimensions of sustainability (environmental, technical, social), and do not take into consideration broader scales of analysis beyond the WWTP itself (to include for example the impacts of the WWTP's function on the watershed or the subcatchment). We posit that a sustainability assessment must be multi-scalar (considering several territorial scales or spatial boundaries in one same study) and multi-dimensional (considering the different dimensions of sustainability).

We therefore developed a method to describe baseline conditions of WWMS and determine the degree of sustainability by (1) constructing a comprehensive and adaptable dataset framework and (2) applying a "distance-to-target" approach (further described in the methods section).

The method is underpinned by an emphasis on participation and transdisciplinarity. Scientists in the field of Integrated Water Resources Management highlight that participation can have positive effects on finding integrated solutions, e.g., by gathering and exchanging knowledge between vital stakeholders [18,19]. In terms of specific WASH-related problems, participation can help identify acceptable solutions on the ground. Based on this knowledge, practitioners and especially international donor organisations, apply participatory approaches in various contexts [20,21].

A research approach in which scientific and non-scientific actors collaborate in a participatory manner with the aim of creating scientific knowledge meant to address practical problems is here understood as transdisciplinary research (e.g., Reference [22]). 'Transdisciplinary' generally refers to an intensive inclusion of practitioners in the research process. To conceptualize transdisciplinary research, research provides a set of design criteria that are likely to have an impact on addressing complex problems in practice. These design criteria refer to (i) the type of actors involved, (ii) the stage of the research process where these stakeholders are involved, (iii) the degree of their involved, and (iv) the respective methodology [23]. Hence, various actors have been involved at different stages of the research process, from the design of research projects, via the implementation of the research projects, up to the evaluation of research results. In doing so, research questions, methods, and results are possibly better adapted to local needs, accepted, and thus also implemented [22,24]. Transferred to the field of wastewater management, the involvement of different scientific disciplines and practitioners from different realms may enable an ecologically, economically, environmentally and socially sustainable treatment of wastewater.

Participation is however no panacea for successful solutions. To achieve the potential benefits of participation, the thoughtful design of participatory processes is essential, including the right mix of actors (e.g., households, farmers, public authorities), degrees of participation (e.g., information sharing or co-decision-making), at the right scale (e.g., local or basin scale) [22,25].

In brief, in order to codesign sustainable options for the WWMS at the pilot sites, we built a method to first assess baseline sustainability, considering different territorial scales and the environmental, technical, economic and social dimensions. To broaden the possibility of accurate understanding of the issue and successful outcomes of the project, we worked in a transdisciplinary manner, i.e., in a diverse scientific team which closely worked with stakeholders and local partners, in every stage of research.

2. Materials and Methods

The method consists of four 'building blocks': (1) A thorough understanding of baseline conditions, which are then assessed under three different but converging perspectives: (2) Sustainability Assessment (SA), (3) Stakeholder Analysis and (4) Wickedness Analysis (WA). Blocks 1 and 2 are consecutive, i.e., number one is needed to perform number two. Blocks 3 and 4 are carried out separately. The assessment is made more thorough and comprehensive by bringing in the specific knowledge of each building block. This facilitates the understanding of bottlenecks and pathways towards sustainability, and as a final outcome, makes it possible to envision and evaluate solution options (Figure 1).

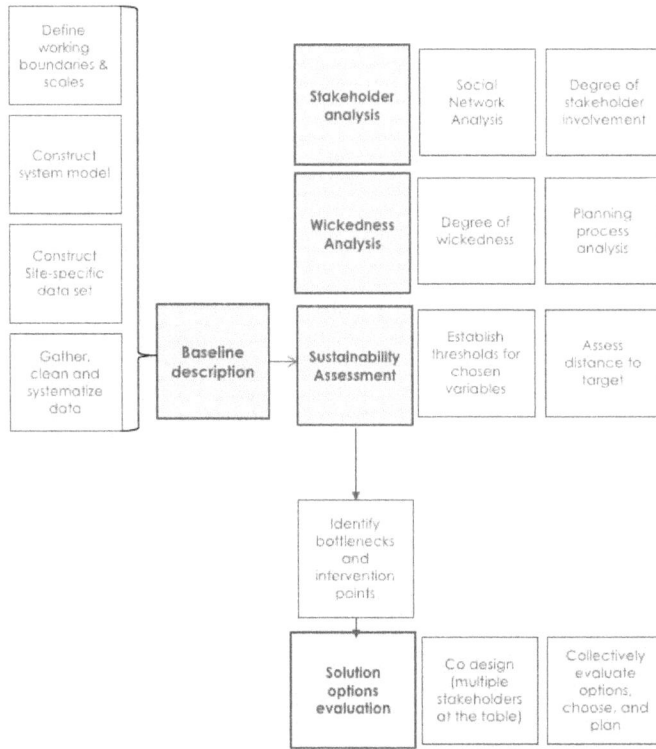

Figure 1. The general method used in this research project. Highlighted blocks are the four building blocks in our method. This paper deals in detail with two blocks: Baseline Description and Sustainability Assessment.

This paper describes the first two building blocks in detail, while the remaining two are the object of future publications.

2.1. Pilot Sites

Pilot sites (Figure 2) were chosen by local project partners based on their knowledge of the reality on the ground.

2.1.1. Panajachel Site description

At Panajachel, Guatemala, the pilot site is the Cebollales WWTP, an extended aeration, activated sludge plant built in 2013. The plant is operated by the municipality, with its financing sources being 100% public. The design flow is 37 liters per second (lps), and the current average flow is ~25 lps. It discharges into the San Francisco River, which, 200 m further downstream, feeds the Atitlan Lake. In the lake's endorheic basin, 55% of households are connected to a sewage system, while the remaining 45% use latrines, septic tanks, or soak latrines. 45,500 m^3 of wastewater is generated every day in the basin, and only approximately 20% receives treatment. Moreover, in the existing WWTPs, poor removal of pathogens and nutrients is a crucial challenge. These WWTPs face, among others, operation and maintenance problems.

2.1.2. Tepeji Site Description

At Tepeji, Mexico, the pilot site is the Tlaxinacalapan WWTP (built in 2017, started operations in January 2018). The WWTP has two treatment steps: a train of plastic anaerobic digestors built on site, followed by constructed wetlands. The design flow is 1.5 lps, and the current average flow is ~0.4 lps. It discharges into a tank from where water is taken to irrigate a football field and agricultural plots.

Figure 2. Location of the pilot sites.

2.2. Dataset Framework

2.2.1. Preliminary Step. System Model: Boundaries and Scales of Analysis

Wastewater management is a wicked problem: a complex network of components, often interlinked in non-linear relationship and expanding across different territories. In addressing sustainability problems, systems approaches have been widely recognized to enable researchers to describe and understand reality more accurately, shedding light on a phenomenon's structure and function [26–28], helping reveal otherwise "hidden" flows [29] and promoting the integrative thinking and interdisciplinary knowledge synthesis needed for sustainability [27,30]. System models are a key tool of systems approaches [27] and are widely used in cybernetics, physics, ecology and other fields where it is necessary to visually represent the complexity of real life networks and processes, in order to grasp the performance and behaviour patterns of systems. Our approach builds on systems thinking by using a system model as a fundamental research tool.

In building a system model, it is important to remember that, although WWMS are bound to human settlements, the sourcing of their inputs and the effluent and other outputs may have consequences well beyond their immediate geographical setting. Therefore, defining relevant scales of analysis and tracing analytical boundaries of the system is crucial. The choice of scales can determine the accuracy of diagnosis, and the effectiveness of projects [10,31]. Spatial resolution determines the visibility of objects and relations. If a model's boundaries are too small, important factors influencing the model may be missed, whereas if they are too large, detail on specific processes may be lost. Avellán et al. [32] postulate that 'the boundaries of the [Water-Soil-Waste Nexus] systems need to be (a) wide enough (to avoid microanalyses of plot levels as in some cases of INRM [Integrated Natural Resource Management]), (b) clear (to avoid confusion as in the WEF [Water-Energy-Food] Nexus), and (c) flexible enough to accommodate varying needs (to avoid geographic constrictions as is the case of the basin discussions in IWRM [Integrated Water Resources Management])'. By mixing in and contrasting different perspectives, a multi-scalar approach provides for more comprehensive analyses, which can lead to reduce biases caused by the use of a single "viewing-point" [10,33].

Different types of boundaries were identified: administrative (municipality, department, state, etc.), biophysical (catchments, geological, soil, etc.), and technical (treatment system, canal network, etc.). The relevance of each of these different spatial definitions was evaluated (Figure 3a), and four working scales were decided upon: 01 WWTP, 02 Municipality, 03 Subcatchment, 04 Watershed (Figure 3b). We argue that these scales together exhibit the needed specificity of the actual problem of wastewater treatment on the one hand but also enough scope to determine the impact that the system has on its surroundings.

A system model for the WWMS was drafted for each study site, using these chosen boundaries. System components (stocks) are represented in boxes and relations between them (flows) with lines (Figure 3c). The first versions of the system model were refined with participation from stakeholders during an assessment workshop held in Panajachel, Guatemala, in March 2018 [34]. Figure 3c shows the final version of the system model for the Panajachel site, resulting from the participative work at the workshop.

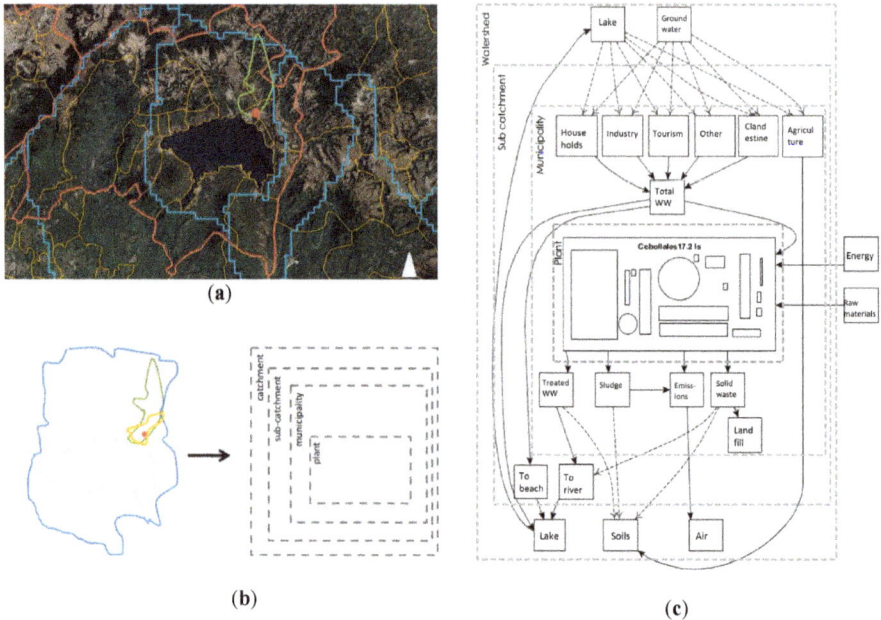

Figure 3. (**a**) A first boundary explorations map for the Panajachel case, showing different scales of analysis that were initially found to be of interest: the plant scale (red dot), the subcatchment (light green), the municipality (yellow), the watershed (blue) and the province (orange). The large water body in the center is the Atitlan Lake. (**b**) The abstraction of "real-world" boundaries into boundaries for the modeling process. (**c**) The system model for the Panajachel case, showing the systems components in the scale they (mostly) operate in.

2.2.2. Constructing the Dataset Framework

2.2.2.1. Extended Dataset Framework

We created the framework for a dataset that allows for a deep and holistic understanding of baseline conditions and sustainability performance, across scales and across the dimensions—environmental, and social, technical, economic—of the nexus problem of wastewater management, specifically in Latin America. To do so, we iterated between a top-down method (literature reviews) and a bottom-up method (working directly on the pilot sites, e.g., analysing the system model, asking stakeholders what sort of data is relevant to them) (Table 1). The result is an extended dataset framework (EF), which is described in more detail in the results section.

Table 1. Steps followed in the construction of the extended dataset framework.

	Bottom Up	Top Down
1	System model analysis	Research literature review
2	Stakeholder input (assessment workshop) on locally relevant data items and indicators	Policies and regulations review
3	-	Technical guidelines review

2.2.2.2. Site-Specific Dataset Framework

The EF was edited into a smaller, site-specific dataset framework for each site. This was necessary in order to respond to local data needs as expressed by stakeholders, as not all data items on the set were relevant for the specific sites. Additionally, to respond to the research priorities as established by the research team after assessing data availability and incorporating stakeholder input.

To edit the EF (with 492 variables) into the site-specific dataset frameworks (with 195 and 218 variables), we classified and prioritised each item on the EF according to the criteria in Table 2. Note that these criteria were chosen for these two project-specific pilot sites, but they could easily be applicable generically for other WWMS.

Table 2. Criteria used to prioritize the data items in the extended dataset framework and to create a site-specific dataset framework.

	Criteria	Priority
1a	Stakeholders chose the item during Assessment Workshop PLUS (+)	P1
1b	Literature on wastewater management mentions it	
2	Locally applicable regulation calls for the parameter	P2
3	Thresholds to compare current value against are available	P3

Priority 1 (P1) was given to an item if two conditions are met: (a) that stakeholders had chosen it during the Assessment Workshop held in March 2018 in Panajachel, and (b) that the item had been found in wastewater management guidelines or other relevant literature during our literature review. Priority 2 (P2) was given when the data item is included in relevant local regulation, e.g., monitoring standards. Priority 3 (P3) was given when a threshold to compare current values to could be identified. Thresholds were found looking in:

a.   Local legislation (region, state, basin).
b.   National legislation.
c.   Legislation valid for the other case study of this project (in this case Mexico or Guatemala).
d.   International organisations (not legally binding but accepted as guidelines or recommendations).

In some cases, a data item is a "yes or no" question, and a threshold can be established with relative ease; for example, the existence of an operation manual for the plant, for which the threshold is "yes", since that would be the desirable situation.

After the data gathering phase (see Section 2.4), these dataset frameworks were "filled in" with data, allowing to understand baseline conditions and perform a sustainability assessment.

2.3. Data Gathering

2.3.1. Identifying Data Holders

Possible data sources were identified through (1) an Assessment Workshop and (2) deskwork. The Assessment Workshop took place in March 2018 in Panajachel, Guatemala. Stakeholders from the Mexican cases were also present. More than 80 local stakeholders were invited, of which a total of 39 participated. The represented stakeholder groups were coming from Academia (43%),

Federal officials (21%), Non-Governmental Oirganisations (NGOs) (20%), Municipal officials (8%), and Private enterprise (8%). Through the participatory activities crucial input needed to refine both the technical and the social assessment components of the project framework was obtained. A thorough comprehension of the current problem structure, made possible by including the views of key stakeholders in a very interactive and participatory manner [34]. Participants drafted lists with institutions and experts who they thought could have the information needed (or access to it) for each data item. The data holder lists made by the participants were screened, refined and complemented through desk work. The final list of data holders consulted or interviewed can be seen in Appendix C.

2.3.2. Data Collection

Part of the data needed was collected on the field. Fieldwork was carried out for two weeks at each site in August 2018, and included meetings with experts, practitioners and local authorities identified as data holders, as well as sampling and laboratory analysis. During the meetings, the interviewer (L. Benavides) went through the dataset with the stakeholder, who provided the answers he or she had available. Data holders were always asked to provide supporting documentation, but this was rarely available. In some cases, stakeholders did not have information at hand, but committed to sending it via email after the meeting.

For water quality parameters, sampling and laboratory analysis were carried out (Appendix D). Sampling and analysis were done in accordance to the norms in each country, in collaboration with certified local laboratories. In both cases, a composite sample of both the plant's inflow and outflow was taken during a 24-h period. At the Panajachel site, sludge was also sampled. The sludge available had been stabilized on a covered drying yard for 28 days and was then piled up outdoors (i.e., under sun and rain) for at least two months prior to our visit. In Tepeji it was not possible to sample sludge, as according to the managers, the plant had not produced any in the 8 months of operation.

Further data was obtained from the revision of literature and documents produced by local and national authorities, which were made available to the research team during field work.

2.4. Sustainability Assessment

Sustainability Assessment (SA) processes aim at guiding decision-making towards sustainability [35] using different evaluative techniques [36] and definitions. 'Sustainability Assessment' can be considered a broad name tag for a series of methods and approaches: e.g., Sustainability Appraisal, Integrated Assessment, Integrated Sustainability Assessment, Sustainability Impact Assessment, Triple Bottom-Line Assessment, 3-E Integrated Assessment and Extended Integrated Assessment [35–37]. Methodologies found for SA include multicriteria approaches, systems analysis, life cycle analysis, economic analysis (cost-benefit analysis, life-cycle costing, etc.), weighting methods (exergy analysis, entropic weighing method), distance-to-target approaches, among others [15,38].

To determine the level of sustainability, we used a "distance-to-target" approach, comparing the current value of a variable with the threshold previously identified (see Section 2.2.2.2). The availability of a threshold finally defines whether a data item could be used in the sustainability assessment or not. Even though data for an item is available, if there is no appropriate threshold to compare it with, it is impossible to profit from this already existing data. Appendix E lists the variables for which thresholds could be identified and the thresholds values used to evaluate each variable of the site-specific dataset framework.

The "distance-to-target" was evaluated by adopting the "traffic light" method [39], where a variable is coded with green if it meets the threshold (good performance), with yellow when its performance does not meet ideal standards but is not far away from doing so, and red when it is performing sub-optimally. Table 3 discloses the quantitative criteria for each colour. Each variable was evaluated following these criteria. The result is a colour-coding of the data set (Appendix F).

Table 3. Colour ranking in Sustainability Assessment.

Data Type	Criteria	Ranking		
		Red	Yellow	Green
Real number	10% tolerance	MV > TH × 1.1	TH < MV ≤ TH × 1.1	MV ≤ TH
Percentage	Range divided into 3 equal parts (33% each)	MV < 33% or 67% ≤ MV	33% ≤ MV < 67%	67% ≤ MV or MV < 33%
Absolute values (e.g., yes/no questions)	No yellow range, unless mentioned otherwise	YES/NO Present/Absent Outside pH range	-	YES/NO Present/Absent Within pH range
Social variables (dataset IIb)	Scale 1 to 4	1 ≤ MV < 2	2 ≤ MV < 3	3 ≤ MV ≤ 4

MV: measured value; TH: threshold (numeric thresholds where normally defining a maximum, not a minimum).

Once the colour ranking was calculated for each variable, a colour ranking was also calculated for each of the three dimensions into which the variables are grouped in the sets: technical-environmental, economic, and social. To do this, we also followed the method described in Bertanza et al. (2016) [39], where a numeric value is assigned to each colour:

1. Green = 1
2. Yellow = 0
3. Red = −1
4. The colour-values are added, and a simple average in each category is calculated.
5. The results are later presented again using the "traffic light" colour-coding for the performance of each dimension of sustainability, as follows: (see results section)

a. Green: >0.33
b. Yellow: between −0.33 and 0.33
c. Red: ≤−0.33.

Although as stated in the introduction we believe a multi-scalar approach is necessary for a wide-enough perspective and an accurate understanding of a WWMS, due to the limited time scope of the SludgeTec project and the prolonged waiting periods to obtain data from data holders, it was only possible to perform a sustainability assessment on the first scale (WWTP, grey shaded areas on, and to include a multi-scalar social assessment of participation and social acceptance in the region where the WWTP operates (dataset IIb on Tables 4 and 5).

3. Results

3.1. Dataset Framework

3.1.1. Extended Dataset Framework

The iterative collection process of data items to describe the multi-scalar WWMS resulted in a large dataset framework with 492 data items (for an overview see Table 4, for the full content see SM 1). This comprehensive or "extended" dataset framework contains data items useful for the transdisciplinary study of WWMS (environmental and technical, economic, and social factors). It is organised into three datasets, namely: Dataset 0 which describes generic context data, Dataset I containing technical and environmental data, and Dataset II, containing socio-economic data. All datasets contain information across the four different spatial scales identified in Section 2.3.1. (WWTP, municipality, subcatchment, watershed) (see Table 4).

Table 4. Extended dataset framework, overview of subsets and number of data items in each.

Subset	Description		Scales		Number of Data Items
Dataset 0 Context indicators	Understanding of context: geographical location and characteristics, poverty and employment indicators, etc.	50 data items for 4 scales	01 02 03 04	WWTP Municipal Subcatchment Watershed	7 18 13 12
Dataset I Technical-Environmental	Technical and environmental variables (e.g., population served, chemical parameters of water bodies and of effluents, WWTP management)	380 data items across 4 scales	01 02 03 04	WWTP Municipal Subcatchment Watershed	211 31 70 68
Dataset II Socio-Economical	Economic, financial, budget variables. Dataset IIb useful to understand the social acceptance of the system	IIa. 52 data items for 4 scales	01 02 03 04	WWTP Municipal Subcatchment Watershed	16 17 7 12
		IIb. 10 data items, across scales		Social space (cross-scale)	10
Total data items					492

3.1.2. Site-Specific Dataset Framework

The EF proved too extensive to be used for the assessment of the sites, as time was a limiting factor, and also because not all variables on the set were necessarily a priority or the data for needed for all was not available at the different sites. Therefore, from the 492 data items in the EF, a site-specific dataset framework was created for the Panajachel pilot site with 218 data items, and for the Tepeji pilot site with 195 data items (Table 5). The full site-specific dataset frameworks can be found in Appendix A for Panajachel and in Appendix B for Tepeji.

Table 5. Site-specific dataset frameworks for both pilot sites, after prioritizing the EF.

Tepeji Dataset Framework			Panajachel Dataset Framework		
Dataset	**Scale**	**Number of Items**	**Dataset**	**Scale**	**Number of Items**
Dataset 0 Context	01	3	Dataset 0 Context	01	1
	02	3		02	0
	03	4		03	0
	04	5		04	0
	Total	15		Total	1
Dataset I Technical Environmental	01	107	Dataset I Technical Environmental	01	98
	02	15		02	15
	03	15		03	55
	04	18		04	18
	Total	155		Total	186
Dataset IIa Social-Economic	01	7	Dataset IIa Social-Economic	01	8
	02	5		02	8
	03	0		03	0
	04	3		04	5
	Total	15		Total	20
Data IIb Multi-scalar Social	Total	10	Data IIb Multi-scalar Social	Total	10
Total items in framework		195	Total items in framework		218
Grey shaded areas indicate the data that used in sustainability assessment					

3.2. Data Gathering

Figure 4a,b show the distribution of sources from which the data came from.

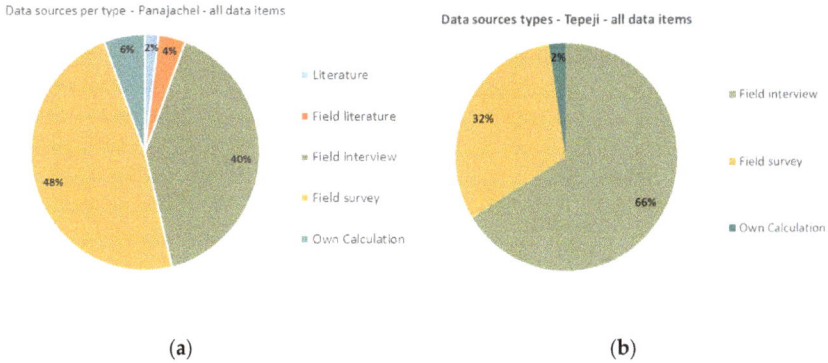

Figure 4. Data sources per type. (**a**) Panajachel pilot site; (**b**) Tepeji pilot site.

About 77% of all data items in Scale 01 could be gathered for the Panajachel site, and ~76% for Tepeji (Tables 6 and 7). However, out of the data that was gathered, only a fraction was of use, as can be seen in the last column of Tables 6 and 7. The reasons why some of the data had to be discarded were:

1. Data quality. Stakeholders sometimes provided no supporting facts or documentation for the data they provided, or there was a considerable difference between data found for the same item from various sources, with no straight-forward way to choose amongst them.
2. No existing threshold. The data could be obtained but no threshold was found, and therefore the data was not used further.

Table 6. Data gathered for all scales, data gathered specifically for Scale 01 and data finally computed into the sustainability assessment: Panajachel pilot site.

	All Scales: Data Found			Scale 01: Data Found			Scale 01: Data Found and Useful	
	Total Items	Items Found	% Found	Total Items	Items Found	%	Number of Items Found and Useful	%
Dataset 0	1	1	100.00	1	1	100	- *	- *
Dataset I	186	88	47.31	98	73	74.49	52	71.23
Dataset II	31	23	74.19	18	16	88.89	10	62.50
Total	218	112	51.38	117	90	76.92	62	68.89

* NOTE: Dataset 0 contains context data and was not used directly in the sustainability assessment.

Table 7. Data gathered for all scales, data gathered specifically for Scale 01 and data finally computed into the sustainability assessment: Tepeji pilot site.

	All Scales: Data Found			Scale 01: Data Found			Scale 01: Data Found and Useful	
	Total Items	Items Found	% Found	Total Items	Items Found	%	Number of Items Found and Useful	%
Dataset 0	15	10	66.67	3	3	100.00	- *	- *
Dataset I	155	93	60.00	107	81	75.70	48	59.26
Dataset II	25	18	72.00	17	12	70.59	7	58.33
Total	195	121	62.05	127	96	75.59	55	57.29

* NOTE: Dataset 0 contains context data and was not used directly in the sustainability assessment.

This filtering process removed ~32% of the data for Panajachel and ~43% of the data gathered for Tepeji. With the remaining variables (62 variables for Panajachel, 55 variables for Tepeji) the sustainability assessment was performed.

3.3. Sustainability Assessment

3.3.1. Panajachel

Sustainability at the *Cebollales* WWTP in Panajachel was assessed with 62 variables: 52 in the technical-environmental dimension, three in the economic and seven in the social. All dimensions show a medium performance (yellow) except for the economic dimension, where the assessment is "poor" (coded with red). Just about half of the variables are performing relatively well (23 variables coded with green) and about half are coded in red (27), with two variables coded in yellow. Therefore, overall sustainability performance can be classified as medium to low (See Table 8. To see performance per variable, see Appendix F).

Table 8. Sustainability performance per dimension: Panajachel site.

Dimension	Variables Per Category				% Variables Per Category				Dimension Average	
	R *	Y	G	Total	R	Y	G	Total	Value	Colour
Technical-Environmental (TE)	27	2	23	52	52%	4%	44%	100%	−0.08	Y
Economic (Ec)	3	0	0	3	100%	0%	0%	100%	−1.00	R
Social (S)	2	1	4	7	29%	14%	57%	100%	0.29	Y
Total or Average	32	3	27	62	60%	6%	34%	100%	−0.26	Y

* R = Red. Y = Yellow. G = Green. ND = No Data.

In the technical environmental dimension variables that performed well included heavy metal concentrations in the plant's water outflow and sludge which were all found to comply with norms at the moment of sampling (except for Arsenic in the sludge). Additionally, the Sampling frequency is complied with. The plant is sampled with regularity. However, the results do not make it to on-the-ground stakeholders (plant operator, for example).

In contrast, variables performing sub-optimally include nutrients and organics: variables such as Total Nitrogen (TN), Total Phosphorus (TP), Biological Oxygen Demand (BOD), Chemical Oxygen Demand (COD), and Total Coliforms. The plant is not able to meet treated water outflow standards. This situation is likely partially driven by the fact that the inflow to the plant (municipal sewage) is already non-compliant as per regulations on discharges into the public sewage system. In addition, strong odours were detected while visiting and were also reported by local stakeholders. Lastly, maintenance is very irregular to non-existing; salaries are irregularly paid; no operation manual exists on site, the operators lack training and equipment. The risks that the WWTP and the treated water discharge into the nearby San Francisco river poses to health or to environment are unknown, as no risk assessment has been carried out, either for health or for ecosystems.

In the economic dimension there is no compliance. For example, the per capita cost of treatment is higher than the WHO illustrative value for activated sludge plants (upper limit set by WHO is 8 USD per capita per year. Using data provided by the municipality, we calculated 9.7 USD). The budget deficit is constant, i.e., the operating entity practically never has access to enough resources to cover operating costs or deliver worker's salaries on time. There is also no valorisation of by products (biogas, sludge), nor has a plan for this purpose been outlined by managers.

In the social dimension stakeholders are generally aware and interested in wastewater-related issues and see opportunities for their suggestions to be heard. They however do no not perceive the solution(s) currently in place as acceptable, nor do they perceive that others accept them.

3.3.2. Tepeji

Sustainability at the Tlaxinacalpan WWTP in Tepeji was assessed with 55 variables: 48 in the technical-environmental dimension and 7 in the social dimension. No economic data was available from the WWTP managers at the time of data gathering, and therefore this dimension could not be evaluated. In the two dimensions evaluated, it shows a moderate to good performance (Table 9).

Table 9. Sustainability performance per dimension: Tepeji site.

Dimension	Variables Per Category				% Variables Per Category				Dimension Average	
	R *	Y	G	Total	R	Y	G	Total	Value	Colour
Technical-Environmental (TE)	15	0	33	48	31%	0%	69%	100%	0.38	G
Economic (Ec)	0	0	0	0	ND	ND	ND	0%	ND	ND
Social (S)	2	2	3	7	29%	29%	43%	100%	0.14	Y
Total or Average	17	2	36	55	ND	ND	ND	ND	ND	ND

* R = Red. Y = Yellow. G = Green. ND = No Data.

The technical-environmental dimension performance' falls just above the border between medium and good performance, with 33 out of 48 being coded with green. Variables that perform well include compliance with heavy metal concentrations in the outflow (as established in local regulations, i.e., Norma Official Mexicana (NOM) 001), except for Cadmium (0.02 mg/L, which is double the allowed value). Additionally, all physical parameters are complied with (Total Suspended Solids (TSS), conductivity, colour, floating matter, grease and oils).

Variables that perform sub-optimally include nutrients and organics: TN, Faecal coliforms, pH, are not performing satisfactorily, neither when compared with the local norm (NOM 001) or with WHO standards for the use of treated wastewater use in agriculture. Odours were detected while visiting and were also reported by local stakeholders. No operation manual is available to key stakeholders such as the operator himself. No regular sampling seems to be occurring on the plant's outflow, as although some interviewed stakeholders assured sampling has been done, no results were provided to us. Operators and managers lack adequate training on anaerobic plant operation. Standard design and operation practices are not being followed (such as an initial inoculation of the system with appropriate bacteria at the start of operations, assurance condition of air-tight conditions within anaerobic digestion tanks). Finally, the risks that the WWTP poses to health or to environment have not been studied, either prior to construction or once in operation, by any of the possibly interested parties.

In the social dimension stakeholders indicated that they are interested in and aware of wastewater related problems. They however do not feel that there is enough information available or opportunities to participate in decision making or to give recommendations to decision makers and managers. The current wastewater management system is generally not accepted or perceived as being accepted by interviewed stakeholders.

4. Discussion

4.1. Dataset Framework for Describing Wastewater Management Systems

We designed a transdisciplinary approach to assess baseline conditions and sustainability performance of wastewater management systems in Latin America, building on methods from both the social and the natural sciences (Figure 1), and with a heavy emphasis on stakeholder involvement and the understanding of baseline conditions. The approach was designed along with the development of a research project, in an iterative process between academic knowledge and the real experiences of what was possible to achieve within the conditions on the field.

We created an Extended dataset framework (EF, see Section 3.1), which we propose to be useful as a general guidance for data item selection for WWMS. It can be used as a sort of repertoire that can be "curated" or edited, choosing the items that are relevant to a specific site or research question, and thus creating a site-specific data framework.

4.1.1. Methodological Issues

The approach calls for not only a transdisciplinary but also a multi-scalar assessment. We attempted to simultaneously look at local scales within technical and administrative boundaries (the WWTP and the municipality), and ecological scales within hydrological boundaries (subcatchment and watershed). However, gathering, evaluating and processing the data required for a multi-scalar assessment was impracticable within our time scope. We therefore implemented the approach only on the scale of the WWTP (Scale 01) and were able to gather enough data to describe baseline conditions and to assess sustainability across all dimensions of sustainability in Panajachel and two out of three in Tepeji.

The approach proved to be practicable at one scale, with the strength of being able to incorporate local needs and conditions through the site-specific editing of the extended dataset framework (Table S1, see Section 2.2.2.2). The resulting datasets are useful as snapshots of the current status quo, and the data items can be used as a guideline for future data generation and periodical evaluation.

Building on a systems perspective, this approach calls for the construction of a system model as a tool: to identify important data items or variables to be investigated, or to localise "invisible" parts of the system, such as stakeholders, boundaries or legal frameworks. The tool proved useful not only for our own research process, but was also helpful during stakeholder involvement activities, where it helped structure the discussion. One example is that when discussing key stakeholders and responsibilities ("who is involved and who is responsible for what?"), the system model clearly depicts that, because the WWTP's outflow eventually reaches the lake at the bottom of the basin, basin authorities (federal), river authorities (provincial) and tourists visiting the lake are involved, albeit to different degrees, in the problem. In other words, by explicitly linking upstream sewage system users to downstream fishermen affected, for example, the visual representations appeal strongly to very different stakeholders sitting around a discussion table, promoting a holistic and inclusive understanding of issues.

The same can be said for the boundaries discussion, also guided by the system model. By clearly illustrating which components fall into which boundary (e.g., the whole of the WWTP falls within the municipal boundary, but its outflow, ten meters ahead, falls into the river and thus provincial jurisdiction, while some of its inputs, such as pump parts, come from a different continent) administrative responsibilities can be made clearer and better understood; conflicting interests or overlapping mandates are made visually explicit and can therefore be more easily comprehended, and complexity is more easily grasped.

Finally, the combination of a technical-environmental assessment (Dataset I) with an economic (Dataset IIa) and a social (Dataset IIb) assessment proved not only enriching but allowed for insight into the drivers of the technical and environmental results. The technical-environmental variables provide an answer to the question *"How is the system behaving?"* while the social and economic data provide perspective into *"Why the system is behaving so?"* making the method better poised to identify bottlenecks and point to solution pathways. We see a challenge but a promising opportunity to improve sustainability thought and its tools in a more thorough transdisciplinary integration in the future. Overall, the approach showed potential for investigating the sustainability of WWMS. We see areas of improvement in, for example, reducing data intensity, systematising thresholds, and operationalizing the multi-scalar approach.

4.1.2. Data Availability

As shown in Tables 6 and 7, roughly over 50% of the data we originally set out to gather for the multi-scalar site-specific dataset framework(s) was available with certain ease of access. Once we decided to focus on a single scale (WWTP), this proportion grew to ~75%, of which around a third had to be discarded due to quality issues.

Basic information such as monthly or yearly budgets and expenditures records, technical drawings and plans of the WWTP were for example not available for the Mexican case. In the Guatemalan site, non-continuous time series for monthly expenditures, inflow and outflow measurements were finally obtained via email after a waiting period following a stakeholder interview. Although indeed useful, the time series were neither long, nor gap-free, and data was not easily made available. In general, we found that stakeholders who, in theory, should have information (operating facilities, government bodies, WWTP managers) may be able to provide verbal answers in an interview, because of their empirical knowledge. The however very often lack supporting documentation, written records and systematic registries. In other cases, they lack the willingness or permission to share information. This is true mostly at the municipal and state levels, while federal agencies, particularly in the Mexican case, usually have well integrated and functional databases. The scale of federal-level data is however often not fine-grained or detailed enough to study a single treatment plant or even a municipal-level WWMS.

It is clear that large efforts are needed in terms of data generation, systematisation, sharing, and transparency. Examples would be digitising written records, using same standards throughout the region, information sharing between institutions, researchers and stakeholders or placing documents and data on the internet. Good starting points already exist, such as the National System for Water Information, kept by the National Water Commission in Mexico (CONAGUA), where geo-referenced information on water quality and quantity, irrigation, and watersheds is disclosed. We suggest that an immediate area of work should be furthering the capacity of key stakeholders, such as municipal and state or provincial governments to generate data, and the integration of all data generators into more detailed and/or numerous data bases, or conversely the creation of citizen-led observatories that foster awareness raising and demand and contribute to regular environmental and economic monitoring.

A significant issue to meaningfully assess sustainability stems from the still incipient integration of social indicators into sustainability. In the extended dataset framework 10 indicators (versus 380 for the environmental-technical dimension) across all scales could be identified from the literature or the stakeholder discussions. They are often linked to information that is not readily available but has to be generated via questionnaires and on-site interviews, coded analysis and other qualitative research methods. In order to strengthen future sustainability assessments of WWMS it is imperative to continue work on the integration of social indicators and methods to streamline the collection and analysis of these.

4.2. Sustainability of the Pilot Wastewater Management Systems

Our analyses show that the wastewater treatment systems currently in place in Panajachel, Guatemala, and Tepeji, Mexico, although performing well in various selected parameters, cannot be considered sustainable when looked at in a multidimensional manner, i.e., in terms of technical, environmental, economic, and social factors.

4.2.1. Technical-Environmental Issues

Both plants treat a municipal wastewater flow of domestic origin, with low to negligible heavy metal and metalloids content. In both cases, the quality of the inflowing wastewater is already below locally applicable standards for discharges into the public sewage network (Total Nitrogen and Total Coliforms in both cases, and Total Phosphorus, Biological Oxygen Demand, Chemical Oxygen Demand, and Total Suspended Solids (TSS) in Guatemala as well) meaning the plants are receiving a low-quality inflow from the start (see results per variable in Appendix F).

Once within the WWTPs, processes show different efficiency levels. Omitting metals, in which both plants practically fully comply (arguably because of an original low-metal content), the WWTP in Panajachel does not comply with virtually any of the examined physical, chemical or biological parameters, while the Tepeji WWTP performs slightly better, complying with half of them. Both plants, however, are performing poorly in the treatment of faecal coliforms, a crucial variable in terms of human and ecosystem health. In the particular case of Tepeji, where the water is being used for irrigation and the de-centralised, small-scale technology is being introduced to the community, a low quality and potentially risky outflow is not only a health risk, but an important hinderance to the success of the de-centralised treatment project, which has the aim of fostering wastewater use for agricultural irrigation. Social acceptance is key to the success of such new technology, and thus trust among the community has to be gained by the promoting entities.

Technical efficiency and environmental compliance are a major issue in both plants. Although visibly more critical in the Mexican case, lack of training of the operating and management personnel is a shared issue that is contributing to the situation. Systematic data generation and environmental monitoring, particularly in the Mexican case, are a challenge. Guatemalan managers were keeping a more detailed track of the WWTP performance. This may have to do with the fact that the WWTP in Guatemala has been in operation for a longer period (5 years versus 8 months), and also that it is operated by the municipality (vs a Trust).

4.2.2. Management Issues

In management-related issues both WWTPs perform the same, with only 2 out 7 management variables evaluated as positive, even though the systems operate at different scales (design flows of 1.5 lps in Tepeji vs 37 lps in Panajachel on average). Both plants lack operation manuals accessible to the operator, personnel lack training and capacities and laboratory analyses are not accessible and hassle-free in any of the cases (although in Panajachel sampling is carried out with norm-compliant frequency by a federal authority, key stakeholders—such as the plant's operator—do not have access to the results and have therefore no feedback on their work).

In terms of risk and safety, operators in both pilot sites lack appropriate working conditions, (clothing, equipment, adequate-hand-washing-facilities). In neither of the cases had the risks posed to the environment or surrounding populations by malfunctioning of the WWTP been studied. Panajachel stakeholders manifested that environmental risk assessment is relatively new in public administration, and that they hope it will be integrated along with, for example, Environmental Impact Assessment (EIA), soon.

In Latin America in general, investments are often made to build an infrastructure project, but the funding for its long-term operation and maintenance (including equipping operators, performing routine samplings, etc.) is not secured, and nor are income generating options (resource recovery, for instance) duly considered [40]. Although this is a known issue, new infrastructure is being built as we write in Panajachel, while in Tepeji funds are being sought for the building of a large scale WWTP, still without a clear idea of how current infrastructure will continue to be financed or maintenance challenges faced (e.g., equipment repairs, salaries). Without a change towards adequate financial planning it is likely that both existing and new WWTPs at the studied sites will continue to operate sub-optimally.

4.2.3. Social Issues

The overarching recommendation, applicable to both sites, is to facilitate stakeholders the access to the information about their own social network. A common understanding of the problem itself is lacking. Who should be contacted with which need, or as formulated by Reed et al. "who is in and why?" [23] is a key question with a high degree of influence on the social development in both pilot sites. A common understanding of the problem is the basis for facilitating social interaction among the involved stakeholders. Economic and human resources should be provided to conduct an in-depth Stakeholder and Social Network Analysis in both pilot sites.

5. Conclusions

To advance towards sustainability in the urgent topic of wastewater management in the Americas, data scarcity and scatteredness must be overcome to allow for precise understanding of current or baseline sustainability performance. From such an understanding, bottlenecks can be made visible, and pathways towards sustainability can be envisioned. To increase the accuracy of the assessment and the adequacy of proposed solutions, research should go beyond one single perspective. To this end, we have proposed a multi-scalar data framework that includes variables for four different territorial scales: the WWTP, the municipality, the subcatchment and the watershed. Other scales could be chosen in other projects, what we propose is the multiscalar approach, not necessarily the scales themselves.

Additionally, we propose to assess sustainability across four dimensions (environmental, technical, economic and social), and to incorporate other strands of scientific practice into the assessment (stakeholder analysis and wickedness analysis).

Transdisciplinarity is also a tool for improved success of research projects in this topic (see introduction). Throughout this project, we worked closely with local stakeholders and non-scientific practitioners. Their input was crucial in tailoring the framework to be locally relevant (see Section 2.2.2.2), and in the process of envisioning and evaluating solution options.

In this paper we present the method itself (Section 2) and partial results of its application in two pilot sites (Section 3). We also discuss the benefits and limitations of the method, and point to ideas for its future improvement and further application (Section 4.1).

As to the method itself, we found the multiscalar approach to enrich assessment and to allow to make visible issues that are not shown by single scale analysis, namely the interconnections of the technical system (WWTP) with ecological systems (watershed, riparian areas) and social systems (government, public administration, community dynamics, social perception). Shedding light on these interconnections, bottlenecks and obstacles to achieve sustainability are understood in a deeper and more detailed way, as many of the bottlenecks would be invisible when looking only at one scale or one dimension. The main limitations of the method are data and time intensity. Good planning, working closely with engaged local partners and performing a preliminary screening of data availability and data holders is recommended.

As to the results of the assessment presented here, Sustainability Assessment showed that technical and environmental variables tend in general to perform medianly to well, with microbiological parameters performing below the norms in both cases. Social and economic variables are the weakest spot of both of the WWMS analysed (Section 3.3). The results of the other two components of the method (stakeholder analysis and wickedness analysis) will be the object of future publications.

Supplementary Materials: The following are available online at http://www.mdpi.com/2073-4441/11/2/249/s1, Table S1: Extended Framework.

Author Contributions: Conceptualization, L.B. and T.A.; Data curation, A.M.; Formal analysis, L.B. and A.M.; Funding acquisition, T.A. and S.C.; Investigation, L.B. and A.H.; Methodology, L.B., T.A., S.C., A.H., S.K. and A.M.; Project administration, T.A., S.C. and A.H.; Supervision, T.A.; Visualization, L.B.; Writing—original draft, L.B.; Writing—review & editing, L.B., T.A., A.H. and A.M. With the exception of the first author, alphabetical order was applied for the remaining authorship's order.

Funding: This research was funded by the German Bundesministerium für Bildung und Forschung, (BMBF) under the grant number 01DF17001. The Guatemalan Consejo Nacional de Ciencia y Tecnología provided support to the researchers of the University of San Carlos de Guatemala and financed, in part, the Assessment Workshop in Panajachel in March 2018. The Fideicomiso de Infraestructura Ambiental para los Valles de Hidalgo (FIAVHI) provided support to the participation of Mexican stakeholders to the Assessment Workshop and access to their premises. The APC was funded by the German Bundesministerium für Bildung und Forschung, (BMBF) under the grant number 01DF17001.

Acknowledgments: The authors wish to acknowledge the project's partners: Ing. Jorge Cifuentes (USAC), and his students, Ing. Carlos Paillés and staff at FIAVHI, and our partners from the Technische Universität Dresden. Many thanks to AMSCLAE, the municipal authorities in Panajachel and Tepeji who kindly supported our research, and all stakeholders who accepted to be interviewed. Special thanks to Enrique Cosenza and Thelma Lopez for making the field in Panajachel possible. Finally, we wish to thank Laura Ferrans, Leon Zimmermann and Nestor de la Paz for their preliminary work into pilot sites, as well as Guido Bartolini for his support in the preparation of the final manuscript.

Conflicts of Interest: The authors declare no conflict of interest. The funders had no role in the design of the study; in the collection, analyses, or interpretation of data; in the writing of the manuscript, or in the decision to publish the results.

Appendix A

Prioritised (Site-Specific) Dataset Framework—Panajachel

Total Data Items	218
PS = Prioritised by stakeholders	LI = Data Item comes from the literature
RG = Included in Guatemala regulation	RM = Included in Mexican regulation
	The numbers in the ID column refer to those of the extended set.

DATASET 0—Context Data—WWTP Scale

Category	ID	PS	LI	RG	RM	Data Item	Item Description	Notes
GEOGRAPHY A	A0.003			1		Map	Cartography at the adequate scale to understand the location of the plant in relation to nearest population settlement, water resources and other relevant features.	All non-domestic wastewater generators have to prepare a technical study including this item. Acuerdo Gubernativo 12-2011, article 5 and 6

DATASET I.01—Technical Environmental Data—WWTP Scale

Category	ID	PS	LI	RG	RM	Data Item	Item Description	Notes
GENERAL A	A0.001	x	x	1		Technology used	Technical procedure with which the plan treats wastewater. Note any relevant particularities. If needed, include a diagram of the process in an annex.	All non-domestic wastewater generators have to prepare a technical study including this item. Acuerdo Gubernativo 12-2011, article 5 and 6
	A0.005	x	x			Number of people served		
INPUTS B	B0.001	x	x			Design inflow	Flow capacity that the plant was originally designed for.	
68	B0.002	x	x			Volume wastewater input	Total volume of water entering the plant in the reporting year	

			Average plant capacity utilization	Percent of design capacity being used, on average, during the reporting year
	B0.005	x	x	
	B0.006	x	x	Volumetric Efficiency	Total wastewater entering the plant/Treated Wastewater (100)
Inflow quality parameters	B1.001	x	x	Temperature	
B1	B1.002	x	x	BOD	Biological Oxygen demand
	B1.003	x	x	COD	Chemical oxygen demand
Inflow Nutrients	B1.004	x	x	Total Nitrogen	
	B1.008	x	x	Total Phosphorus	
	B1.015		x	Faecal coliforms	
Pathogens inflow	B1.016	x	x	E.Coli	
	B1.021		x	TSS	Total suspended solids
	B1.023		x	pH	
Other inputs B2	B2.001	x	x	Raw materials used	Raw materials as inputs necessary for the plant to function (e.g., machine oils, fuel, chemicals for the flocculation phase or other stages of the process, etc.), as well as office supplies and such. When data available is in other units, make sure to note so in the units column. Tonnes per year is a recommended unit.
	B2.003	x	x	Total energy consumed	Energy consumed in the reporting year, all energy carriers together and all energy uses considered.

OUTPUTS C	Code	Parameter	1	Total volume Treated Water produced	Total Outflow of wastewater from the plant, in yearly total average.
	C0.001			x	x
	C1.001	Temperature	1	x	x
	C1.002	BOD (Biological Oxygen demand)	1	x	x
	C1.003	COD (Chemical oxygen demand)	1	x	x
	C1.004	Total Nitrogen	1	x	x
	C1.008	Total Phosphorus	1	x	x
Pathogens in outflow	C1.015	Faecal coliforms	1	x	x
	C1.016	E.coli		x	x
	C1.017	Helminths		x	x
	C1.019	Organic Matter		x	x
	C1.021	Sedimentable solids	1	x	x
	C1.022	TSS	1	x	x
	C1.023	Turbidity		x	x
	C1.024	pH	1	x	x
Metals, metalloids and trace elements in outflow	C1.025	Al		x	x
	C1.026	As	1	x	x
	C1.027	Cd		x	x
	C1.028	Cyanide (CN)	1	x	x
	C1.029	Co		x	x
	C1.030	Cr	1	x	x
	C1.031	Cu	1	x	x
	C1.032	Fe		x	x
	C1.033	Mn		x	x

Code	Section	Parameter					Description
C1.034		Ni	1	1	x	x	
C1.035		Ti		1	x	x	
C1.036		Zn	1	1	x	x	
C1.037		Hg	1	1	x	x	
C1.038		Pb	1	1	x	x	
C1.039		Se		1	x	x	
C1.040		B		1	x	x	
C1.041		Mo		1	x	x	
C1.043		Grease and oils	1	1		x	
C1.044		Floating matter	1	1		x	
C1.045		Colour	1	1			
C2.001	Wastewater Reuse C2	Percentage of wastewater output being recycled or reused			x	x	
C3.001	Sludge C3	Total Sludge produced yearly			x	x	Total amount of sludge produced in the reporting year.
C3.002	Sludge Quality parameters	Al	1		x	x	
C3.003	Metals, metalloids and trace elements in sludge	As	1	1	x	x	
C3.004		Cd	1	1	x	x	
C3.005		Co		1	x	x	
C3.006		Cr	1	1	x	x	
C3.007		Cu	1	1	x	x	
C3.008		Fe		1	x	x	
C3.009		Mn		1	x	x	
C3.010		Ni	1	1	x	x	

Group	Code	Parameter				Description
	C3.011	Ti	x		x	
	C3.012	Zn	x	1	x	
	C3.013	Hg	x	1	x	
	C3.014	Pb	x	1	x	
	C3.015	Se	x		x	
	C3.016	B	x		x	
	C3.017	Mo	x		x	
	C3.030	Calorific value	x		x	
Pathogens in sludge	C3.031	Helminths	x	1	x	
	C3.032	Total coliforms	x	1	x	
	C3.033	E.coli	x	1	x	
	C3.034	Salmonella sp.		1	x	
Organics	C3.035	Organic Matter	x		x	
Sludge use C4	C4.001	Scope of sludge management	x		x	% of sludge that is managed, including treatment in different ways, such as use in agriculture, thermal disposal, landfills, etc. As proposed by Popovic & Kraslawski (2018).
	C4.002	Current use/management of sludge	x		x	What is done with sludge once it is dried at the plant?
	C4.004	Potential sludge users	x		x	
Emissions C5	C5.001	Total Biogas production	x		x	How much biogas was produced in the reporting year?

Category	Code			Indicator	Description
	C5.005	x	x	GHG emissions	Can be divided into GHG emissions linked to plant operation and maintenance, and emissions produced by the wastewater itself. Specify and disclose method for Calculations performed in an annex. The online tool ECAM (wacclim.org/ecam) is an option for estimation.
Management D2	D0.001	x	x	Number of operators	
Staff D0	D0.003	x	x	Employee/inhabitant ratio	Number of employees per 1000 inhabitants served by the plant.
Management D1	D1.001	x	x	Existence Operation manual	Does a clear, up to date operations manual exist on site, and available to all people operating the plant?
	D1.002	x	x	Regularity of maintenance	
Capacities D2	D2.001	x	x	Capacity sufficiency	Does all the personnel involved have the knowledge and skills they need to have?
	D2.003	x	x	Accessible sampling and processing equipment	Does the plant have its own equipment or easy and hassle-free access to sample and analyse incoming wastewater, treated water and by-products quality?
Compliance and certification D3	D3.001		x	Discharge standards compliance	Percent of time that the plant's outflow complies with applicable regulations. State the regulations are being considered.

Category	ID	PS	LI	Data Item	Item Description	Notes
	D3.002	x		Analysis frequency compliance	Ratio of number of effluent samplings per month to number of effluent sampling per month required by law of wastewater treatment policy (as proposed by Popovic & Kraslawski (2018).	1
	D3.003	x		Certification	Does the plant have some quality certification (ISO, or other national/international standards)?	
RISK E1	E0.001	x	x		Has a health risk assessment related to wastewater been performed at the site?	
	E0.002	x	x		Are health risks being managed?	
Health E0	E0.003	x	x		Do the operators have the necessary health and safety equipment?	
	E1.001				Has a natural hazard risk assessment been performed at the facility?	
	E1.002				Are natural hazard risks being managed?	
	E1.003				Has an environmental impact study relating wastewater with ecosystem health been performed at the site?	
Other hazards E1	E1.004	x	x		What efforts are being made to reduce or manage environmental impacts?	
	E1.005				Presence or risk of groundwater pollution	
	E1.006				Presence or risk of surface water pollution	

DATASET IIA.01—Economic Data—WWTP Scale

Category	ID	PS	LI	Data Item	Item Description	Notes
Costs A0	A0.002	x	x	Cost per m³ of water treated	Cost of producing one cubic meter of water	
	A0.003		x	Cost per inhabitant served		
	A0.006	x	x	Proportion of costs: maintenance and repairs	What proportion of the total expenses corresponds to energy?	

Category	ID	PS	IL	Data Item	Item Description	Notes
	A0.009			Proportion of costs: training, capacity building	What proportion of the total expenses corresponds to energy?	
Income A1	A1.001		x	Total plant income	Total income of the plant yearly. Specify currency used under 'units'	
	A1.002		x	Real financial availability per inhabitant served		
	A1.003			Budget deficit		
	A1.006		x	Valorisation of by products	Are products of the plant being valorised (sold, recycled, etc.)	

DATASET IIB.01—Social Acceptance—Multi-Scalar

Category	ID	PS	IL	Data Item	Item Description	Notes
SOCIAL B	B0.001				Personal interest in wastewater management problems	
Inclusion/Participation	B0.002				Personal awareness of wastewater management problems	
	B0.003				Willingness to be informed about the wastewater management problems	
	B0.004				Accessibility to information	
	B0.005				Possibilities for providing a recommendation	
	B0.006				Recommendations are considered?	
	B0.007				Willingness to participate in decision-making	
	B0.008				Participative decision-making	
	B0.009				Personal acceptance of the current wastewater management	
	B0.010				Perception of social acceptance of the current wastewater management	

Appendix B

Prioritised (Site-Specific) Dataset Framework—Tepeji

Total Data Items	195

PS = Prioritised by stakeholders
RG = Included in Guatemala regulation
LI = Data Item comes from the literature
RM = Included in Mexican regulation
The numbers in the ID column refer to those of the extended set.

DATASET 0.1—Context Data—WWTP Scale

Category	ID	PS	LI	RG	RM	Data Item	Item Description	Notes
GEOGRAPHY A	A0.003			1		Map	Cartography at the adequate scale to understand the location of the plant in relation to nearest population settlement, water resources and other relevant features.	All non-domestic wastewater generators have to prepare a technical study including this item. Acuerdo Gubernativo 12-2011, article 5 and 6.
	A0.006		x			Land uses in 1 km radius		
	A0.007		x			Distance to nearest house		

DATASET 1.01—Technical Environmental Data—WWTP Scale

Category	ID	PS	LI	RG	RM	Data Item	Item Description	Notes
GENERAL A	A0.001		x			Technology used	Technical procedure with which the plan treats wastewater. Note any relevant particularities. If needed, include a diagram of the process in an annex.	All non-domestic wastewater generators have to prepare a technical study including this item. Acuerdo Gubernativo 12-2011, article 5 and 6
	A0.002		x			Construction year	Year of construction. When construction lasted more than one year, state ending year.	

	A0.005	Number of people served	x
INPUTS B	B0.001	Design inflow	x	Flow capacity that the plant was originally designed for.
	B0.002	Volume wastewater input	x	Total volume of water entering the plant in the reporting year
Inflow B0	B0.003	Average inflow (AF)		Average flow (in a year) of wastewater into WWTP.
	B0.005	Average plant capacity utilization	x	Percent of design capacity being used, on average, during the reporting year
	B0.006	Volumetric Efficiency	x	Total incoming wastewater/total treated water
Inflow quality parameters	B1.001	Temperature	x	
B1	B1.002	BOD	x	Biological oxygen demand
	B1.003	COD	x	Chemical oxygen demand
Inflow Nutrients	B1.004	Total Nitrogen	x	
	B1.008	Total Phosphorus	x	
Salts inflow	B1.009	K	x	
	B1.010	Ca	x	
	B1.011	Mg	x	
	B1.012	Na	x	
	B1.014	Electric conductivity	x	Useful when data for Na and other related parameters is not available, as general guidance of salts contents.
	B1.015	Faecal coliforms	x	

105

	Code	Name	Description	
Pathogens inflow	B1.016	E.coli		x
	B1.021	TSS	Total suspended solids	x
	B1.023	pH		x
	B1.025	As		x
	B1.026	Cd		x
	B1.028	Cr		x
	B1.029	Cu		x
	B1.030	Fe		x
	B1.031	Mn		x
	B1.032	Ni		x
	B1.033	Ti		x
	B1.034	Zn		x
	B1.035	Hg		x
	B1.036	Pb		x
	B1.037	Se		x
	B1.038	B		x
	B1.039	Mo		x
Others	B1.040	Residual chlorine		x
	B1.041	Grease and oils		
	B1.042	Floating matter		
	B1.043	Colour		
	B2.003	Total energy consumed	Energy consumed in the reporting year, all energy carriers together and all energy uses considered.	x
	B2.004	Energy/m^3 treated water		x

	Code		Parameter	Description
OUTPUTS C	C0.001	x	Total volume Treated Water produced	Total Outflow of wastewater from the plant, in yearly total average.
	C1.001	x	Temperature	
	C1.002	x	BOD	Biological oxygen demand
	C1.003	x	COD	Chemical oxygen demand
	C1.004	x	Total Nitrogen	
Nutrients in outflow	C1.006	x	Nitrates	
	C1.007	x	Nitrites	
	C1.008	x	Total Phosphorus	
Salts in outflow	C1.009	x	K	
	C1.010	x	Ca	
	C1.011	x	Mg	
	C1.012	x	Na	
	C1.014	x	Electric conductivity	Useful when data for Na and other related parameters is not available, as general guidance of salts contents.
Pathogens in outflow	C1.015	x	Faecal coliforms	
	C1.016	x	E.coli	
	C1.017	x	Helminths	
	C1.021		Sedimentable solids	
	C1.022	x	TSS	Total suspended solids
	C1.024	x	pH	
	C1.026	x	As	
	C1.027	x	Cd	
	C1.028		Cyanide (CN)	

	Code	Parameter		Note
	C1.030	Cr	x	
	C1.031	Cu	x	
	C1.034	Ni	x	
	C1.036	Zn	x	
	C1.037	Hg	x	
	C1.038	Pb	x	
	C1.043	Grease and oils		
	C1.044	Floating matter		
	C1.045	Colour		
Wastewater Reuse C2	C2.001	Percentage of wastewater output being recycled or reused	x	
Sludge C3	C3.001	Total Sludge produced yearly	x	Total amount of sludge produced in the reporting year.
Metals, metalloids and trace elements in sludge	C3.003	As	x	
	C3.004	Cd	x	
	C3.006	Cr	x	
	C3.007	Cu	x	
	C3.010	Ni	x	
	C3.012	Zn	x	
	C3.013	Hg	x	
	C3.014	Pb	x	
Pathogens in sludge	C3.031	Helminths	x	
	C3.032	Total coliforms	x	
	C3.034	Salmonella sp.		

Category	Code	Indicator	Description	
sludge use C4	C4.001	Scope of sludge management	% of sludge that is managed, including treatment in different ways, such as use in agriculture, thermal disposal, landfills, etc. As proposed by Popovic & Kraslawski (2018)	x
GHG Emissions	C5.006	Are there complaints regarding odours?	E.g., neighbours	x
	C5.007	Strength of odour in the treated wastewater	high, medium, low	x
Solid Waste	C6.002	Solid waste sustainable management plan	Is there a waste management programme in place that considers reuse and/or recycling of solid waste, and/or plans to reduce waste or eliminate it, e.g. by changing inputs?	x
Staff D0	D0.003	Employee/inhabitant ratio	Number of employees per 1000 inhabitants served by the plant.	x
Management D1	D1.001	Existence Operation manual	Does a clear, up to date operations manual exist on site, and available to all people operating the plant?	x
	D1.002	Regularity of maintenance		x
Capacities D2	D2.001	Capacity sufficiency	Does all the personnel involved have the knowledge and skills they need to have?	x
	D2.003	Accessible Sampling and processing equipment	Does the plant have its own equipment or easy and hassle-free access to sampling and analysis to monitor wastewater, treated water and by-products quality?	x

Category	ID	PS	LI	Data Item	Item Description	Notes
Compliance and certification D3	D3.001	x		Discharge standards compliance	Percent of time that the plant's outflow complies with applicable regulations. State which regulations are being considered	
	D3.002	x		Analysis frequency compliance	Ratio between the number of effluent samplings per month and number of effluent sampling per month required by law of wastewater treatment policy (as proposed by Popovic & Kraslawski (2018))	
	D3.003			Certification	Does the plant have some quality certification (ISO, or other national/international standards)	
RISK E1	E0.001	x			Has a health risk assessment related to wastewater been performed at the site?	
	E0.002	x			Are health risks being managed?	
Health E0	E0.003	x			Do the operators have the necessary health and safety equipment?	
	E1.001				Has a natural hazard risk assessment been performed at the facility?	
	E1.002				Are natural hazard risks being managed?	
	E1.003				Has an environmental impact study relating wastewater with ecosystem health been performed at the site?	
Other hazards E1	E1.004	x			What efforts are being made to reduce or manage environmental impacts?	
	E1.005				Presence or risk of groundwater pollution	
	E1.006				Presence or risk of surface water pollution	

DATASET IIA.01—Social Economic Data—WWTP Scale

Category	ID	PS	LI	Data Item	Item Description	Notes
Costs A0	A0.002		x	Cost per m³ of water treated	Cost of producing one cubic meter of water	

Category	ID	PS	LI	Data item	Item description	Notes
16	A0.003		x	Cost per inhabitant served		
	A0.009			Proportion of costs: training, capacity building		What proportion of the total expenses corresponds to energy?
Income A1	A1.001		x	Total plant income		Total income of the plant yearly. Specify currency used under 'units'
	A1.002		x	Real financial availability per inhabitant served		
	A1.003			Budget deficit		
	A1.006		x	Valorisation of by products		Are products of the plant being valorised (sold, recycled, etc.)

DATASET IIB.01—Social Acceptance—Multi-Scalar

Category	ID	PS	LI	Data item	Item description	Notes
SOCIAL B	B0.001				Personal interest in wastewater management problems	
Inclusion/Participation	B0.002				Personal awareness of wastewater management problems	
	B0.003				Willingness to be informed about the wastewater management problems	
	B0.004				Accessibility to information	
	B0.005				Possibilities for providing a recommendation	
	B0.006				Recommendations are considered?	
	B0.007				Willingness to participate in decision-making	
	B0.008				Participative decision-making	
	B0.009				Personal acceptance of the current wastewater management	
	B0.010				Perception of social acceptance of the current wastewater management	

Appendix C

Dataholders for the Panajachel Study Site—Final List

	1—Stakeholder Local/Municipality		2—Stakeholder Provincial or National		3—Own Calculations		4—Scientist Interview or Scientific Literature		5—NGO Interview or Report
1	Plant operator Julio Pablo de León	1	AMSCLAE interviews	1	Sampling and analysis	1	UVG—CEA	1	Amigos del Lago
2	Encargado de la planta Cebollales Ing. Genaro Umul	2	AMSCLAE reports	2	Calculations	2	Laura Ferrans	2	Mancomunidad (Mankatitlán). Delvin Rolón, gerente
3	Environmental office (oficina municipal del medio ambiente)/DIGAM	3	NE			3	ERIS	3	Proyecto ProAtitlán
4	Reports, monographs, other documentation published by municipality DGP—Planning	4	MARN—provincial delegation at Sololá			4	Elisandra Hernandez USAC	4	ANACAFE
5	authority at the municipality. Oficina Municipal de Agua	5	Ministerio de Salud					5	Puravida
6	Agua	6	MAGA—Ministerio de agricultura y ganadería					6	Vivamos mejor
		7	Instituto Nacional de Estadística						
		8	Energuate						

Dataholders for the Tepeji Study Site—Final List

1—Stakeholder Local/Municipality		2—Stakeholder Provincial or National		3—Own Calculations		4—Scientist Interview or Scientific Literature		5—NGO Interview or Report
1	CAAMTROH director	1	CONAGUA at state capital Pachuca	1	Sampling and analysis	1	Research by UNAM	
2	CAAMTROH/Field personnel	2	CONAGUA central office Mexico City	2	Calculations			
3	Dirección de ecología municipal	3	INEGI					
4	FIAVHI director							
5	FIAVHI technical staff							
6	Plant operator							
7	Urban development office at the municipality							
8	Owner of agricultural field who will receive treated WW							

Appendix D

Water Quality Parameters Analyzed in Panajachel—Field Campaign 08.2018

Raw (WW) and Treated Wastewater (TWW)		Sludge	
1	Temperature	1	Fecal coliforms
2	pH	2	Helminth eggs
3	Grease and oils	3	Al
4	Floating matter	4	As
5	BOD	5	Ca
6	COD	6	Cd
7	TSS	7	Co
8	Total Nitrogen	8	Cr
9	Total Phosphorus	9	Cu
10	Fecal coliforms	10	Fe
11	Apparent Color	11	Hg
12	Al	12	K
13	As	13	Mn
14	Ca	14	Na
15	Cd	15	Ni
16	Co	16	P
17	Cr	17	Pb
18	Cu	18	Se
19	Fe	19	Zn
20	Hg		
21	K		
22	Mn		
23	Na		
24	Ni		
25	P		
26	Pb		
27	Se		
28	Zn		

Water Quality Parameters Analyzed in Tepeji—Field Campaign 08.2018

Raw and Treated Wastewater

1	Grease and oils
2	Floating matter
3	BOD
4	COD
5	Suspended solids
6	TN
7	TP
8	pH
9	Fecal coliforms
10	Apparent color
11	Al
12	As
13	Ca
14	Cd
15	Co
16	Cr
17	Cu
18	Fe
19	Hg
20	K
21	Mn
22	Na
23	Ni
24	P
25	Pb
26	Se
27	Zn
27	Cn
28	Sedimentable solids
29	Nitrites
30	Nitrates

Appendix E

Variables and threshold values considered for the Sustainability Assessment at the two study sites (Panajachel, Guatemala and Tepeji, Mexico). This table discloses the values and sources of the thresholds used in the Sustainability Assessment.

Variables and Thresholds for SA in Panajachel

Gt: Guatemala Regulation Mx: Mexican Regulation ST-Team: SludgeTec Team WHO (2006): Guidelines for SUWA—Vol 2

No.	Code (ID)	Variable	Unit	Threshold Value	Source	Red	Yellow	Green
1	TE7B	Temperature—WW	°C	40	AG 12-2011 Art. 14 (p.10) Gt	>44	>40 and ≤44	≤40
2	TE8B	Biological Oxygen Demand (BOD)—WW	mg/L	100	AG 12-2011 Art. 14 (p.10) Gt	>110	>100 and ≤110	≤100
3	TE9B	Chemical Oxygen Demand (COD)—WW	mg/L	200	AG 12-2011 Art. 14 (p.10) Gt	>220	>200 and ≤220	≤200
4	TE10B	Total Nitrogen—WW	mg/L	20	AG 12-2011 Art. 14 (p.10) Gt	>22	>20 and ≤22	≤20
5	TE11B	Total Phosphorus—WW	mg/L	10	AG 12-2011 Art. 14 (p.10) Gt	>11	>10 and ≤11	≤10
6	TE12B	Faecal coliforms—WW	MPN/100 mL	100,000	AG 12-2011 Art. 14 (p.10) Gt	>110,000	>100,000 and ≤110,000	≤100,000
7	TE14B	Total Suspended Solids (TSS)—WW	mg/L	125	AG 12-2011 Art. 14 (p.10) Gt	>137.5	>125 and ≤137.5	≤125
8	TE15B	pH—WW	pH unit	between 6–9	AG 12-2011 Art. 14 (p.10) Gt	<6 and >9	-	≥6 and ≤9
9	TE19C	Temperature—TWW	°C	TRWB ±3	AG 12-2011 Art. 11 (p.7) Gt	<20 and >26	-	≥20 and ≤26
10	TE20C	Biological Oxygen Demand (BOD)—TWW	mg/L	30	AG 12-2011 Art. 11 (p.7) Gt	>33	>30 and ≤33	≤30
11	TE21C	Chemical Oxygen Demand (COD)—TWW	mg/L	60	AG 12-2011 Art. 11 (p.7) Gt	>66	>60 and ≤66	≤60
12	TE22C	Total Nitrogen—TWW	mg/L	5	AG 12-2011 Art. 11 (p.7) Gt	>5.5	>5 and ≤5.5	≤5
13	TE23C	Total Phosphorus—TWW	mg/L	3	AG 12-2011 Art. 11 (p.7) Gt	>3.3	>3 and ≤3.3	≤3
14	TE24C	Faecal coliforms—TWW	MPN/100 mL	500	AG 12-2011 Art. 11 (p.7) Gt	>550	>500 and ≤550	≤500
15	TE26C	Helminths—TWW	-	5	NOM-003-SEMARNAT-1997 Mx	>5.5	>5 and ≤5.5	≤5
16	TE29C	Total Suspended Solids (TSS)—TWW	mg/L	40	AG 12-2011 Art. 11 (p.7) Gt	>44	>40 and ≤44	≤40
17	TE31C	pH—TWW	pH units	between 6–9	AG 12-2011 Art. 11 (p.7) Gt	<6 and >9	-	≥6 and ≤9
18	TE33C	Arsenic (As)—TWW	mg/L	0.1	AG 12-2011 Art. 11 (p.10) Gt	>0.11	>0.1 and ≤0.11	≤0.1
19	TE34C	Cadmium (Cd)—TWW	mg/L	0.1	AG 12-2011 Art. 11 (p.10) Gt	>0.11	>0.1 and ≤0.11	≤0.1
20	TE37C	Chromium (Cr)—TWW	mg/L	0.1	AG 12-2011 Art. 11 (p.10) Gt	>0.11	>0.1 and ≤0.11	≤0.1

Variables and Thresholds for SA in Panajachel

Gt: Guatemala Regulation Mx: Mexican Regulation ST-Team: SludgeTec Team WHO (2006): Guidelines for SUWA—Vol 2

No.	Code (ID)	Variable	Unit	Threshold Value	Source	Red	Yellow	Green
21	TE38C	Copper (Cu)—TWW	mg/L	0.5	AG 12-2011 Art. 11 (p.10) Gt	>0.55	>0.5 and ≤0.55	≤0.5
22	TE41C	Nickel (Ni)—TWW	mg/L	0.5	AG 12-2011 Art. 11 (p.10) Gt	>0.55	>0.5 and ≤0.55	≤0.5
23	TE43C	Zinc (Zn)—TWW	mg/L	1	AG 12-2011 Art. 11 (p.10) Gt	>1.1	>1 and ≤1.1	≤1
24	TE44C	Mercury (Hg)—TWW	mg/L	0.01	AG 12-2011 Art. 11 (p.10) Gt	>0.011	>0.01 and ≤0.01	≤0.01
25	TE45C	Lead (Pb)—TWW	mg/L	0.1	AG 12-2011 Art. 11 (p.10) Gt	>0.11	>0.1 and ≤0.11	≤0.1
26	TE49C	Grease and oils—TWW	mg/L	15	NOM-001-SEMARNAT-1996 (p.15) Mx	>16.5	>15 and ≤16.5	≤15
27	TE50C	Floating matter—TWW	Present-Absent	Present-Absent	AG 12-2011 Art. 11 (p.10) Gt	Present	-	Absent
28	TE51C	Colour—TWW	PCU	400	AG 12-2011 Art. 11 (p.10) Gt	>440	>400 and ≤440	≤400
29	TE52C	Water reuse	YES-NO	YES-NO	ST team	NO	-	YES
30	TE55C	Arsenic (As)—Sludge	mg/kg dry matter (104 °C)	50	AG 236-2006 para lodos—Application in soil Gt	>55	>50 and ≤55	≤50
31	TE56C	Cadmium (Cd)—Sludge	mg/kg dry matter (104 °C)	50	AG 236-2006 para lodos Gt	>55	>50 and ≤55	≤50
32	TE58C	Chromium (Cr)—Sludge	mg/kg dry matter (104 °C)	1500	AG 236-2006 para lodos Gt	>1650	>1500 and ≤1650	≤1500
33	TE59C	Copper (Cu)—Sludge	mg/kg (dry weight)	1500	NOM-004-SEMARNAT-2002 (p.6)—Excellent Biosolid Mx	>1650	>1500 and ≤1650	≤1500
34	TE62C	Nickel (Ni)—Sludge	mg/kg (dry weight)	420	NOM-004-SEMARNAT-2002 (p.6)—Excellent Biosolid Mx	>462	>420 and ≤462	≤420
35	TE64C	Zinc (Zn)—Sludge	mg/kg (dry weight)	2800	NOM-004-SEMARNAT-2002 (p.6)—Excellent Biosolid Mx	>3080	>2800 and ≤3080	≤2800
36	TE65C	Mercury (Hg)—Sludge	mg/kg dry matter (104 °C)	25	AG 236-2006 para lodos—Application in soil Gt	>27.5	>25 and ≤27.5	≤25
37	TE66C	Lead (Pb)—Sludge	mg/kg dry matter (104 °C)	500	AG 236-2006 para lodos—Application in soil Gt	>550	>500 and ≤550	≤500
38	TE71C	Helminths—Sludge	egg/g (dry weight)	10	NOM-004-SEMARNAT-2002 (p.6) Mx	>11	>10 and ≤11	≤10
39	TE72C	Total coliforms—Sludge	MPN/g (dry weight)	1000	NOM-004-SEMARNAT-2002 (p.6) Mx	>1100	>1000 and ≤1100	≤1000
40	TE74C	Salmonella—Sludge	-	300	NOM-004-SEMARNAT-2002 (p.6) Mx	>330	>300 and ≤330	≤300
41	TE76C	Scope of sludge management	%	100	ST team	<33.33	≥33.33 and <66.67	≥66.67 and ≤100
42	TE78C	Identification of potential sludge consumers/users	YES-NO	YES-NO	ST team	NO	-	YES
43	TE80C	Quantification of GHG emissions	YES-NO	YES-NO	ST team	NO	-	YES

Variables and Thresholds for SA in Panajachel

Gt: Guatemala Regulation Mx: Mexican Regulation ST-Team: SludgeTec Team WHO (2006): Guidelines for SUWA—Vol 2

No.	Code (ID)	Variable	Unit	Threshold Value	Source	Red	Yellow	Green
44	TE83D	Operation Manual	YES-NO	YES-NO	ST team	NO	-	YES
45	TE84D	Regular maintenance	YES-NO	YES-NO	ST team	NO	-	YES
46	TE85D	Capacity sufficiency	YES-NO	YES-NO	ST team	NO	-	YES
47	TE86D	Accessible Sampling and processing equipment	YES-NO	YES-NO	ST team	NO	-	YES
48	TE87D	Discharge standards compliance	YES-NO	YES-NO	ST team	NO	-	YES
49	TE88D	Analysis frequency compliance—water	samples/year	2	AG 236-2006 para lodos Gt	<2	-	≥2
50	TE89D	Analysis frequency compliance—sludge	samples/year	2	AG 236-2006 para lodos Gt	<2	-	≥2
51	TE90D	Certification	YES-NO	YES-NO	ST team	NO	-	YES
52	TE91D	Health risk assessment	YES-NO	YES-NO	ST team	NO	-	YES
53	TE92E	Current management of health risks	YES-NO	YES-NO	ST team	NO	-	YES
54	TE93E	Health and safety equipment	YES-NO	YES-NO	ST team	NO	-	YES
55	TE94E	Performance of risk assessment	YES-NO	YES-NO	ST team	NO	-	YES
56	TE95E	Current management of risks	YES-NO	YES-NO	ST team	NO	-	YES
57	TE96E	Environmental impact assessment (EIA)	YES-NO	YES-NO	ST team	NO	-	YES
58	TE97E	Efforts to reduce or manage environmental impacts	YES-NO	YES-NO	ST team	NO	-	YES
59	TE98E	Presence or risk of groundwater pollution	YES-NO	YES-NO	ST team	YES	-	NO
60	TE99E	Presence or risk of surface water pollution	YES-NO	YES-NO	ST team	YES	-	NO
61	Ec2A	Per capita cost of WWT	USD/hab (inhabitants)/year	4–8	WHO	>8.8	>8 and ≤8.8	≤8
62	Ec7A	Budget deficit	YES-NO	YES-NO	ST team	YES	-	NO

Variables and Thresholds for SA in Panajachel

Gt: Guatemala Regulation Mx: Mexican Regulation ST-Team: SludgeTec Team WHO (2006): Guidelines for SUWA—Vol 2

No.	Code (ID)	Variable	Unit	Threshold Value	Source	Red	Yellow	Green
63	Ec8A	Valorisation of by-products	YES-NO	YES-NO	ST team	NO	-	YES
64	S1B	Personal interest in wastewater management problems	scale 1–4	between 1–4	ST team	≥ 1 and <2	≥ 2 and <3	≥ 3 and ≤ 4
65	S2B	Personal awareness of wastewater management problems	scale 1–4	between 1–4	ST team	≥ 1 and <2	≥ 2 and <3	≥ 3 and ≤ 4
66	S3B	Willingness to be informed about the wastewater management problems	scale 1–4	between 1–4	ST team	≥ 1 and <2	≥ 2 and <3	≥ 3 and ≤ 4
67	S4B	Accessibility to information	scale 1–4	between 1–4	ST team	≥ 1 and <2	≥ 2 and <3	≥ 3 and ≤ 4
68	S5B	Possibilities for providing a recommendation	scale 1–4	between 1–4	ST team	≥ 1 and <2	≥ 2 and <3	≥ 3 and ≤ 4
69	S9B	Personal acceptance of the current wastewater management	scale 1–4	between 1–4	ST team	≥ 1 and <2	≥ 2 and <3	≥ 3 and ≤ 4
70	S10B	Perception of social acceptance of the current wastewater management	scale 1–4	between 1–4	ST team	≥ 1 and <2	≥ 2 and <3	≥ 3 and ≤ 4

Variables and Thresholds for SA in Tepeji

Gt: Guatemala Regulation Mx: Mexican Regulation ST-Team: SludgeTec Team WHO (2006): Guidelines for SUWA—Vol 2

No.	Code (ID)	Variable	Unit	Threshold Value	Source	Red	Yellow	Green
1	TE9B	Temperature—WW	°C	40	AG 236-2006 Art. 28 Gt	>44	>40 and ≤44	≤40
2	TE12B	Total Nitrogen—WW	mg/L	80	AG 236-2006 Art. 28 Gt	>88	>80 and ≤88	≤80
3	TE13B	Total Phosphorus—WW	mg/L	20	AG 236-2006 Art. 28 Gt	>22	>20 and ≤22	≤20
4	TE19B	Faecal coliforms—WW	MPN/100 mL	10,000	AG 236-2006 Art. 28 Gt	>1100	>1000 and ≤1100	≤1000
5	TE22B	pH—WW	pH unit	between 6-9	AG 236-2006 Art. 28 Gt	<6 and >9	-	≥6 and ≤9
6	TE23B	Arsenic (As)—WW	mg/L	0.5	NOM-002-SEMARNAT-1996 (p.41) Mx	>0.55	>0.5 and ≤0.55	≤0.5
7	TE24B	Cadmium (Cd)—WW	mg/L	0.5	NOM-002-SEMARNAT-1996 (p.41) Mx	>0.55	>0.5 and ≤0.55	≤0.5
8	TE25B	Chromium (Cr)—WW	mg/L	0.5	NOM-002-SEMARNAT-1996 (p.41) Mx	>0.55	>0.5 and ≤0.55	≤0.5
9	TE26B	Copper (Cu)—WW	mg/L	10	NOM-002-SEMARNAT-1996 (p.41) Mx	>11	>10 and ≤11	≤10
10	TE29B	Nickel (Ni)—WW	mg/L	4	NOM-002-SEMARNAT-1996 (p.41) Mx	>4.4	>4 and ≤4.4	≤4
11	TE31B	Zinc (Zn)—WW	mg/L	6	NOM-002-SEMARNAT-1996 (p.41) Mx	>6.6	>6 and ≤6.6	≤6
12	TE32B	Mercury (Hg)—WW	mg/L	0.01	NOM-002-SEMARNAT-1996 (p.41) Mx	>0.011	>0.01 and ≤0.01	≤0.01
13	TE33B	Lead (Pb)—WW	mg/L	1	NOM-002-SEMARNAT-1996 (p.41) Mx	>1.1	>1 and ≤1.1	≤1
14	TE38B	Grease and oils—WW	mg/L	50	NOM-002-SEMARNAT-1996 (p.41) Mx	>55	>50 and ≤55	≤50
15	TE39B	Floating matter—WW	Absent-Present	Absent	AG 236-2006 Art. 28 Gt	Present	-	Absent
16	TE40B	Colour—WW	PCU	500	AG 236-2006 Art. 28 Gt	>550	>500 and ≤550	≤500
17	TE47C	Total Nitrogen—TWW	mg/L	30	WHO	>33	>30 and ≤33	≤30
18	TE54C	Sodium (Na)—TWW	meq/l	9	WHO	>9.9	>9 and ≤9.9	≤9
19	TE55C	Electric conductivity—TWW	µS/cm	30	WHO	>33	>30 and ≤33	≤30
20	TE56C	Faecal coliforms—TWW	MPN/100 mL	1000	NOM-003-SEMARNAT-1997 Mx	>1100	>1000 and ≤1100	≤1000
21	TE58C	Helminths—TWW	egg/L	5	NOM-003-SEMARNAT-1997 Mx	>5.5	>5 and ≤5.5	≤5
22	TE60C	Total Suspended Solids (TSS)—TWW	mg/L	100	WHO	>110	>100 and ≤110	≤100
23	TE61C	pH—TWW	pH units	between 6.5-8	WHO	<6.5 and >8	-	≥6.5 and ≤8
24	TE62C	Arsenic (As)—TWW	mg/L	0.1	WHO	>0.11	>0.1 and ≤0.11	≤0.1
25	TE63C	Cadmium (Cd)—TWW	mg/L	0.01	WHO	>0.011	>0.01 and ≤0.01	≤0.01
26	TE64C	Cyanide (CN)—TWW	mg/L	2	NOM-001-SEMARNAT-1996 (p.14) Mx	>2.2	>2 and ≤2.2	≤2
27	TE65C	Chromium (Cr)—TWW	mg/L	0.1	WHO	>0.11	>0.1 and ≤0.11	≤0.1
28	TE66C	Cupper (Cu)—TWW	mg/L	0.2	WHO	>0.11	>0.1 and ≤0.11	≤0.1
29	TE67C	Nickel (Ni)—TWW	mg/L	0.2	WHO	>0.11	>0.1 and ≤0.11	≤0.1
30	TE68C	Zinc (Zn)—TWW	mg/L	2	WHO	>2.2	>2 and ≤2.2	≤2
31	TE69C	Mercury (Hg)—TWW	mg/L	0.005	NOM-001-SEMARNAT-1996 (p.14) Mx	>0.0055	>0.01 and ≤0.01	≤0.01
32	TE70C	Lead (Pb)—TWW	mg/L	5	WHO	>5.5	>5 and ≤5.5	≤5

Variables and Thresholds for SA in Tepeji

Gt: Guatemala Regulation Mx: Mexican Regulation ST-Team: SludgeTec Team WHO (2006): Guidelines for SUWA—Vol 2

No.	Code (ID)	Variable	Unit	Threshold Value	Source	Red	Yellow	Green
33	TE71C	Grease and oils—TWW	mg/L	15	NOM-001-SEMARNAT-1996 (p.14) Mx	>16.5	>15 and ≤16.5	≤15
34	TE72C	Floating matter—TWW	Absent-Present	Absent	NOM-001-SEMARNAT-1996 (p.14) Mx	Present	-	Absent
35	TE73C	Colour—TWW	PCU	400	AG 12-2011 Art. 11 (p.10) Gt	>440	>400 and ≤440	≤400
36	TE74C	Water reuse	%	between 0–100	ST team	<33.33	≥33.33 and <66.67	≥66.67 and ≤100
37	TE88C	Odours	YES-NO	YES-NO	ST team	YES	-	NO
38	TE89C	Solid waste management	-	YES-NO	ST team	NO	-	YES
39	TE91C	Operation Manual	YES-NO	YES-NO	ST team	NO	-	YES
40	TE92C	Regular Maintenance	YES-NO	YES-NO	ST team	NO	-	YES
41	TE93C	Capacity sufficiency	YES-NO	YES-NO	ST team	NO	-	YES
42	TE94C	Accessible Sampling and processing equipment	YES-NO	YES-NO	ST team	NO	-	YES
43	TE95C	Discharge standards compliance	YES-NO	YES-NO	ST team	NO	-	YES
44	TE96C	Analysis frequency compliance—water	YES-NO	YES-NO	ST team	NO	-	YES
45	TE98C	Certification	YES-NO	YES-NO	ST team	NO	-	YES
46	TE99C	Health risk assessment	YES-NO	YES-NO	ST team	0	-	-
47	TE100C	Current management of health risks	YES-NO	YES-NO	ST team	NO	-	YES
48	TE101C	Health and safety equipment	YES-NO	YES-NO	ST team	NO	-	YES
49	TE102C	Performance of risk assessment	YES-NO	YES-NO	ST team	NO	-	YES
50	TE103C	Current management of risks	YES-NO	YES-NO	ST team	0	-	-
51	TE104C	Environmental impact assessment (EIA)	YES-NO	YES-NO	ST team	NO	-	YES
52	TE105C	Efforts to reduce or manage environmental impacts	YES-NO	YES-NO	ST team	0	-	-
53	TE106C	Presence or risk of groundwater pollution	YES-NO	YES-NO	ST team	0	-	-
54	TE107C	Presence or risk of surface water pollution	YES-NO	YES-NO	ST team	YES	-	NO
55	Ec2A	Per capita cost of WWT	USD/hab/year	1-1.5	WHO	>8.8	>8 and ≤8.8	≤1.5
56	Ec6A	Budget deficit	YES-NO	YES-NO	ST team	YES	-	NO

Variables and Thresholds for SA in Tepeji

Gt: Guatemala Regulation Mx: Mexican Regulation ST-Team: SludgeTec Team WHO (2006): Guidelines for SUWA—Vol 2

No.	Code (ID)	Variable	Unit	Threshold Value	Source	Red	Yellow	Green
57	Ec7A	Valorisation of by-products	YES-NO	YES-NO	ST team	NO	-	YES
58	S1B	Personal interest in wastewater management problems	scale 1-4	between 1-4	ST team	≥ 1 and <2	≥ 2 and <3	≥ 3 and ≤ 4
59	S2B	Personal awareness of wastewater management problems	scale 1-4	between 1-4	ST team	≥ 1 and <2	≥ 2 and <3	≥ 3 and ≤ 4
60	S3B	Willingness to be informed about the wastewater management problems	scale 1-4	between 1-4	ST team	≥ 1 and <2	≥ 2 and <3	≥ 3 and ≤ 4
61	S4B	Accessibility to information	scale 1-4	between 1-4	ST team	≥ 1 and <2	≥ 2 and <3	≥ 3 and ≤ 4
62	S5B	Possibilities for providing a recommendation	scale 1-4	between 1-4	ST team	≥ 1 and <2	≥ 2 and <3	≥ 3 and ≤ 4
63	S9B	Personal acceptance of the current wastewater management	scale 1-4	between 1-4	ST team	≥ 1 and <2	≥ 2 and <3	≥ 3 and ≤ 4
64	S10B	Perception of social acceptance of the current wastewater management	scale 1-4	between 1-4	ST team	≥ 1 and <2	≥ 2 and <3	≥ 3 and ≤ 4

Appendix F

Sustainability Assessment Results Per Variable (Panajachel, Guatemala)

R: Red Y: Yellow G: Green

No.	Code (ID)	Variable	Unit	Measured/Gathered Data	Category
1	TE7B	Temperature—WW	°C	23.50	G
2	TE8B	Biological Oxygen Demand (BOD)—WW	mg/L	1060.00	R
3	TE9B	Chemical Oxygen Demand (COD)—WW	mg/L	1150.00	R
4	TE10B	Total Nitrogen—WW	mg/L	33.05	R
5	TE11B	Total Phosphorus—WW	mg/L	26.65	R
6	TE12B	Faecal coliforms—WW	MPN/100 mL	2.75×10^{15}	R
7	TE14B	Total Suspended Solids (TSS)—WW	mg/L	610.00	R
8	TE15B	pH—WW	pH unit	7.27	G
9	TE19C	Temperature—TWW	°C	22.68	G
10	TE20C	Biological Oxygen Demand (BOD)—TWW	mg/L	287.50	R
11	TE21C	Chemical Oxygen Demand (COD)—TWW	mg/L	224.00	R
12	TE22C	Total Nitrogen—TWW	mg/L	33.50	R
13	TE23C	Total Phosphorus—TWW	mg/L	16.19	R
14	TE24C	Faecal coliforms—TWW	MPN/100 mL	1.32×10^{11}	R
15	TE29C	Total Suspended Solids (TSS)—TWW	mg/L	565.00	R
16	TE31C	pH—TWW	pH units	6.80	G
17	TE33C	Arsenic (As)—TWW	mg/L	Not detectable	G
18	TE34C	Cadmium (Cd)—TWW	mg/L	Not detectable	G
19	TE37C	Chromium (Cr)—TWW	mg/L	0.10	G
20	TE38C	Copper (Cu)—TWW	mg/L	0.01	G
21	TE41C	Nickel (Ni)—TWW	mg/L	Not detectable	G
22	TE43C	Zinc (Zn)—TWW	mg/L	0.12	G
23	TE44C	Mercury (Hg)—TWW	mg/L	Not detectable	G
24	TE45C	Lead (Pb)—TWW	mg/L	Not detectable	G
25	TE49C	Grease and oils—TWW	mg/L	367.50	R
26	TE50C	Floating matter—TWW	Present-Absent	Present	R

Sustainability Assessment Results Per Variable (Panajachel, Guatemala)

R: Red Y: Yellow G: Green

No.	Code (ID)	Variable	Unit	Measured/Gathered Data	Category
27	TE51C	Colour—TWW	PCU	648.00	R
28	TE52C	Water reuse	YES-NO	NO	R
29	TE55C	Arsenic (As)—Sludge	mg/kg dry matter (104 °C)	53.00	Y
30	TE56C	Cadmium (Cd)—Sludge	mg/kg dry matter (104 °C)	1.00	G
31	TE58C	Chromium (Cr)—Sludge	mg/kg dry matter (104 °C)	60.00	G
32	TE59C	Copper (Cu)—Sludge	mg/kg (dry weight)	100.00	G
33	TE62C	Nickel (Ni)—Sludge	mg/kg (dry weight)	21.00	G
34	TE64C	Zinc (Zn)—Sludge	mg/kg (dry weight)	0.15	G
35	TE65C	Mercury (Hg)—Sludge	mg/kg dry matter (104 °C)	Not detectable	G
36	TE66C	Lead (Pb)—Sludge	mg/kg dry matter (104 °C)	61.00	G
37	TE71C	Helminths—Sludge	egg/g (dry weight)	9.00	G
38	TE72C	Total coliforms—Sludge	MPN/g (dry weight)	9×10^{13}	R
39	TE76C	Scope of sludge management	%	Negligible	R
40	TE78C	Identification of potential sludge consumers/users	YES-NO	0.00	G
41	TE83D	Operation Manual	YES-NO	NO	Y
42	TE84D	Regular maintenance	YES-NO	NO	R
43	TE85D	Capacity sufficiency	YES-NO	NO	R
44	TE86D	Accessible Sampling and processing equipment	YES-NO	NO	R
45	TE88D	Analysis frequency compliance—water	samples/year	2	G
46	TE89D	Analysis frequency compliance—sludge	samples/year	2	G
47	TE90D	Certification	YES-NO	NO	R
48	TE91D	Health risk assessment	YES-NO	NO	R
49	TE93E	Health and safety equipment	YES-NO	NO	R
50	TE94E	Performance of risk assessment	YES-NO	NO	R
51	TE96E	Environmental impact assessment (EIA)	YES-NO	NO	R

Sustainability Assessment Results Per Variable (Panajachel, Guatemala)

R: Red Y: Yellow G: Green

No.	Code (ID)	Variable	Unit	Measured/Gathered Data	Category
52	TE99E	Presence or risk of surface water pollution	YES-NO	YES	R
53	Ec2A	Per capita cost of WWT	USD/hab/year	1.00	R
54	Ec7A	Budget deficit	YES-NO	Yes	R
55	Ec8A	Valorisation of by-products	YES-NO	No	R
56	S1B	Personal interest in wastewater management problems	scale 1–4	4.00	G
57	S2B	Personal awareness of wastewater management problems	scale 1–4	4.00	G
58	S3B	Willingness to be informed about the wastewater management problems	scale 1–4	3.60	G
59	S4B	Accessibility to information	scale 1–4	2.40	Y
60	S5B	Possibilities for providing a recommendation	scale 1–4	3.40	G
61	S9B	Personal acceptance of the current wastewater management	scale 1–4	1.20	R
62	S10B	Perception of social acceptance of the current wastewater management	scale 1–4	1.30	R

Sustainability Assessment Results Per variable (Tepeji, Mexico)

R: Red Y: Yellow G: Green

No.	Code (ID)	Variable	Unit	Measured/Gathered Data	Category
1	TE9B	Temperature—WW	°C	21.00	G
2	TE12B	Total Nitrogen—WW	mg/L	115.38	R
3	TE13B	Total Phosphorus—WW	mg/L	4.71	G
4	TE19B	Faecal coliforms—WW	MPN/100 mL	2.40×10^3	R
5	TE22B	pH —WW	pH unit	8.85	G
6	TE23B	Arsenic (As)—WW	mg/L	0.00	G
7	TE24B	Cadmium (Cd)—WW	mg/L	0.02	G
8	TE25B	Chromium (Cr)—WW	mg/L	0.05	G
9	TE26B	Cupper (Cu)—WW	mg/L	0.02	G
10	TE29B	Nickel (Ni)—WW	mg/L	0.05	G
11	TE31B	Zinc (Zn)—WW	mg/L	0.02	G
12	TE32B	Mercury (Hg)—WW	mg/L	0.00	G
13	TE33B	Lead (Pb)—WW	mg/L	0.00	G
14	TE38B	Grease and oils—WW	mg/L	5.41	G
15	TE39B	Floating matter—WW	Absent-Present	Absent	G
16	TE40B	Colour—WW	PCU	100.00	G
17	TE47C	Total Nitrogen—TWW	mg/L	120.62	R
18	TE55C	Electric conductivity—TWW	µS/cm	1.84	G
19	TE56C	Faecal coliforms—TWW	MPN/100 mL	2400.00	R
20	TE60C	Total Suspended Solids (TSS)—TWW	mg/L	46.00	G
21	TE61C	pH —TWW	pH units	8.32	R
22	TE62C	Arsenic (As)—TWW	mg/L	0.00	G
23	TE63C	Cadmium (Cd)—TWW	mg/L	0.02	R
24	TE64C	Cyanide (CN)—TWW	mg/L	0.64	G
25	TE65C	Chromium (Cr)—TWW	mg/L	0.05	G
26	TE66C	Cupper (Cu)—TWW	mg/L	0.02	G
27	TE67C	Nickel (Ni)—TWW	mg/L	0.05	G
28	TE68C	Zinc (Zn)—TWW	mg/L	0.02	G
29	TE69C	Mercury (Hg)—TWW	mg/L	0.00	G
30	TE70C	Lead (Pb)—TWW	mg/L	0.10	G
31	TE71C	Grease and oils—TWW	mg/L	5.00	G
32	TE72C	Floating matter—TWW	Absent-Present	Absent	G

Sustainability Assessment Results Per variable (Tepeji, Mexico)

R: Red Y: Yellow G: Green

No.	Code (ID)	Variable	Unit	Measured/Gathered Data	Category
33	TE73C	Colour—TWW	PCU	100.00	G
34	TE74C	Water reuse	%	100.00	G
35	TE88C	Odours	YES-NO	YES	R
36	TE89C	Solid waste management	-	NO	R
37	TE91C	Operation Manual	YES-NO	NO	R
38	TE92C	Regular Maintenance	YES-NO	Daily	G
39	TE93C	Capacity sufficiency	YES-NO	NO	R
40	TE94C	Accessible Sampling and processing equipment	YES-NO	NO	R
41	TE95C	Discharge standards compliance	YES-NO	NO	R
42	TE96C	Analysis frequency compliance—water	YES-NO	NO	R
43	TE98C	Certification	YES-NO	YES	G
44	TE100C	Current management of health risks	YES-NO	YES	G
45	TE101C	Health and safety equipment	YES-NO	YES	G
46	TE102C	Performance of risk assessment	YES-NO	NO	R
47	TE104C	Environmental impact assessment (EIA)	YES-NO	NO	R
48	TE107C	Presence or risk of surface water pollution	YES-NO	NO	G
49	S1B	Personal interest in wastewater management problems	scale 1–4	3.71	G
50	S2B	Personal awareness of wastewater management problems	scale 1–4	3.57	G
51	S3B	Willingness to be informed about the wastewater management problems	scale 1–4	3.29	G
52	S4B	Accessibility to information	scale 1–4	1.86	R
53	S5B	Possibilities for providing a recommendation	scale 1–4	2.71	Y
54	S9B	Personal acceptance of the current wastewater management	scale 1–4	2.64	Y
55	S10B	Perception of social acceptance of the current wastewater management	scale 1–4	1.64	R

References

1. Muga, H.E.; Mihelcic, J.R. Sustainability of wastewater treatment technologies. *J. Environ. Manag.* **2008**, *88*, 437–447. [CrossRef] [PubMed]
2. UN Habitat. *Water and Sanitation in Latin America and the Caribbean*; UN Habitat: Nairobi, Kenya, 2010; ISBN 978-92-1-132300-9.
3. UN Habitat. *Estado de las Ciudades de América Latina y el Caribe*; UN Habitat: Nairobi, Kenya, 2012.
4. Ballestero, M.; Arroyo, V.; Mejía, A. *Inseguridad Económica del Agua en Latinoamérica: de la abundancia a la inseguridad*; CAF: Caracas, Venezuela, 2015.
5. Mateo-Sagasta, J. *Reutilización de Aguas para Agricultura en América Latina y el Caribe: Estado, Principios y Necesidades*; FAO: Santiago, Chile, 2017.
6. Libralato, G.; Volpi Ghirardini, A.; Avezzù, F. To centralise or to decentralise: An overview of the most recent trends in wastewater treatment management. *J. Environ. Manag.* **2012**, *94*, 61–68. [CrossRef] [PubMed]
7. UN General Assembly A/RES/71/256—New Urban Agenda GA Adopted 68th Plenary N1646655-E. Available online: http://habitat3.org/wp-content/uploads/New-Urban-Agenda-GA-Adopted-68th-Plenary-N1646655-E.pdf (accessed on 3 May 2017).
8. UN. *Transforming our World: The 2030 Agenda for Sustainable Development: Sustainable Development Knowledge Platform*; United Nations: New York, NY, USA, 2015.
9. UN-Habitat. *Urbanization and Development: Emerging Futures*; World Cities Report; UN-Habitat: Nairobi, Kenya, 2016; ISBN 978-92-1-132708-3.
10. Alcamo, J.; Bennet, E. *Ecosystems and Human Well-Being: A Framework for Assessment*; Millenium Ecosystem Assessment, World Resources Institute, Eds.; Island: Washington, DC, USA, 2003; ISBN 978-1-55963-402-1.
11. Ghodeif, K. *Baseline Assessment Study for Wastewater Treatment Plant for Al Gozayyera Village, West Kantara City, Ismailia Governorate, Egypt*; Network of Demonstration Activities for Sustainable Integrated Wastewater Treatment and Reuse in the Mediterranean: Cairo, Egypt, 2013.
12. Seecon Sustainable Sanitation and Water Tool 2017. Available online: URL (accessed on 10 May 2018).
13. Deutsche Gesellschaft für Technische Zusammenarbeit (GTZ). *Ministry of Construction-Hanoi Asset Management Guideline for Drainage Sewerage System*; Ministry of Construction: Hanoi, Vietnam, 2009.
14. GIZ. *IWA The Energy Performance and Carbon Emissions Assessment and Monitoring (ECAM) Tool*; Deutsche Gesellschaft für Internationale Zussamenarbeit: Eschborn, Germany, 2017.
15. Balkema, A.J.; Preisig, H.A.; Otterpohl, R.; Lambert, F.J. Indicators for the sustainability assessment of wastewater treatment systems. *Urban Water* **2002**, *4*, 153–161. [CrossRef]
16. Quadros, S.; João Rosa, M.; Alegre, H.; Silva, C. A performance indicators system for urban wastewater treatment plants. *Water Sci. Technol.* **2010**, *62*, 2398–2407. [CrossRef] [PubMed]
17. Lundin, M.; Olofsson, M.; Pettersson, G.; Zetterlund, H. Environmental and economic assessment of sewage sludge handling options. *Resour. Conserv. Recycl.* **2004**, *41*, 255–278. [CrossRef]
18. Özerol, G.; Newig, J. Evaluating the success of public participation in water resources management: Five key constituents. *Water Policy* **2008**, *10*, 639–655. [CrossRef]
19. Mostert, E. The challenge of public participation. *Water Policy* **2003**, *5*, 179–197. [CrossRef]
20. Millington, P. *Integrated River Basin Management: From Concepts to Good Practice*; The World Bank: Washington, DC, USA, 2006; pp. 1–20.
21. Agarwal, A. *Integrated Water Resources Management*; Global Water Partnership: Stockholm, Sweden, 2000; ISBN 978-91-630-9229-9.
22. Lang, D.J.; Wiek, A.; Bergmann, M.; Stauffacher, M.; Martens, P.; Moll, P.; Swilling, M.; Thomas, C.J. Transdisciplinary research in sustainability science: Practice, principles, and challenges. *Sustain. Sci.* **2012**, *7*, 25–43. [CrossRef]
23. Reed, M.S.; Graves, A.; Dandy, N.; Posthumus, H.; Hubacek, K.; Morris, J.; Prell, C.; Quinn, C.H.; Stringer, L.C. Who's in and why? A typology of stakeholder analysis methods for natural resource management. *J. Environ. Manag.* **2009**, *90*, 1933–1949. [CrossRef] [PubMed]
24. Zscheischler, J.; Rogga, S. Transdisciplinarity in land use science—A review of concepts, empirical findings and current practices. *Futures* **2015**, *65*, 28–44. [CrossRef]

25. Von Korff, Y.; d'Aquino, P.; Daniell, K.A.; Bijlsma, R. Designing Participation Processes for Water Management and Beyond. Synthesis, part of a special feature on implementing participatory water management: Recent advances in theory, practice and evaluation. *Ecol. Soc.* **2010**, *15*. [CrossRef]

26. Grant, W.E.; Grant, W.E.; Pedersen, E.K.; Marin, S.L.; Marín, S.L. *Ecology and Natural Resource Management: Systems Analysis and Simulation*; John Wiley & Sons: New York, NY, USA, 1997; ISBN 978-0-471-13786-3.

27. Grant, W.E. Ecology and natural resource management: reflections from a systems perspective. *Ecol. Model.* **1998**, *108*, 67–76. [CrossRef]

28. Odum, H.T. *General and Ecological Systems*; University Press of Colorado: Niwot, CO, USA, 1980.

29. Schmidt-Bleek, F. *The Factor 10/MIPS-Concept: Bridging Ecological, Economic, and Social Dimensions with Sustainability Indicators*; ZEF: Tokyo, Japan; Berlin, Germany, 1999.

30. Ehrenfeld, J.R. The roots of sustainability. *MIT Sloan Manag. Rev.* **2005**, *46*, 23.

31. Kissinger, M.; Rees, W.E. An interregional ecological approach for modelling sustainability in a globalizing world—Reviewing existing approaches and emerging directions. *Ecol. Model.* **2010**, *221*, 2615–2623. [CrossRef]

32. Avellan, T.; Koester, K.; Selvam, V.; Brueggemann, K.; Ruiz, N. Understanding the Water-Soil-Waste Nexus. In Proceedings of the Constructed Wetland Knowledge Platform Meeting 2017, Dresden, Germany, 10 November 2017.

33. Loiseau, E.; Junqua, G.; Roux, P.; Bellon-Maurel, V. Environmental assessment of a territory: An overview of existing tools and methods. *J. Environ. Manag.* **2012**, *112*, 213–225. [CrossRef] [PubMed]

34. UNU-FLORES. *Sustainability of Wastewater Systems: Current and Future Perspectives—An Assessment Workshop*; United Nations University Institute for Integrated Management of Material Fluxes and of Resources (UNU-FLORES): Dresden, Germany, 2018.

35. Hacking, T.; Guthrie, P. A framework for clarifying the meaning of Triple Bottom-Line, Integrated, and Sustainability Assessment. *Environ. Impact Assess. Rev.* **2008**, *28*, 73–89. [CrossRef]

36. Pope, J.; Bond, A.; Hugé, J.; Morrison-Saunders, A. Reconceptualising sustainability assessment. *Environ. Impact Assess. Rev.* **2017**, *62*, 205–215. [CrossRef]

37. Pope, J.; Annandale, D.; Morrison-Saunders, A. Conceptualising sustainability assessment. *Environ. Impact Assess. Rev.* **2004**, *24*, 595–616. [CrossRef]

38. Toumi, O.; Le Gallo, J.; Ben Rejeb, J. Assessment of Latin American sustainability. *Renew. Sustain. Energy Rev.* **2017**, *78*, 878–885. [CrossRef]

39. Bertanza, G.; Baroni, P.; Canato, M. Ranking sewage sludge management strategies by means of Decision Support Systems: A case study. *Resour. Conserv. Recycl.* **2016**, *110*, 1–15. [CrossRef]

40. Fay, M.; Andres, L.A.; Fox, C.; Narloch, U.; Straub, S.; Slawson, M. *Rethinking Infrastructure in Latin America and the Caribbean*; International Bank for Reconstruction and Development/The World Bank: Washington, DC, USA, 2017.

water [MDPI]

Article

Analysis of Greenhouse Gas Emissions in Centralized and Decentralized Water Reclamation with Resource Recovery Strategies in Leh Town, Ladakh, India, and Potential for Their Reduction in Context of the Water–Energy–Food Nexus

Mounia Lahmouri, Jörg E. Drewes and Daphne Gondhalekar *

Chair of Urban Water Systems Engineering, Department of Civil, Geo and Environmental Engineering, Technical University of Munich, 80333 München, Germany; mounia.lahmouri@tum.de (M.L.); jdrewes@tum.de (J.E.D.)
* Correspondence: d.gondhalekar@tum.de

Received: 2 February 2019; Accepted: 10 April 2019; Published: 29 April 2019

Abstract: With the constant increase of population and urbanization worldwide, stress on water, energy, and food resources is growing. Climate change constitutes a source of vulnerability, raising the importance of implementing actions to mitigate it. Within this, the water and wastewater sector represents an important source of greenhouse gas (GHG) emissions, during both the construction and operation phase. The scope of this study is to analyze the GHG emissions from the current and future water supply scheme, as well as to draw a comparison between possible water reclamation with resource recovery scenarios in the town Leh in India: a centralized scheme, a partly centralized combined with a decentralized scheme, and a household level approach. Precise values of emission factors, based on the IPCC Guidelines for National Greenhouse Gas Inventories, previous studies, and Ecoinvent database, have been adopted to quantify the different emissions. Potential sources of reduction of GHG emissions through sludge and biogas utilization have been identified and quantified to seize their ability to mitigate the carbon footprint of the water and wastewater sector. The results show that the future water supply scheme will lead to a significant increase of the GHG emissions during its operation. Further, it is shown that decentralizing wastewater management in Leh town has the least carbon footprint during both construction and operation phases. These results have implications for cities worldwide.

Keywords: greenhouse gas emissions; decentralized water reclamation with resource recovery; Water-Energy-Food Nexus; climate change mitigation

---

1. Introduction

Today, more than half of the world's population lives in urban areas, and by 2050, 66% of the world's population is projected to be urban [1]. Apart from being the centers of demographic growth, cities and urbanized regions are considered to be centers of economic activity and increased consumption of resources. It has been reported that a global "business as usual" scenario will lead to an increase of 30–50% of water, energy, and food demand by the year 2030 [2]. As a consequence, rapid urbanization requires a rigid integrated approach able to foster urgent synergies between sectors to limit adverse consequences of urbanization, mitigate trade-offs, and ensure sustainable urban development [3].

In light of the dialogues within the international community, many regional and local governments are committed to addressing global challenges through the Sustainable Development Goals (SDGs) defined by the United Nations (UN), such as ensuring equitable drinking water and adequate sanitation to all (SDG 6), achieving sustainable food production and resilient agricultural practices (SDG 2), reaching universal access to affordable and clean energy (SDG 7), as well as strengthening resilient and adaptive capacities to climate related disasters worldwide (SDG 13). However, this support started before the support of the SDGs in the framework of the 2030 Agenda For Sustainable Development—many countries have supported the UN global Agenda 21, 1991 [4], the Habitat Agenda, 1997, the establishment of the Cities Alliance, 1999, and the New Urban Agenda, which was adopted during the UN Conference on Housing and Sustainable Urban Development (Habitat III) in Quito, Ecuador, in 2016. The water sector was a core theme of this last conference, which promoted "the conservation and sustainable use of water by rehabilitating water resources within the urban, peri-urban, and rural areas, reducing and treating wastewater" [5].

The importance of the current challenges accompanying rapid urbanization are understood and supported in the international development agendas. Nevertheless, cities and urbanized areas should also be recognized as being sources of solutions to, rather than only the cause of, these challenges. If an urban area is well-managed and well-planned, urbanization could be a powerful tool for sustainable development for both developing and developed countries.

The Urban Water–Energy–Food (WEF) Nexus approach [6] responds to the urgent need for policies and implementation mechanisms, which are equipped with an approach to counter "silo-thinking" in vital urban sectors. This approach addresses the crucial metropolitan sectors of Water, Energy, and Food (WEF), to reach a more efficient and effective use of resource cycles in urban and peri-urban areas. A key factor for the success of the Urban WEF Nexus approach is its methodology for both vertically- and horizontally-integrated governance, and building upon existing urban development strategies for enhanced coordination. In doing so, the Urban WEF Nexus offers customized solutions to sustainably govern rural-urban linkages and resource interdependencies through comprehensive spatial perspectives. It provides salient future-oriented solutions in the context of the Habitat III debate, as well as for the shaping of the New Urban Agenda.

1.1. Water Reclamation with Resource Recovery as Key Nexus Potential

As mentioned previously, the water and wastewater sector will be facing important challenges in the future, due to the increase of population and resulting increase of water demand. Further, water pollution is becoming an important issue, which could partly be due to the unsafe handling of wastewater, resulting in a higher incidence of waterborne diseases [7]. Another further source of vulnerability is climate change, as the observed effects already show water availability declination, shifts in precipitation patterns, as well as the deterioration of water quality [8]. This puts an extra adaptation burden on water and wastewater utilities against predicted climate change impacts.

At the same time, ensuring safe drinking water and safe wastewater treatment and disposal also contribute to greenhouse gas (GHG) emissions, since the different water utilities, which are the companies or institutions responsible for managing water and wastewater, are particularly energy intensive [9]. Water losses can imply even higher energy consumption, and methane or nitrous oxide emissions from wastewater treatment plants can increase the carbon footprint of the water sector [10].

Energy consumption, which is necessary at different levels in the water and wastewater sector, is witnessing an increasing demand, since the water supplied and wastewater to be treated is increasing. That increase implies a simultaneous increase in the GHG emissions, contrary to their aims of having a smaller carbon footprint. Some studies demonstrate the important share of GHG emissions in the water sector. For instance, the GHG emissions from the water sector in the United States (U.S.) has a percentage of 5% of total GHG emissions of the country, in comparison to a higher value in the United Kingdom (U.K.), due to end uses of water in heating [11]. In 2015, the share of GHG emissions from wastewater treatment and discharge in both the United Kingdom and United States was around 0.3% [12].

By drawing special attention to the synergies between climate change mitigation and adaptation measures, especially in the water sector, an integrated approach may support the effective action of cities on climate change and the implementation of the SDGs. The opportunities that this sector offers are multiple in terms of recovery of water, energy, nutrients, and other materials from the water reclamation with resource recovery. Investing in more efficient technologies can lead the water utilities to decrease their need and expenses related to energy use, as well as improve the carbon balance of water and wastewater companies and contribute to climate change mitigation. This is important in developing economies, which are developing urban infrastructure systems, as well as developed economies, where infrastructure systems are very water- and-energy intensive and may need modifications in view of future climate change challenges.

In this paper, we discuss carbon footprint analysis for the developing economy context of India. Centralized and decentralized water reclamation with resource recovery strategies were compared using the case study of Leh Town, Ladakh, India, and highlight potential for greenhouse gas emission reductions in context of the WEF Nexus.

1.2. GHG Emissions in India

The Intergovernmental Panel on Climate Change (IPCC) was created in 1988 by the World Meteorological Organization (WMO) and the United Nations Environment Program (UNEP), to allow assessments on all aspects of climate change and its impacts and propose realistic strategies for response. The scientific proof brought by the IPCC Assessment Report of 1990 [13] played a decisive role in the creation of the United Nations Framework Convention for Climate Change (UNFCCC), which is considered to be the key international treaty to mitigate global warming and adapt to the outcomes of climate change.

UNFCCC divides countries into three main groups. Annex I Parties to the Convention mainly includes industrialized countries that are members of the Organization for Economic Co-operation and Development (OECD), plus countries with economies in transition (EIT parties), such as the Russian Federation, the Baltic States, and several Central and Eastern European States. Annex II Parties consist of the OECD members of Annex I, but not the EIT Parties. These are required to provide financial resources to support developing countries to undertake measures against climate change. Finally, Non-Annex I parties are mainly developing countries [14].

In accordance with Articles 4 and 12 of the Convention and the relevant decisions of the Conference of the Parties (COP), Annex I parties are required to submit yearly to the UNFCCC secretariat national GHG inventories of anthropogenic emissions by sources and removals by sinks of GHG not under control by the Montreal Protocol: this is an international treaty that entered into force in 1989, and is designed to protect the ozone layer by eliminating the production of different substances that are responsible for ozone depletion [15]. These inventories are subject to an annual technical review process. In addition, these parties submit information on the implementation of the Convention, counting national activities to mitigate and adapt to climate change [16]. All non-Annex I Parties are also required to submit a national inventory of emissions by sources, to the extent its capacities permit.

The IPCC provides practical guidelines for countries to prepare GHG inventories. The 2006 IPCC Guidelines for National Greenhouse Gas Inventories includes the reporting methodology of emissions from the sectors of Energy, Industrial Processes and Product Use, Agriculture, Forestry and Other Land Use, Waste, and others. According to India's inventory on GHG emissions [17], the wastewater handling percentage increased from 47.4% in 1994 to 78.5% in 2010 of the GHG emissions of the waste sector, which had a share of 2% in 1994 and 4% in 2010 of the total GHG emissions of the country. In comparison, Germany demonstrates a stable value of GHG emissions from the wastewater treatment and discharge sector, with 9% in 1990 in comparison to 10% in 2015. The high percentage of GHG emissions from the wastewater sector in India proves the importance of developing studies targeting the wastewater sector, in order to identify sources of GHG emissions and potential for climate change mitigation.

This paper hypothesizes that in the context of a city in a developing economy, where existing water infrastructures are inadequate to meet basic needs and services, implementing a decentralized approach for wastewater treatment leads to less GHG emissions than a centralized wastewater treatment strategy. The study uses the town of Leh in India as a case study and hypothesizes that this analysis can lead to propose future opportunities of energy, water, and GHG saving potentials. The objectives of the study in Leh are to undertake the following:

- Study the current and planned wastewater management;
- Investigate and analyze different scenarios of possible water reclamation with resource recovery;
- Propose opportunities of improvement that can lead to energy, water, and GHG emission saving potential in the water and wastewater sector.

1.3. State-of-the-Art on GHG Emissions Estimation in the Wastewater Sector

The ongoing global warming trend is strongly related to the ongoing emissions of CO_2 and other GHG to the atmosphere, such as Nitrous Oxide (N_2O) and Methane (CH_4) [18]. The Global Warming Potential (GWP) is the index widely being used to compare the greenhouse effect of different gases. CO_2 has, by definition, a GWP of 1 [19]. CH_4 contributes significantly to the greenhouse effect. Its effect has a considerable lifespan of 12 years and a GWP of around 34 times higher than CO_2 [20]. This potent gas is responsible for 20% of anticipated warming. Similarly, and due to increasing human activities and their impact on the global environment, N_2O is increasing in the atmosphere at a rate of around 0.3% per year. With a GWP of 298 CO_2-equivalent over a 100-year time horizon and a lifetime of 120 years, N_2O is responsible for 6% of anticipated warming [21].

Water and wastewater management is facing different challenges in the coming years, as the energy use in the water sector is growing [11]. An important source of GHG is the electricity used, which depends on the process of production (from renewable sources or burning of fossil fuels). Depending on the energy mix of the country, electricity production can emit a more or less important amount of CO_2. India's electricity emissions increased from 901.7 g CO_2/kWh in 2005 to 926 g CO_2/kWh in 2012, which are values higher than the global averages [22]. A reason for that is the dependence of India on its large coal reserves for energy production and electricity generation. [23] proposes three impact categories, namely embodied energy, carbon footprint, and eutrophication potential, and key contributors to these environmental impact categories, to assess the environmental sustainability of wastewater treatment plants (WWTPs) during construction and operation [23].

The IPCC (2006) suggested a method to calculate GHG from wastewater treatment while stressing the importance of choosing country specific data, when available, and default values in the opposite case. [24] developed an energy consumption model to quantify the energy consumption and the relative GHG emissions from sewage sludge treatment processes, while taking into account the data from sewage sludge treatment plants in Osaka, Japan. Other studies use life cycle assessment methods for this aim. However, no general calculation method, besides the one developed by the IPCC, has yet been set up to compute the GHG emissions from the wastewater sector. Generally there is a lack of a proper and generalized methodology to study the interlinkages between water and energy use and intensity and GHG emissions in the wastewater sector, which poses a hindrance in comparing different water infrastructure systems across various regions [25]. This study aims to address this research gap.

Emissions can be distinguished into those occurring during the construction phase and the ones resulting during the operation. For the purposes of this paper, they are differentiated as follows into sources of methane (CH_4) and nitrous oxide (N_2O) emissions.

1.3.1. Sources of CH_4 Emissions

The common sources of CH_4 emissions in wastewater management are from sewers and from WWTPs. CH_4 emissions from sewers occur during conversion of organic carbon by methanogenic archaea, which are microorganisms that can reduce CO_2 with H_2 to CH_4 under anaerobic conditions, released to the atmosphere through manholes and atmospheric discharge points. Both the CH_4 emissions from gravity [26] and pressurized sewers [27] have been studied. However, the latter is considered to have a higher risk of production, because no air/water interface to diffuse oxygen into the liquid phase and promote aerobic conditions is present. In general, unless air or oxygen is used to control sulphide, anaerobic conditions are typically the predominant ones. Hence, sewers are optimal conditions for methanogenesis. Even if the IPCC, 2006, indicates that closed underground sewers, which are predominant in urban water infrastructure, do not imply important CH_4 emissions, studies have shown the contrary. [27] relate that the sewage CH_4 can contribute to an extra share ranging between 12–100% of GHG emissions to the ones from a WWTP. However, conventional methods for estimating these emissions are lacking. [27] also demonstrate a strong positive correlation between methane and sulphide production. The longer the sewers, the more production of methane and sulphide [27].

CH_4 produced in WWTP is approximated to 0.85% of the influent COD [20]. While [28] reveal a CH_4 percentage of 75% as CO_2 equivalent (CO_2 eq) in the total GHG emissions from wastewater management, [29] find that the proportion of CH_4 generation can constitute 12% of the total GHG emissions of the WWTP, three-quarters of which was found to be from the anaerobic digestion of the primary and secondary sludge. CH_4 sources from WWTP could include:

1.   Dissolved CH_4 that is produced and transported from the collection system and that is then stripped a the WWTP headworks or in the aerobic reactors;
2.   Dissolved CH_4 that is produced from anaerobic digestion and is left in the reject water that is recycled to the aerobic tanks, where a fraction of the dissolved CH_4 is ultimately stripped;
3. CH_4 gas produced in anaerobic digestion that escapes via gas piping leaks;
4. CH_4 gas produced in anaerobic digestion that is not fully combusted in cogeneration [30] or thermally destructed by flaring;
5. CH_4 gas escaping from digested sludge storage facilities [29]; and
6.   Anaerobic lagoon treatment systems.

1.3.2. Sources of N_2O Emissions

The N_2O emissions can occur from different sources. N_2O emissions from sewers: [31] tried to fill in this gap knowledge by providing a presumptive emission factor for gravity sewage systems around 1.4–1.8 g N_2O person^{-1} year^{-1}. N_2O emissions can also emanate from biological wastewater treatment. The proportion of N_2O contribution to a WWTP's total GHG emissions has been estimated to account for 78% [29]. During biological wastewater treatment, N_2O emissions can be the result of processes, including nitrification and denitrification through the following pathways [32]: (1) Autotrophic nitrification: (NH_2OH) oxidation in the conversion of ammonia (NH_3) to nitrite (NO2) and during the reduction of nitric oxide (NO) produced through ammonia oxidizing bacteria (AOB) denitrification; and (2) heterotrophic denitrification during the oxidation of nitrate NO_3- to N_2. IPCC includes a default emission factor for N_2O from wastewater treatment, which only takes into consideration the population, despite the various studies directly linking the N_2O emissions and the operational conditions of the wastewater treatment plant [33]. They should be accounted in the optimization strategies of GHG emission minimization. N_2O emissions from effluent discharge in receiving waters: N_2O can also be indirectly emitted from the conversion of the nitrogen in the effluent of a WWTP by nitrifying and denitrifying bacteria, depending on the nitrogen removal in the treatment plant [20].

1.3.3. Other CO_2, CH_4, and N_2O Emissions

Aside from CO_2, CH_4, and N_2O emissions from biological conversions, other sources of these emissions can include: (1) Onsite stationary fossil fuel combustion sources, which encompass engine generators, pumping equipment and driving processes at water treatment and pumping facilities; (2) electricity use; and (3) sludge disposal off-site due to vehicle fuel consumption for transport.

1.4. Water Reclamation with Resource Recovery

With the aim of reducing energy and costs, decreasing the stress on natural resources, and achieving a more circular economy, recovering resources from treated wastewater can be helpful. Water reuse practices over the world have proved that resource recovery is far beyond just a matter of technology. The products of this process have to meet some requirements, such as the ability to be sold and to fit to the purpose of the usage [34]. The recovery of resources from wastewater can be categorized into three main groups: water, energy, and nutrients, which are described in the following in detail.

1.4.1. Water

Reclaimed water is defined as the effluent from a WWTP, which has undergone a set of physical, chemical, and biological treatments to remove suspended solids, dissolved solids, pathogens, organic substance, nutrients, and metals [35]. It can be exploited for different purposes. Agricultural irrigation is the most common water reuse [36]. In the urban water use context, many successful pilot projects all over the world have implemented water reclamation and water reuse. In the industrial sector, reclaimed water can be used for cooling. Groundwater recharge and potable reuse have also received much interest [37].

In order to achieve the quality standards of the treated water for reuse, microbial and chemical contaminants can be removed by combinations of advanced oxidation processes, activated carbon and biofiltration, or integrated membranes, such as ultrafiltration followed by reverse osmosis [37].

Water and energy interaction are an important component to consider in designing reuse schemes. Decentralized reclamation can, in fact, reduce the energy needed for pumping reclaimed water over a long distance and tailor water qualities to the identified purposes [38].

The need for water reclamation in India is recognized through political frameworks. However, detailed treatment standards and types of water reuse applications have not been developed yet. The Ministry of Urban Development has addressed the issues in several advisories in recent years for recycling and reuse of wastewater. However, it was not until the last revision and update of the Manual on Sewerage and Sewage Treatment Systems [39] that guidelines were specified for the water quality for the treated water based on its intended reuse [40]. CPHEEO defines quality parameters for the non-potable water reuse in India. Different standards are defined for water reuse for toilet flushing, fire protection, vehicle washing, non-contact storage, horticulture, and golf courses [39].

1.4.2. Energy

Energy plays an important role in resource recovery, since electricity is used for the treatment of wastewater and it can also be recovered from the treatment process. Similarly, heat can also be recovered and used on site. It can sometimes exceed the energy needs of the plant. Hence, the discussions about energy use in the treatment processes are now moving from achieving energy efficiency to achieving energy neutrality or realizing a production which surpasses energy demand [34].

Meda et al., [41] state that the used water counts different forms of energy: potential energy, chemically-bound energy, and thermal energy. The two most important energy recovery paths from the water reclamation process are [34] anaerobic digestion and heat recovery–anaerobic digestion (AD) of the sludge, which can be burnt in a cogeneration unit. The resulting heat and electricity can be utilized for specific needs of the plant. AD typically converts the readily biodegradable portion of the solids. This portion can be enhanced through pre-treatment and co-digestion. The first opens the cells of the bacteria in the activated sludge, which allows for the release of the contents of the cells and makes them available to the anaerobic bacteria for conversion to biogas. Thermal hydrolysis, sonication, mechanical disintegration, and electrical pulse treatment are all technologies used in the pre-treatment of sludge. Co-digestion, on the other hand, is based on the addition of readily biodegradable feedstock, such as fats, oils, and grease, which can be locally collected, to the digester. These are afterwards co-digested with the sludge, increasing the biogas yield. In terms of heat recovery, [42] state that recovered heat from used water treatment plants can be used either for district heating, sludge drying, or thermophilic heating. While the implementation of these options depends greatly on the current price of carbon; district heating is considered to be the most promising in terms of carbon reduction potential [42]. Thermal energy can be concentrated using heat pumps, which can be characterized by different thermal capacities, to be used on-site in the process or off-site [43]. Other energy recovery paths include microbial fuel cells, a technology that uses bacteria as the catalysts to ensure the oxidation of organic and inorganic matter and generate a current.

1.4.3. Nutrients

The isolated management of the WEF resources has caused unsustainable and open nutrient loops. Synthetic fertilizers are produced from finite resources and energy intensive processes to ensure the bioavailable nutrients for agricultural soils. Following fertilizer production and food consumption, humans concentrate these nutrients in wastewater. However, in place of being reused and recycled, effluent of WWTPs is regularly discharged to freshwater sources, contributing simultaneously to eutrophication [44]. Therefore, WWTPs should be seen as reliable sources of nutrients and water used for agriculture through irrigation.

Phosphorus use is primarily recurrent in agricultural fertilizers. Almost 83% of the world's reserves are concentrated in Morocco, China, South Africa, and the United States [45]. Although phosphorus production is predicted to enter a long decline once it reaches its peak in 2033, its consumption level is predicted to be increasing [46]. Nitrogen is another important component used in agriculture. Its production is energy intensive and produces consequent GHG emissions [47].

These nutrients can be recovered from raw wastewater sources, semi-treated wastewater effluents, and biosolids from treatment byproducts. The latter can be applied to the land after treatment, leading to soil conditioning and synthetic fertilizer use reduction. The most important concerns of this recovery path are mainly health and safety issues, as well as odor and public acceptance [48]. Urine separation is another possible path of recovery that ensures nutrient loads of WWTPs are reduced significantly. This maximizes nutrient recovery from wastewater, as up to 80% of nitrogen and 50% of phosphorous of domestic water are contained in urine [49]. The major challenge relative to this technology is the requirement of intensive support and engagement from local communities, which is sometimes difficult to handle [50]. Other uncommon methods for resource recovery include controlled struvite crystallization, which is based on extracting struvite from sludge digester liquors because of its high concentrations of phosphorus, ammonium, and magnesium. This path can be economically feasible, however it can block valves, pumps, and pipes [51].

1.5. Decentralized Water Reclamation with Resource Recovery

Centralized wastewater management strategies face many different challenges to cover the needs of communities in terms of collecting and safely managing the generated wastewater, especially in low-income countries [52]. They are very water and energy intensive and either do not function well, or at all, with reduced availability of water and energy. They are also very capital intensive. As an alternative, decentralized water reclamation with resource recovery can contribute to reducing drinking water consumption, and is more cost-efficient than centralized systems [53], especially in less dense areas [54].

A decentralized approach for water reclamation with resource recovery offers many opportunities, including water reuse. Decentralized systems are defined as the collection, treatment, and distribution of water and wastewater near the point of use or generation [55]. They are more likely to reduce collection transport distances and use a smaller piping system installed at a shallower depth [56], all of which leads hypothetically to less use of energy and construction material, simultaneously reducing the relative GHG emissions in comparison to a centralized approach. In this section, different decentralized water reclamation with resource recovery options alongside potential GHG emissions are discussed.

"Decentralized Wastewater Treatment Systems" (DEWATS) refers to a modular systems approach, which provides treatment for wastewater flows ranging from 1 m^3 and 1000 m^3 per day per unit. They can also offer the possibility of achieving energy recovery for cooking, lighting, or power generation. These systems include 4 steps, encompassing the sedimentation, the anaerobic digestion, the aerobic and facultative decomposition, and the post treatment [57].

1.5.1. Constructed Wetlands

Constructed wetlands (CW) systems are systems that are designed and built to utilize the natural process for wastewater treatment, in addition to multiple ecosystem services (flood control, climate regulation, supporting biological diversity, and others [58,59]. This technology is able to treat to a good quality domestic, industrial, and municipal wastewater, storm water, and agricultural land runoff, as well as landfill leachate [60]. There are two types of CWs: Free Water Surface (FWS) or Subsurface Flow (SF). From this latter, Vertical Subsurface Flow CWs (VSSF) and Horizontal Subsurface Flow CWs (HSSF) can be distinguished. Due to the different removal efficiencies of these types of treatment, they are usually combined. A common design consists of two steps [58]:

- Several VSSF beds to remove organics and suspended solids and ensure nitrification; and
- Two or three HSSF beds to ensure denitrification and the further removal of organic and suspended solids. The pre-treatment step is necessary before the CW to avoid clogging, which is an obstruction of the free pore spaces due to the accumulation of solids. Conventional treatment settling time requires a retention time extended up to 4 h [61].

1.5.2. Fecal Sludge Management

Fecal sludge (FS) includes all liquid and semi-liquid components of pit and vaults, which accumulate in on-site sanitation installations [62]. These latter are systems of sanitation within the immediate surroundings of the occupier, such as septic tanks, private latrines or toilets, and aqua privies. In some cases, the Fecal Sludge Management (FSM) can be conducted on site and by extended in-pit consolidation and storage. In other cases, such as septic tanks, single pit, or vault installation, the sludge has to be collected and treated off site [63]. FS is known to be more concentrated in suspended and dissolved solids. It can be treated in a separate system, such as a waste stabilization pond, aerated pond, upflow anaerobic sludge blanket, activated sludge reactor, anaerobic biogas reactor [62], or co-treated with the sludge produced from a WWTP. It can constitute undigested or partially digested slurry or solids resulting from storage or treatment of blackwater or excreta. FSM encounter many challenges in terms of (1) sludge collection due to cost intensive mechanized vehicles and narrow streets, (2) health hazard for handlers and the public nearby in case of traditional and uncontrolled FSM, and (3) potential lack of treatment facilities, which leads to environmental damage. FSM involves the activities of collection, treatment, and reuse or storage.

2. Case Study Area Leh Town, Ladakh, and Basis for GHG Emission

Leh town, located on the banks of the upper Indus River, is the capital of the Ladakh district in the most northern Indian state, Jammu and Kashmir. Being in the Indian Himalayas, Ladakh is characterized by a cold high-altitude desert with precipitation averaging around 61 mm/annum, occurring in the format of both rain and snow [64].

Ladakh's valley is weighted down with well-maintained latera and terminal moraines. Therefore, in the event the melting of the glacier or during rain, water drains in the valley and is considered a primordial source of drinking water and irrigation water. Groundwater is available as an unconfinement condition in the deposits of gravel mixed with salt and sand material. However, groundwater is facing high risk of contamination due inefficient or absent wastewater disposal strategies [65].

A cold arid climate characterizes the desert of Ladakh. Severe cold is typical during dry winters, whereas during summer, moderate hot and dry weather distinguishes the climate in Leh.

The temperature can drop to −40 °C during winter and escalate to 35 °C during the summer season [66], with an annual daily average of 7.3 °C. A slight but noticeable warming trend has been signaled over the years while analyzing temperature datasets. This development was accompanied during the last decades by an inverse relationship regarding the precipitations patterns. The change in climatic pattern is predicted to have an important impact on the ecology, vegetation, the hydrology, as well as the human activities, such as agriculture [67].

Leh town is divided into 21 administrative wards and is classified as a Class III Urban Agglomeration and has a Nagar Panchayat or a Notified Area Council (NAC), which is an urban political unit in India comparable to a municipality. The population as per the Indian census increased from 28,639 in 2001 to 30,870 in 2011, resulting in a growth rate of 7.8%. In addition to the permanent resident population, a significant number of floating population exists. The tourism industry and army forces are the two main reasons for the latter, which is constituted of migrant laborers, who are mainly present in the short summer seasons from June to September [64].

Historically, agriculture has been an important part of the economy of the Ladakh region. Trading dairy, wool, and pashmina goat fiber, as well as farming barley and wheat, has been sufficient for subsistence for about 8800 years for the local communities [68]. The economy in Leh has, however, gradually shifted during the last decades. The opening to tourism in 1974 has generated many job opportunities and contributed to 50% of the region's Gross Domestic Product (GDP) as of 2010 [69]. The number of tourists has drastically increased by almost 280% from 2007 to 2017 [70]. This number is highly fluctuant yearly and seasonally, as tourists are mainly present during the summer, which almost matches the agricultural activity of the area [68].

To ensure the water supply to Leh town inhabitants, significant amounts of energy are being used to pump water from groundwater aquifers to the city. Water is for instance extracted from the Indus River Aquifer and is pumped over 300 m in vertical distance upwards and several kilometers horizontally to be stored in reservoirs, and later distributed by gravity in Leh town [71].

Due to the remote location of Leh town, it is, similarly to the whole Ladakh region, not connected to the Indian grid. Currently, the Ladakh region is supplied power from diesel generators (about 12 MW) of the army and paramilitary forces and micro hydroelectricity units (about 14 MW). Therefore, a large amount of diesel is still being used for supplying Leh-Ladakh's daily energy demands. This amount is estimated currently at 8000 L of diesel a day to cover the power needs of Ladakh [72]. The used fuel is imported from the outside, which outweighs both the expenses of the region, and the resulting GHG emissions from fuel combustion. In fact, on a regional level, Ladakh's power requirement is increasing by an annual rate of 7%. [72] predicts this to be around 140.5 MW by 2025.

The increasing energy demand has led the authorities to build a hydro power plant, Nimoo Bazgo Hydroelectric Plant, commissioned in 2014 with a capacity of 45 MW. Leh town is facing repeated power cuts despite this. Solar radiation is one of the most abundant energy resources in Ladakh, at around 7–7.5 kWh m^{-2} day^{-1}. This causes higher than the average irradiance on horizontal surfaces in India (5.6 kWh m^{-2} day^{-1}) [72]. In Leh, the solar radiation available is of the same order (5.54 kWh m^{-2} day^{-1}). Currently the full potential of this resource is yet to be exploited and can constitute a suitable off-grid and household scale technology [73]. Direct applications can be to:

- Lift the irrigation water from Indus river tributaries;
- Distribute drinking water to remote areas;
- Ensure heating and hot water for buildings.

A further resource for energy is the geothermal energy. [74] stress the existence of geothermal resource potential at suitable depths, which in the Ladakh region can be explored and developed.

Due to the lack of information for the GHG emissions inventory at a region or a town level, emissions from the state Jammu and Kashmir are related to in the following. The 2013–2014 Emissions Inventory of CO_2 reports emission in in Jammu and Kashmir as 0.68% of India's annual emissions [75], where the most important emitting sector is the energy sector, and the least is the waste sector.

2.1. Wastewater Management Situation in Leh

A sanitation system can be defined as to be able "to protect and promote human health by providing a clean environment and breaking the cycle of disease" [76] highlights the importance that these systems are sustainable: economically viable, accepted in the society, and suitable technically and institutionally.

The World Health Organization (WHO)/United Nations Children's Fund (UNICEF)'s Joint Monitoring Programme (JMP) for water supply, sanitation, and hygiene, allows one to access open source country specific data online [77]. It shows the evolution of access to sanitation between 2000 and 2015 in the rural and urban areas of India, analyzed by the service level provided. The trend shows an overall increase of access to sanitation through use of improved latrines, septic tanks, and sewer systems. Septic tanks are the most used type of sanitation in rural and urban areas, whereas improved latrine coverage has declined in urban areas. JMP also enables benchmarking and comparison of the sanitation service levels across countries. In order to reach safely managed sanitation service (SDG 6.2), improved sanitation facilities should be ensured, so that the excreta produced is either [78]:

- Treated and disposed in situ;
- Stored temporarily and then emptied and transported to treatment off-site; or
- Transported through a sewer with wastewater and then treated off-site.

On the other hand, the share open defecation has substantially decreased both on a rural and urban level in India from 2000 to 2015, with the increase of safely managed facilities or basic sanitation facilities. Open defecation is, however, still very recurrent in urban areas, with a share of over 50% in 2015.

According to [79], the access to sanitation in Leh is mainly based on "Ladakhi dry toilets", a traditional form of pit toilets, with a percentage of 74%. Flush or pour toilets are the second most important sanitation type with a share of 14%. The number of toilets according to the 2011 census data is around 4000 for a population of 30,870. However, 4% of the households still are not equipped with private toilets and rely on open defecation or public toilets. Districts like Leh Ladakh, Kargil, Ramban, and Jammu, where more than 60% of the households are dependent upon on-site sanitation facilities, need priority interventions. Also, in extreme cold climates, the fecal sludge and septage tends to freeze inside the containments and the decomposition processes comes to a halt below 0 °C. Since the septage will not decompose, the volume inside the tank or pit will be higher and it becomes necessary to desludge septic tanks in cold areas, such as Leh and Kargil, where temperatures drop as low as −20 °C to −40 °C in winter nights [80].

Currently, Leh encounters a lack of wastewater management structure to manage the increasing wastewater amounts. Many houses in the town have soak pits and septic tanks that gather the wastewater. However, this infrastructure is poorly managed and rarely emptied. Therefore, it can be assumed that seepage occurs from these pits and tanks infiltrating and polluting the upper stretches of groundwater in Leh [71]. A higher density of septic tanks is more likely to increase the probability of pollution [81].

Overflowing septic tanks may cause leeching into the ground, and septic tanks or soak pits that are poorly designed can increase the risk of contamination of groundwater. The Municipal Committee of Leh (MCL) realized the need to improve the management of the septic tanks through yearly desludging, as well as building a fecal sludge treatment plant (FSTP) [82]. The FSTP was designed by BORDA and CDD society and realized by MCL and the Blue Water Company. It targets the hotels that would not be connected to the proposed centralized treatment plant in Leh. The collection of the sewage is through a vacuum truck, which runs depending on clients need. The FSTP of Leh, which was inaugurated in August 2017, is a DEWATS solution that is based on the Planted Drying Bed Technology. It has a capacity of 12 m^3/day within a built up area of 60 m^2/m^3. The FSTP must run optimally in Leh's extreme climatic conditions, especially during the cold winter temperatures and low sludge inflow [82]. The collected sludge, after passing through a screen chamber, flows by gravity to one of the 10 Planted

Drying Beds. Here, the solid and liquid separation occurs and the digestion of a fraction of the solids is completed. The treated liquid is then directed to one of the two Horizontal Planted Gravel Filters. The final step is disinfection, where the water is places in a Polishing Pond. The sludge retained in the first drying bed accumulates and is dried. Once it reaches 0.9 m it can be removed and used as an organic soil conditioner.

The authorities in Leh have also started the construction of a centralized sewerage system that will service around 50% of Leh town, with a WWTP planned near the banks of the River Indus. A consulting firm was commissioned in 2009 to design the sewage line, main pumping station, and the WWTP, with the aim to collect sewage, transport it to a treatment plant, treat it, and dispose it without causing any health and environmental problems. The treated water is to be reused for agricultural irrigation partially and the rest discharged to the River Indus. The technology chosen for the water treatment is the activated sludge followed by a settling tank [83]. However, no disinfection step, which is necessary for water reuse for agricultural irrigation, is planned in the project. The excess produced sludge is to be dried using vast drying beds with no further valorization.

The estimated population for the design of the centralized WWTP in Leh was based on computing the average of five population estimation methods, while taking the census data from 1921 to 2001 and using the arithmetic progression method, geometrical increase method, incremental increase method, state urban method, and the geometrical method [83]. However, this has led to an overestimation of the estimated population for the year 2001, at 35,496, in comparison to 30,870 stated in the last census [79].

2.2. Scope of This Paper

This study undertakes a comparative analysis of GHG emissions emitted from the operation of the water supply scheme and the potential scenarios of water reclamation with resource recovery strategies in the town of Leh, India, in order to identify potential pathways of GHG emissions reduction. The carbon footprint and emissions assessment constitute primordial fundamentals in low-carbon research [84]. However, the ubiquitous term "carbon footprint", which is widely used, seems to not have a clear definition [85]. According to the Parliamentary Office of Science and Technology, 2006, calculating carbon footprint is summing up the total amount of CO_2 and other greenhouse gases emitted during the life cycle of a process.

In this study, the total carbon footprint is the sum of individual GHGs, in which carbon dioxide (CO_2), methane (CH_4), and nitrous oxide (N_2O), are expressed in CO_2 equivalent (CO_2 eq). The conversion of CH_4 and N_2O emissions to CO_2 eq will be realized by using their GWP, according to [86]. The GHG emissions are estimated for the water scheme, and in the proposed future water reclamation with resource recovery schemes in Leh during an assessment period of one year. Measuring the carbon footprint takes into consideration the difference between the wastewater to be generated and handled during the winter and the summer.

The GHG emissions from the water supply scheme included in this study are those from the abstraction of the water, treatment of the water, and distribution of drinking water. The GHG emitted from the operation of the wastewater sector and accounted for in the context of this study are those from the collection, treatment, and discharge of the wastewater, and transport and sludge disposal when applicable.

The GHG emissions estimated are the ones from the current water supply scheme run by the Public Health Engineering Department (PHE), which is the governmental institution responsible for ensuring water supply and wastewater management of the town of Leh, during the assessment period of one year. The analysis will be based on previous research results of current and future fuel consumption related by [87]. The GHG emitted from the operation of the wastewater sector is evaluated from three proposed scenarios of future water reclamation with resource recovery schemes in Leh, which take into consideration the complete or partial coverage of the proposed sewage treatment plant, and which are described as follows:

Scenario 1: Water Reclamation with Resource Recovery through a Centralized WWTP

The final design of the centralized wastewater collection and treatment solution is aimed at the year 2040, with a design population that has been recognized to be overestimated [65,83]. As the available data about the running costs of the WWTP are related to the final design year of 2040, the GHG emissions will be computed for the design year and population in 2040 (sub-scenario 1). A more realistic population estimation will be calculated for the year 2040, alongside the relative GHG emissions (sub-scenario 2). For the two sub-scenarios, the population connected will be the estimated population living in households during the year, as well as the floating population during the summer.

Different methods of population estimation exist, which take into consideration various parameters of population growth, leading to different results [88]. In order to highlight a more realistic estimation of the design population related by [83], a new population estimation will be computed as the average of the arithmetic progression method, the geometrical increase method, the incremental increase method, the state urban method, and the geometrical method, while taking as reference the census years from 1921 to 2011. It is the same method proposed by Tetra Tech, however it converges.

Scenario 2: Partial Management of the Wastewater from the Areas that can be Connected to the WWTP, Combined with Decentralized Water Reclamation with Resource Recovery

This scenario supposes a more realistic approach, where only some parts of Leh will be serviced through the centralized scheme. The unconnected parts of the city are subdivided into pockets where wastewater is treated using a decentralized water reclamation solution. The chosen treatment technology is Subsurface CWs due to their economic, technical, and climatic appropriateness.

Scenario 3: Partial Water Reclamation with Resource Recovery from the Areas that can be Connected to the WWTP, Combined with Household Level Treatment Solutions

Similarly to scenario 2, only a part of the town is connected to the WWTP. The rest of the households are hypothetically equipped with a septic tank, which is desludged once a year. The fecal sludge is transported by vacuum trucks to an FSTP, where it is treated. Wastewater generation is assumed to be similar to scenario 2. The emissions from the connected area to the WWTP will be based on interpolating the population connected to the WWTP with the results of the WWTP's designed population GHG emissions from scenario 1. The three scenarios are depicted in Figure 1 below.

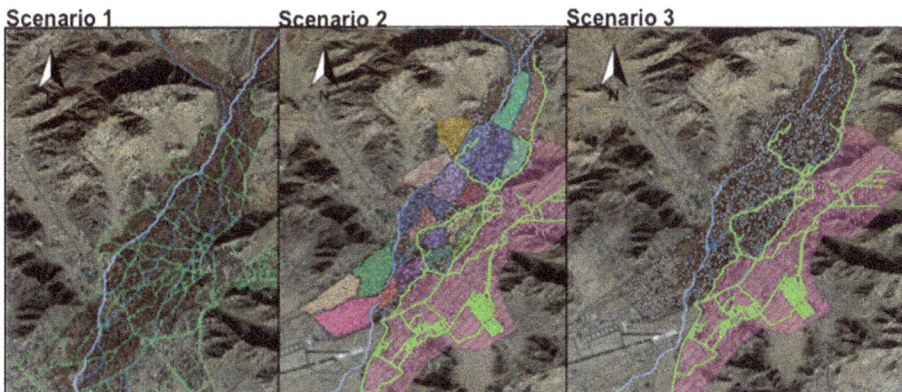

Figure 1. Three scenarios for water reclamation with resource recovery in Leh, India.

It is supposed that 80% of the daily consumed water becomes wastewater. The water use is 135 L/day/capita for the part of Leh connected to the WWTP for scenario 1.84 L/day/HH and 76 L/day/tourist during the summer for scenarios 2 and 3, according to [89]. All the hotels and guesthouses are assumed to be full during the summer. The number of tourists is, hence, equal to the number of beds. The number of beds is taken from the results of a survey completed in 2015. Wastewater generation takes into account the permanent residents during the year, combined with the surplus wastewater production during the summer.

2.3. GHG Emissions Assessment

As seen above, different sources of GHG emissions can occur while ensuring drinking water supply or safely managing the wastewater. The GHG Protocol orders the different emissions in three categories or "scopes", in order to avoid counting the emissions multiple times. Scope 1, or "Direct emissions", are those from sources owned or under the control of the water utility. The second scope, also called "indirect emissions", are the emissions from the consumption of electricity. The third scope encompasses "Other indirect emissions", which are the product of the water utility's activities [90]. Figure 2 includes the emissions in the wastewater sector comprised in the assessment, according to the scope. The ones assessed in the context of this study are highlighted in green, while those that exist, and which are not accounted for in this study, are highlighted in red.

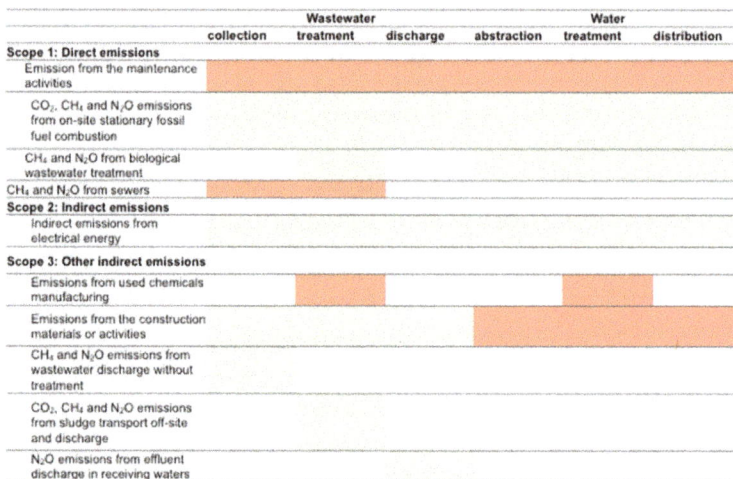

	Wastewater			Water		
	collection	treatment	discharge	abstraction	treatment	distribution
Scope 1: Direct emissions						
Emission from the maintenance activities						
CO_2, CH_4 and N_2O emissions from on-site stationary fossil fuel combustion						
CH_4 and N_2O from biological wastewater treatment						
CH_4 and N_2O from sewers						
Scope 2: Indirect emissions						
Indirect emissions from electrical energy						
Scope 3: Other indirect emissions						
Emissions from used chemicals manufacturing						
Emissions from the construction materials or activities						
CH_4 and N_2O emissions from wastewater discharge without treatment						
CO_2, CH_4 and N_2O emissions from sludge transport off-site and discharge						
N_2O emissions from effluent discharge in receiving waters						

Figure 2. Types of emissions captured in this study.

2.3.1. GHG Emissions from Operation of the Water Reclamation with Resource Recovery Scenarios

The overall methodology of the assessment is based on using emission factors, which quantify the emissions for every potent gas from their different sources of emission. This methodology is the same as used in the 2006 IPCC Guidelines for National Greenhouse Gas Inventories [20]. The chapter "Energy" will mainly be used for GHG emissions from electricity and fossil fuels usage during the collection, treatment, and discharge of the wastewater; or transport of sludge. Chapter 5 "Wastes" of the "2006 IPCC Guidelines for National Greenhouse Gas Inventories", which includes wastewater treatment and discharge, used for the choice of emission factors of GHG emissions related to wastewater treatment and discharge in the WWTP, and the wastewater treatment in septic tanks. The emission factors from wastewater treatment in CWs are derived from [91,92].

2.3.2. GHG Emissions from Construction Works

In order to account for the carbon footprint from the construction works of the different scenarios, datasets from the Ecoinvent 3.3 database have been used [93]. The assessment from the construction phase mainly includes the carbon footprint from excavation using hydraulic diggers and the production of cement and concrete used. In order to analyze the emissions from Scenario 1 or the connected parts of Leh serviced by the WWTP in scenarios 2 and 3, the Energy Performance and Carbon Emissions Assessment and Monitoring (ECAM) tool was used [94]. The ECAM tool is an open access web-based tool, developed by the German International Development Agency (GIZ) project Water and Wastewater Companies for Climate Mitigation (WaCCliM). It enables water utilities to:

- Assess the origin of their GHG emissions and their relative weight;
- Identify potential areas for improvement, regarding energy savings; and
- Forecast the reduction impact of future measures and monitor the results after their implementation.

As the WWTP of Leh is still not built, the objective of using the ECAM tool is to estimate the future GHG emissions from the operation of the planned WWTP, as well as identifying high sources of emissions during the collection, treatment, and discharge of the wastewater, and pointing out the GHG emissions from the current water supply scheme in Leh.

2.4. GHG Reduction Potential

In order to value the potential of reducing GHG emissions from the wastewater sector, valorization sludge and biogas, through electricity production and usage as a cooking fuel, will be discussed in scenario 1 and scenario 2 in the following. Electricity production from biogas is computed through the formula of [95].

The biogas potential related to the chosen serviced population can be estimated for the assessment year, as follows; the biogas is supposed to be produced during the whole year. The production estimation takes into account the availability of organic matter during summer and winter. [96] specify that the biogas requirement per meal in India is 0.07 m^3/meal. If three meals are typically consumed per day, 0.21 m^3 of biogas per person per day is needed; if only two meals are consumed, 0.14 m^3 daily per person is required. However, due to the technical challenges of transporting the biogas from the point of biogas production to the households, losses should be accounted before utilizing it. The losses from small-scale biogas plants can be due to cracks in biogas digesters and tubing or can be done intentionally when the biogas surpasses the needs. Only the losses from the cracks are assumed here to estimate the maximum potential of utilizing biogas. The value of losses is supposed according to [97] for a value of 5%. The cooking fuel used in domestic cookers in Leh is supposed to be natural gas. This estimation is rough, as the use of other types of cooking fuel would lead to different related GHG emissions.

3. Results and Analysis

3.1. Water Supply Sector Emissions

The Public Health and Environment (PHE) institution is responsible for ensuring the drinking water supply to the population of Leh. The water abstraction and distribution system as of 2017 is composed of:

- Four stages of pumping stations extracting around 1.5 MLD of water and using around 560 L of fuel daily [87];
- Spring water [87];
- Six borewells extracting 1.24 MLD of groundwater daily and using around 86 L of diesel daily for running the pumps [87].

PHE also operates 10 water tankers in Leh town during the year. It extracts 80% of the water from tubewells, whereas the rest comes from spring sources. The average distance driven is 70 km per day for all operating water tankers. The resulting daily fuel consumption is then 17.5 L. The future water supply scheme proposed by PHE is designed to supply about 10 MLD of water per day by 2042 to the town from the Indus aquifer, as well as through using the six same borewells. It is reported that the proposed future scheme would lead to a consumption of around 3600 L of diesel per day for the 4 pumping stages, 86 L/day of diesel for the 6 borewells, as well as 17.5 L from operating the water tankers [87].

After abstraction and pumping of the water from the Indus Aquifer, it is conveyed to storage reservoirs. From there, the residents receive the supplied water by gravity through private household connections and public standposts. The water receives no treatment before distribution. Hence, there are no GHG emissions related to the treatment. As the distribution system is combined with the abstraction system, the resulting GHG emissions can be described together.

The water supply is currently the source of 658 t CO_2 eq per year, with the biggest share from the fuel used for pumping the water from the Indus River Aquifer (around 556 t CO_2 eq per year), followed by the water extraction from the borewells (83 t CO_2 eq per year), and finally the water tankers (17 t CO_2 eq per year). The proposed future supply scheme which would allow to supply an amount of 135 L/day/capita for the estimated population by 2042 would imply an increase of 540% of CO_2 eq emitted during the operation of the pumping system, in comparison to the current water supply scheme. This value certainly highlights the need to propose potential possibilities, according to which the carbon footprint from the operation of the water supply scheme can be decreased.

To make a comparison between Leh's current figure in terms of GHG emissions in the water supply scheme and other countries or cities, data were taken from the literature review, were normalized, and compared according to the GHG emissions per m^3 of supplied water. The current scheme in Leh shows the second highest value of GHG emissions per m^3 of supplied drinking water in comparison to China [10], the United Kingdom [98], the United States [99], Durban in South Africa [100], and Oslo in Norway [101]. This could be explained, similarly to Australia, by the energetic mix in both locations, which is based on burning fossil fuels to cover the energy requirements, diesel fuel in the case of Leh, and coal and oil in the case of Australia [102]. This high value can also be justified in the case of Leh by the need to pump the water from the Indus river aquifer south of the town, up to 345 m elevation change to the northern part of the town.

PHE's water supply scheme is seen as inappropriate, since it is allowing supply of water for only a few hours per day. This is not sufficient for the summer season, when tourists increase the stress on the available and supplied water by PHE. Consequently, many households, guesthouses, and hotels own private borewells and use a pumping unit on site to extract groundwater for their daily consumption [89]. If fossil fuels are used for pumping, the GHG emissions from ensuring sufficient water supply is even higher. The absence of regulation for drilling private borewells in Leh does not only lead to a potential pollution of the groundwater due to the proximity of drinking water wells to septic tanks [71], but also these practices constitute a catalyzer to climate change. It is then necessary to suggest possibilities of GHG reduction through utilizing renewable energies or through minimizing the energy needed for pumping in the water supply scheme.

Ref. [101] propose a comparison of electricity usage patterns during pumping and treatment of drinking water in different cities of India. This allows a comparison in terms of GHG emissions from the electricity used, as well as from fuel usage in the case of Leh. Over 95% of the GHG emissions in the case of Hyderabad and Shimla are due to the extraction and pumping of water [103], while it is of 100% in the case of Leh. The GHG emissions from water supply in Leh are in the same range of the ones of Hyderabad, while Shimla, which is situated in a hill station in the Himalayan Mountains, has to source the water from long distances and pump it uphill. This explains the high value of GHG emissions per m^3 supplied.

3.2. Wastewater Sector Emissions

The GHG emissions from the operation of water reclamation with resource recovery schemes in Leh is discussed and compared between three scenarios of:

- Fully centralized scheme
- Combined decentralized and centralized scheme

- Combined household level and centralized scheme

Scenario 1: Fully Centralized Scheme

Computing the new population projection while including the census of 2011 shows an overestimation ranging from 15,000 to 20,000 people by year 2040 in comparison to the design population estimated by Tetra Tech. Consequently, the WWTP is more likely to operate under the designed capacity. The two estimations will be taken into account in the analysis to provide a range of GHG emissions for the centralized water reclamation with resource recovery scheme.

The first scenario is based on hypothetically connecting the total population of Leh town to the WWTP. It is also supposed that the electricity needed for running electromechanical engines, for pumping purposes, and for aerating the activated sludge tanks being supplied directly from a local electricity supplier through a connection to the power grid.

According to the Tetra Tech report, the estimated population used to design the system is at 120,836, while the new estimated population computed in this study is 52,313 during the year and 68,007 during the summer, including the floating population. The discrepancy between the two population estimations could be explained by the fact that the neighboring villages of Choglamsar, Saboo, and Stok, with a total population of 26,471 as of 2001 census [79], will most probably not be able to connect to the sewerage line.

The results show that the step that contributes most to the GHG emissions is the wastewater treatment step. This could mainly be due to the high electrical energy requirement for aerating the activated sludge tank. It has been reported by [104] that the aeration processes in activated sludge can consume up to 60% of the total plant power requirements. Further, around 80% of GHG emissions are owing to the electrical energy that would be used on site or for collecting the wastewater, followed by N_2O emissions emitted during biological treatment and discharge of treated wastewater.

The results in terms of GHG emissions from the two population estimation scenarios show a big discrepancy, since these are in the function of the inflow wastewater to the plant and can greatly affect the results. It is shown that if more people are connected to the WWTP, the GHG emissions will be higher. On the other hand, the design capacity of the WWTP (population estimation 1) is much higher than the projected one from the new population estimation 2. The lower value of wastewater inflow could lead to operation issues, especially if the pumps run dry or operate near to the shut-off head. This could actually cause overheating and degradation of the quality of the pumps [39]. For this reason and to avoid issues from running the WWTP under capacity, it can be suggested to allow hotels to connect to the plant and simultaneously increase the wastewater inflow to the plant, or to decrease the capacity of the planned WWTP.

Ref. [105] state that a typical Biological Nutrient Removal activated sludge WWTP for 100,000 PE may be the source of around 5300–7400 t CO_2/year of GHG emissions. Their study took in fact into account plant power requirements and energy recovery, as well as energy embodied in the chemicals consumed and transport requirement or disposal of biosolids produced. This is roughly comparable to the value of 4600–7500 t CO_2/year of population equivalent to 90,000 and 120,000 in this study, where energy recovery and embodied chemical energy are not accounted for.

3.2.1. Sludge Disposal

The design of the planned WWTP in Leh does not include a valorization of the produced sludge. It is only proposed to stabilize it through drying it in solar drying beds [83]. Here, different possibilities of sludge disposal will be discussed, as well as their impact on the carbon footprint of the plant.

The discussion will be based on the sludge production from the design population estimation (92,951 during the winter and 120,836 during the summer) to have a maximum value of the possible GHG emissions. The results of the inventory include the sum of the transport of the sludge, as well as the disposal method. The discussed methods are transport of the sludge and disposal of it to a landfill, and transport of the sludge and application of it to the land.

The transport is assumed to be using a truck of a capacity of 5000 kg. The estimated dry weight of the sludge to be sent to the disposal site is at 26,859 kg for the sub-scenario 1. The landfill is estimated to be at a distance of 7.6 km from the WWTP. The land application is supposed to be in the agricultural areas of Leh. The distance from the WWTP to the agricultural area is roughly 10 km. The need to apply the sludge is assumed to be during the summer months only, hence the transport of the sludge to the land is supposed to be completely fulfilled in 6 trips.

After analyzing the results, the total carbon footprint from transporting the sludge is less than 1% from both disposal scenarios and can be supposed as negligible. Land application is a better alternative to manage the sludge produced from the WWTP, as an alternative to the landfill. Although this last leads to lower N_2O emissions than land application, it produces an important amount of CH_4 due to the anaerobic conditions in the landfill. Hence, it is recommended to valorize the produced sludge, by land application for example, instead of disposing it in a landfill, as it offers a potential of 64,076 kg CO_2 eq saving. Land application of the sludge should only be done after biological, chemical, or thermal treatment of the sludge to reduce its fermentability and health hazards.

In utilizing sludge for land application, the concentration of toxic elements, such as Zinc, Copper, Nickel, Cadium, Lead, and Mercury, must not exceed certain limits [106]. Land application of the sludge can enhance the fertility of soil and increase the crop yield of agricultural lands. It can also be used to regenerate barren lands to productivity or to provide the vegetative layer needed for controlling soil erosion. A significant amount of sewage sludge should be applied to land to provide sufficient nutrients and organic matter for supporting the vegetation until achieving a self-sustainable ecosystem (7 to 450 t dry weight/ha) [107]. The produced sludge can be used in Leh town to recover the fertility of the barren land, located in the proximity of the WWTP. If 7 t dry weight/ha of sludge is applied to the land, almost three ha of barren land can be regenerated.

3.2.2. Biogas Potential and Usage

The estimated biogas production during the period of assessment for the two population estimation scenarios is as follows. The produced biogas, if not valorized or flared, would lead to higher CH_4 emissions in the air. The valorization of the biogas produced from a WWTP can be achieved through different paths [108]. If the electricity is used on site to cover a percentage of electrical energy requirements, this would lead to GHG reduction potential of 907 t CO_2 eq (sub-scenario 1) and 510 t CO_2 eq for (sub-scenario 2).

The production of electricity through an internal combustion engine requires a pre-treatment to remove traces of Hydrogen Sulfide (H_2S) and humidity, which can both lead to damaging the equipment and increasing the heating power of biogas [109]. However, onsite electricity consumption of the produced biogas could cover a share of 13% the electricity needed, allowing savings for costs of the electricity. Alternatively, if the biogas from sub-scenario 1 is used as a cooking fuel, it would cover the need of a range of 7000–8000 persons for preparing three meals a day for one year for sub-scenario 1, or 4000–5000 persons for covering the fuel needed for cooking three meals a day for one year for sub-scenario 2. The biogas produced can be sold to interested clients in the surroundings of the WWTP for daily usage, for instance to the inhabitants of the military area. If the biogas produced is used as a cooking fuel, between 110 and 210 t CO_2 eq will be saved per year, depending on whether two or three meals a day are consumed.

Scenario 2: Combined Decentralized and Centralized Scheme

The parts of Leh that will not be connected to the centralized sewerage system can hypothetically be serviced by defining decentralized pockets with an alternative collection and CW treatment option. The pockets defined in this study can be divided into two groups, according to the number of households in each pocket: 18 small size pockets of less than 60 households as of 2015, and 3 big size pockets of households number higher than 100 as of 2015. In addition to these pockets, the areas that will also be connected to the WWTP encompass 2992 households as of 2015, as well as a bed capacity of 2452. The different size pockets have the following wastewater generation:

between 1000 L/d and 2000 L/d in 11 pockets

between 1000 L/d and 24,000 L/d in 7 pockets

higher than 24,000 L/d in 4 pockets

Two sub-scenarios are studied to draw a comparison between the gravity sewers and vacuum sewers for collecting the wastewater in the different pockets. One of the first differences is the electrical energy requirement for the second, which is estimated by [110] to be in the range of 0.2–0.7 kWh/m^3. In order not to underestimate the GHG emissions from these systems, the highest value was chosen for the inventory. Due to this electrical energy requirement, the vacuum sewer would emit 64 t CO_2 eq, in comparison to zero GHG emissions from operation of the gravity sewer.

The collection step is the largest contributor to the GHG emissions in the case if vacuum sewers are used. If gravity sewers are used, the treatment process is the largest contributor to GHG emissions, similar to scenario 1. The treatment step includes both for HSSF CWs and for VSSF CWs emissions from the biological degradation of organic matter. The VSSF CWs require electrical energy to lift the water during the treatment, but despite this, the emissions from HSSF remain higher than the ones from VSSF. This can be explained by the higher CH_4 emissions factor reported by [92] for the HSSF (a median value of 7.4 mg CH_4–C m^{-2} h^{-1}) in comparison to the ones of VSSF with a value of 2.9 mg CH_4–C m^{-2} h^{-1}. In addition to that, the result is also due to the fact that HSSF needs more area per PE than VSSF [111].

While comparing the emissions per m^3 of treated wastewater in the pockets and in the areas connected to the WWTP, the results of emissions per m^3 during the treatment for the connected area to the WWTP are higher than those of the decentralized pockets. This can again be explained by the high electrical energy requirement of the WWTP.

3.2.3. Biogas Potential and Usage

In terms of biogas potential for the different pockets, the produced biogas if not valorized or flared would lead to a higher CH_4 emission in the air. If all the biogas produced in the pockets is utilized for cooking, it would lead to approximately 90–100 persons having enough cooking fuel for three meals a day for one year, or 135–145 persons for two meals a day for one year. Using biogas instead of LPG can lead to a saving potential of around 3 t CO_2 eq from the same number of people using LPG for covering their energy requirement for cooking fuel for one year. Allowing the use of biogas can be a great incentive for the population to support the installation of a decentralized water reclamation with a resource recovery solution, which can simultaneously produce biogas. The biogas could be utilized at hotels or guesthouses, for example, in return for a higher share of investment for installing the decentralized solution or for covering the operation and maintenance costs.

Scenario 3: Combined Household Level and Centralized Scheme

The water reclamation with resource recovery scheme in this scenario is based on combining a household level on-site wastewater treatment approach with a centralized FSTP that treats and stabilizes the fecal sludge, alongside the centralized treatment through the WWTP of the similarly connected areas, as in scenario 2. Every household and hotel is hypothetically equipped with a septic tank.

The results of GHG emissions from the population with on-site sanitation shows that the sludge disposal is the largest contributor to GWP with a value of 331 t CO_2 eq, in comparison to the on-site treatment of the generated wastewater (260 t CO_2 eq) and the sludge transport to the FSTP (260 t CO_2 eq). The CH_4 emissions from septic tanks are higher (92% of total emissions) than N_2O emissions, which account only for 8%. This can be explained by the anaerobic digestion occurring in the septic tank, which results in the production of CH_4 [112]. CH_4 emissions from septic tanks are considered as a big contributor to global warming, since CH_4 has a GWP of 34 over 100 years [86], especially if there is no utilization of the CH_4 produced. A possible strategy for reducing the CH_4 emissions from septic tanks would be to capture it, in order to utilize it as a cooking fuel or simply flare it. Flaring biogas can actually decrease the carbon footprint by converting all CH_4 to CO_2 [101]. However, the flare for a household plant has nearly the same costs as a flare for a large plant of 20,000 inhabitants. Therefore, the specific costs per person are high for flares implemented in small systems at the household level. Since septic tanks lead to CH_4 emissions, and since flares are neither economical nor practical at the small scale, installing septic tanks could have an important negative impact on climate change if this pre-treatment method is adopted [113].

Comparison of the Scenarios

Since the first scenario deals with the future planned centralized sewerage system for the projected population by 2040, while the second and the third scenarios analyze the wastewater management for the population by 2015, the comparison of different emissions from different scenarios will be completed by defining a comparison ratio of the produced emissions divided by the cubic meter of treated wastewater in each scenario. This is also important, since for the first scenario, only the households are supposed to be connected, while in the two others, the wastewater produced from the households, guesthouses, and hotels is managed.

Figure 3 highlights the comparison of the GHG emissions per cubic meter of treated wastewater in scenario 1 (sub-scenario 1), scenario 2 (if the collection is ensured with gravity sewers and the treatment is with VSSF), and scenario 3. The comparison of the 3 scenarios shows that the combination of the proposed centralized system combined with a household-level wastewater treatment solution has the highest ratio of GHG emissions per m^3, in comparison to scenario 1 and scenario 2. Scenario 2 shows the lowest carbon footprint during operation. This could be explained by the fact that for scenario 2 and 3, the wastewater treated by the decentralized treatment solution or by onsite treatment is much lower than the volume of water treated by the WWTP from the connected areas. Hence, it is more appropriate to calculate the ratios of GHG emissions of centralized, decentralized, and household level wastewater management separately.

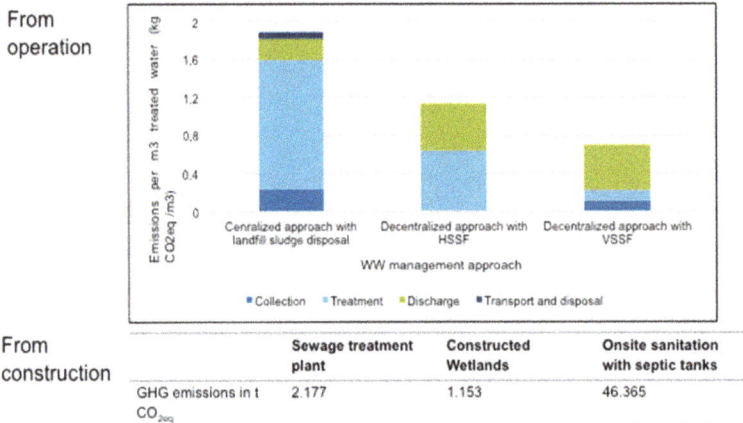

Figure 3. GHG emissions per m^3 of treated water in the 3 scenarios.

When comparing these in terms of GHG emissions per m^3 of treated wastewater, the highest ratio is allocated to the centralized approach, followed by the decentralized approach using CWs. A decentralized approach could lead to a lower carbon footprint during its operation. The household approach has been omitted from this comparison due to the very high values in comparison to the other approaches.

3.2.4. GHG Emissions from Plant Construction

Aside from the carbon footprint of the operation of the different water reclamation with resource recovery solutions, the construction phase can also have a considerable impact on GHG emissions, especially when comparing the construction requirement of a high capacity WWTP in comparison to decentralized or household scale water reclamation with resource recovery alternatives. The analysis of the GHG emissions from the excavation works and production of cement concrete, plum concrete, and mortar concrete in the installation of gravity mains, rising mains, pumping station, and WWTP show an overall value of 2177 t CO_2 eq. With a percentage of more than 20%, the share of installation of the gravity sewage lines account for 625 t CO_2 eq. The excavation activities and the concrete requirement from the WWTP account for over 60% of the total carbon of the centralized sewage scheme, with a value of 1127 t CO_2 eq. This does not include the emissions due to the use of reinforcement steel also used for the construction of the WWTP, which has a share of 33% of the carbon footprint of a WWTP's construction and its dismantling [93]. Taking this into account would lead to an even higher carbon footprint. The same occurs when the transport of materials to the site of construction, and the electricity used for construction are also considered.

On the other hand, if the lifetime of the plant is 30 years, the GHG emissions related to the construction of the WWTP, sewage system, and the main pumping station would account only for 2% of the GHG emissions from the projected electricity used for pumping the wastewater and treating it for 120,000 people (sub-scenario 1). Other studies demonstrate that GHG emissions from the use of raw materials and energy during the construction of two real WWTPs corresponds to 10–20% of the total GHG emissions of the WWTP for a service life of 30 years [114].

The carbon footprint of scenarios 2 and 3 will be even higher if the WWTP is built with the same capacity as scenario 1 but having to treat a smaller share of the generated wastewater. Further, upscaling the construction of septic tanks for a total population as in scenario 1 leads to total emissions of 46,365 t CO_2 eq, whereas upscaling the construction of CWs for 120,000 people will lead to a total of 1153 t CO_2 eq. CWs allow the least carbon footprint during the construction phase, in comparison to the WWTP, as the carbon footprint of installing septic tanks for the same volume of wastewater to treat is by far the highest (46,365 t CO_2 eq). Constructing household sanitation systems is, hence, the largest emitter in comparison to decentralized and fully centralized water reclamation with resource recovery schemes.

4. Discussion

The results of the GHG emissions inventory in the operation of a water supply scheme, and the possible proposed water reclamation with resource recovery approaches presented in this study, show important sources of GHG emissions, as well as opportunities for valorization of byproducts of the water reclamation with resource recovery process. In this chapter, further GHG emission reduction opportunities will be discussed.

As part of the PHE project to ensure drinking water for Leh town, 6 borewells, located in Khankshal, Laödon, Jumabach, Tukcha, Agling, and Murtsey are used to pump water from the groundwater to supply the public stand posts. The borewells pump water from a depth between 6 and 27 m daily and provide a total amount of 1.674 billion L/day. The consumption of the diesel needed for this is estimated to be 31,089 L per year [87], which is equivalent to an emission of 83 t CO_2 eq per year. The energy productivity of diesel fuel pumps can highly be increased by replacing these latter with solar photovoltaic pumps [72]. In fact, as seen above, the availability of solar irradiation in Ladakh can play an important role in substituting the used fuel and its related yearly costs.

Furthermore, solar pumping would allow for a system that operates without any GHG emissions. Solar pumping is already being used in many locations worldwide, due to the economic and environmental implications of using a fossil fuel powered systems [115]. A common use of solar pumping is for irrigation [116]. On a broader scale and based on field experience on multiple installed systems, [117] has proved their technical maturity and reliability. Hence, design of solar pumping systems for the 6 borewells in Leh is proposed.

So far, the carbon footprint from the water supply scheme and the water reclamation with resource recovery schemes have been discussed separately. However, a fully decentralized water reclamation with resource recovery scheme could positively influence the carbon footprint from the water supply scheme—[118] recommends different water supply levels for Indian towns, depending on the type on their sewerage systems. A decentralized water reclamation with resource recovery strategy in Leh, similar to scenarios 2 or 3 in this study, will lead to a requirement of only 70 L/day instead of 135 L/day per capita. According to the information and data received from PHE in 2017, 8 million liters per day will be extracted from the future water supply scheme in Leh town using two pumps, to supply the water for 59,259 people. Afterwards, three other stages, each equipped with two pumps, will run daily to lift the water up. The GHG emissions from using fuel were calculated above, and it is around 3672 t CO_2 eq per year. This is roughly equivalent to a value of 1.26 g CO_2 eq per liter of supplied water. Hence, in case Leh is not provided with a centralized sewerage system, a water supply of 4.2 million liters per day is required. which is equivalent to only 1928 t CO_2 eq per year. A decentralized water reclamation with a resource recovery approach will, thus, allow a saving potential of 1744 t CO_2 eq per year.

4.1. Overcoming the Biogas Challenge

India has been experiencing a rapid increase in installed biogas plants since 2013 [119]. However, in high altitude regions, such as Jammu and Kashmir State, where the town of Leh is located, the trends seems to be unsatisfactory, as the number of plants constitutes only 0.06% (2.739 plants) of the total plants installed in the country, as of March 2013 [119]. The most prominent challenge in biogas technology in high altitude areas is climatic conditions, mainly the temperature, where severe winter conditions can limit the gas production [120]. This is mainly because the methanogenic bacteria, responsible for biogas production, are very sensitive to temperature changes, and these regions are not able to ensure permanent heating provisions and automatic temperature control.

To overcome these challenges, research has led to the development of floating drum biogas plants, while using bricks in the process of fabrication of the digester, insulating around it with straw, enclosing it within a greenhouse, and using hot water to mix the feeding. This set-up has led to a higher biogas production rate of 1.6 to 2.6 m³/day [121]. Another challenge that can endanger the optimal operation for the digester is the low sludge rate production during the winter. However, the sludge for the digester can be collected and conserved during winter and afterwards be used to initiate the digester to boost the anaerobic process. Hence, operationalizing biogas production has very significant energy security potential in Leh.

4.2. Limitations of the Study

The analysis of carbon footprint shows that municipal WWTPs have a small impact at a global scale, constituting 0.45% of the yearly average pro capita CO_2 eq emissions in Europe [114]. Nevertheless, it is a good method to detect sources of GHG emissions and target them to suggest GHG reduction potential, which could help decrease the carbon footprint of urban areas. However, it cannot provide sufficient proof for decision-making, as an ecological evaluation through a life cycle assessment (LCA) offers a more complete method for assessing the impacts of processes on different impact categories [114], such as human toxicity, smog, global warming, and eutrophication [122].

On the other hand, during the construction of the different scenarios, only excavation and cement and concrete production have been taken into consideration, as these were considered to be the largest contributors to the carbon footprint of the construction phase. The resulting calculated carbon footprints are underestimated. However, the results give a very good basis of comparison between the different scenarios.

Further, while emission factors have been assumed to reach an approximation of GHG emissions, these may lack precision and accuracy, since they are not always verified by measurements. While many studies certify that they underestimate the real GHG emissions in the case of wastewater treatment, [26] found that in the context of the GHG emissions study for a WWTP, the measured emissions were almost 36% lower than the estimated emissions based on emission factors. A sensitivity analysis could be helpful to arrive at a range of estimated GHG emissions. Nevertheless, using emission factors allows for a concise comparison between the studied scenarios.

5. Conclusions

This study compared the carbon footprint of the operation of the current and future water supply scheme in Leh town. Further, it analyzed sources of emissions from different water reclamation with resource recovery scenarios in Leh, being fully centralized, or a combination of this with a decentralized treatment scheme or a household level treatment scheme. The results show that the electricity requirement for the planned WWTP will lead to high GHG emissions. In addition, a fully household-level on-site treatment scheme could lead to the highest value of GHG emissions per m^3 during the construction and the operation phase. Thus, this scheme is not encouraged for managing the wastewater in Leh, due to the uncertainty of the safe management of the sludge produced, and due to absence of regulations. Fixing a desludging frequency and operating maintenance activities is important to avoid health and environmental hazards.

The results of this study show that a fully decentralized water reclamation with resource recovery scheme, based on VSSF CWs, can in fact allow the least GHG emissions during operation. This is further supported by the low carbon footprint from the construction of decentralized water reclamation with resource recovery schemes (1153 t CO_2 eq) in comparison to an onsite water reclamation with resource recovery scheme generalized to the whole town (46,365 t CO_2 eq), or a centralized water reclamation with resource recovery option through the WWTP (2177 t CO_2 eq), to manage the wastewater from 120,000 PE in Leh.

The study also highlighted GHG emissions reduction potential, including utilizing the produced biogas and sludge. If electricity is produced and utilized on site from the biogas produced in scenario 1, a range from 500 to 900 t CO_2 eq can be saved from using power grid electricity. Similarly, utilizing the sludge for land application could provide a saving potential of 64 t CO_2 eq in this scenario, in comparison to sending it to a landfill. Alternatively, using biogas as cooking fuel in scenario 2 could allow a saving potential of 3 t CO_2 eq for substituting the LPG need for cooking.

Further, a solar pumping system has been proposed to replace the current diesel fueled pumps to extract groundwater used for water supply, which would allow a saving potential of 83 t CO_2 eq per year. This system is also recommended to be adopted by private hotels and households, as it may imply fuel cost savings and a better water supply reliability. In addition, decentralizing water reclamation with resource recovery can also lead to a saving potential of 1744 t CO_2 eq due to the lower water supply requirements. The results obtained with the available data and upscaled to the treatment capacity of the proposed WWTP or obtained from the normalization of GHG emissions to the cubic meter of treated wastewater should be examined critically due to the uncertainties of emission factors used and to the non-consideration of future urban development patterns. Also, the projected energy consumption for the water supply sector and the planned WWTP are uncertain. However, the results clearly underline the importance to engage in discussions to reevaluate and propose measures to be implemented to achieve more sustainable urban water management in Leh. Supporting decentralized water reclamation with resource recovery pilot projects can demonstrate the necessity of upscaling these types of solutions in the context of Leh.

Author Contributions: In the writing of this paper, M.L. contributed to the conceptualization of the research; design of the research methodology; utilization of the ECAM software; formal analysis and validation; original draft preparation and visualization of the results. J.E.D. and D.G. contributed to conceptualization of the research; design of the research methodology; writing, reviewing and editing the paper; supervision of the research work; and project administration and funding acquisition of the research project under which this research was conducted.

Funding: This research was funded by the Bavarian State Ministry of Environment and Consumer Protection, Munich, Germany, under the "Urban Water-Energy-Food Nexus" research project, grant number 76e0100000217.

Acknowledgments: We gratefully acknowledge the technical support to the paper content provided by Mohammed Al-Azzawi. This work was supported by the German Research Foundation (DFG) and the Technical University of Munich (TUM) in the framework of the Open Access Publishing Program.

Conflicts of Interest: The authors declare no conflict of interest. The funders had no role in the design of the study; in the collection, analyses, or interpretation of data; in the writing of the manuscript, or in the decision to publish the results.

References

1. United Nations. World Urbanization Prospects: The 2014 Revision. Available online: https://esa.un.org/unpd/wup/Publications/Files/WUP2014-Report.pdf (accessed on 20 February 2018).

2. McDonald, R.I.; Weber, K.; Padowski, J.; Flörke, M.; Schneider, C.; Green, P.A.; Montgomery, M. Water on an urban planet: Urbanization and the reach of urban water infrastructure. *Glob. Environ. Chang.* **2014**, *27*, 96–105. [CrossRef]

3. GIZ; ICLEI. Operationalizing the Urban NEXUS Towards Resource-Efficient and Integrated Cities and Metropolitan Regions. Available online: http://www2.giz.de/wbf/4tDx9kw63gma/UrbanNEXUS_Publication_ICLEI-GIZ_2014_kl.pdf (accessed on 20 February 2018).

4. Stakeholder Forum for a Sustainable Future. Review of Implementation of Agenda 21. Available online: https://sustainabledevelopment.un.org/content/documents/1126SD21Agenda21_new.pdf (accessed on 20 February 2018).

5. United Nations. New Urban Agenda, Conference on Housing and Sustainable Urban Development (Habitat III). Available online: http://habitat3.org/wp-content/uploads/NUA-English-With-Index-1.pdf (accessed on 20 February 2018).

6. Hoff, H. Understanding the Nexus. In Proceedings of the 2011 Nexus Conference, Bonn, Germany, 16–18 November 2011.

7. Vörösmarty, C.J.; Green, P.; Salisbury, J.; Lammers, R.B. *Global Water Resources: Vulnerability from Climate Change and Population Growth*. *Science*; American Association for the Advancement of Science: Washington, DC, USA, 2000. [CrossRef]

8. Bates, B.C.; Kundzewicz, Z.W.; Wu, S.; Palutikof, J.P. Climate Change and Water. In Proceedings of the Intergovernmental Panel on Climate Change, Geneva, Switzerland, 1 June 2008.

9. Copeland, C. Energy-Water Nexus: The Water Sector's Energy Use. Available online: https://fas.org/sgp/crs/misc/R43200.pdf (accessed on 20 February 2018).

10. Smith, K.; Liu, S.; Chang, T. Contribution of Urban Water Supply to Greenhouse Gas Emissions in China. *J. Ind. Ecol.* **2016**, *20*, 792–802. [CrossRef]

11. Rothausen, S.G.S.A.; Conway, D. Greenhouse-gas emissions from energy use in the water sector. *Nat. Clim. Change* **2011**, *1*, 210–219. [CrossRef]

12. UNFCCC. Reporting of GHG Data. Available online: http://unfccc.int/ghg_data/new_reporting_requirements/items/9560.php (accessed on 20 February 2018).

13. Houghton, J.T.; Jenkins, G.J.; Ephraums, J.J. *Climate Change: The IPCC Scientific Assessment*; Intergovernmental Panel on Climate Change by Working Group I: Cambridge, UK; New York, NY, USA; Melbourne, Australia, 1990.

14. UNFCCC. United Nations Framework Convention on Climate Change. Available online: https://unfccc.int/resource/docs/convkp/conveng.pdf (accessed on 20 February 2018).

15. UNEP. *Handbook for the Montreal Protocol on Substances that Deplete the Ozone Layer*; United Nations Environment Programme: Nairobi, Kenya, 2006; Available online: http://ozone.unep.org (accessed on 20 February 2018).

16. UNFCC. UNFCCC Data Interface. Available online: http://di.unfccc.int/comparison_by_category (accessed on 20 February 2018).

17. UNFCCC. Guide for Peer Review of National GHG Inventories. Available online: https://unfccc.int/files/national_reports/non-annex_i_natcom/application/pdf/final_guide_for_peer_review_report_final_webupload.pdf (accessed on 20 February 2018).

18. Anderson, T.R.; Hawkins, E.; Jones, P.D. CO_2, the greenhouse effect and global warming: From the pioneering work of Arrhenius and Callendar to today's Earth System Models. *Endeavour* **2016**, *40*, 178–187. [CrossRef] [PubMed]

19. Harvey, L.D.D. A guide to global warming potentials (GWPs). *Energy Policy* **1993**, *21*, 24–34. [CrossRef]

20. IPCC. *2006 IPCC Guidelines for National Greenhouse Gas Inventories, Prepared by the National Greenhouse Gas Inventories Programme*; Eggleston, H.S., Buendia, L., Miwa, K., Ngara, T., Tanabe, K., Eds.; IGES: Hayama/Kanagawa, Japan, 2006.

21. IPCC. *Climate Change: The Physical Science Basis*; Cambridge University Press: Cambridge, UK, 2007.

22. IEA. *World Energy Outlook*; International Energy Agency: Paris, France, 2015.

23. Cornejo, P.K. Environmental Sustainability of Wastewater Treatment Plants Integrated with Resource Recovery: The Impact of Context and Scale. Available online: http://scholarcommons.usf.edu/etd (accessed on 20 February 2018).

24. Soda, S.; Iwai, Y.; Sei, K.; Shimod, Y.; Ike, M. Model analysis of energy consumption and greenhouse gas emissions of sewage sludge treatment systems with different processes and scales. *Water Sci. Technol.* **2010**, *61*, 365. [CrossRef] [PubMed]

25. Nair, S.B.G.; Malano, H.N.; Arora, M.; Nawarathna, B. Water–energy–greenhouse gas nexus of urban water systems: Review of concepts, state-of-art and methods. *Resour. Conserv. Recycl.* **2014**, *89*, 1–10. [CrossRef]

26. De Graaff, M.S.; Zandvoort, M.; Roest, K.; Frijns, J.; Janse, T.; van Loosdrecht, M.C.M. Quantification of Methane and Nitrous Oxide Greenhouse Gas Emissions from the Urban Water Cycle. IWA World Water Congress. Available online: http://livelink.kwrwater.nl/livelink/livelink.exe/open/50222954 (accessed on 20 February 2018).

27. Guisasola, A.; de Haas, D.; Keller, J.; Yuan, Z. Methane formation in sewer systems. *Water Res.* **2008**, *42*, 1421–1430. [CrossRef] [PubMed]

28. Foley, J.; Lant, P.A.; Yuan, Z.; Keller, J.; Donlon, P. Fugitive greenhouse gas emissions from wastewater systems. In Proceedings of the 6th IWA World Water Congress, Vienna, Austria, 7–12 September 2008.

29. Daelman, M.R.J.; De Baets, B.; van Loosdrecht, M.C.M.; Volcke, E.I.P. Influence of sampling strategies on the estimated nitrous oxide emission from wastewater treatment plants. *Water Res.* **2013**, *47*, 3120–3130. [CrossRef] [PubMed]

30. Daelman, M.R.J.; van Voorthuizen, E.M.; van Dongen, U.G.J.M.; Volcke, E.I.P.; van Loosdrecht, M.C.M. Methane emission during municipal wastewater treatment. *Water Res.* **2012**, *46*, 3657–3670. [CrossRef] [PubMed]

31. Short, M.D.; Daikeler, A.; Peters, G.M.; Mann, K.; Ashbolt, N.J.; Stuetz, R.M.; Peirson, W.L. Municipal gravity sewers: An unrecognised source of nitrous oxide. *Sci. Total Environ.* **2014**, *468–469*, 211–218. [CrossRef] [PubMed]

32. Wunderlin, P.; Mohn, J.; Joss, A.; Emmenegger, L.; Siegrist, H. Mechanisms of N_2O production in biological wastewater treatment under nitrifying and denitrifying conditions. *Water Res.* **2012**, *46*, 1027–1037. [CrossRef]

33. Foley, J.; de Haas, D.; Yuan, Z.; Lant, P. Nitrous oxide generation in full-scale biological nutrient removal wastewater treatment plants. *Water Res.* **2010**, *44*, 831–844. [CrossRef]

34. Holmgren, K.E.; Li, H.; Verstraete, W.; Cornel, P. *State of the Art Compendium Report on Resource Recovery from Water*; International Water Association (IWA) Publishers: London, UK, 2014.

35. Jjemba, P.K.; Weinrich, L.A.; Cheng, W.; Giraldo, E.; Lechevallier, M.W. Regrowth of potential opportunistic pathogens and algae in reclaimed-water distribution systems. *Appl. Environ. Microbiol.* **2010**, *76*, 4169–4178. [CrossRef] [PubMed]

36. Lazarova, V.; Bahri, A. *Water Reuse Practices for Agriculture. Water Reuse: An International Survey of Current Practice, Issues and Needs*; IWA publishing: London, UK, 2008; pp. 199–227.

37. Drewes, J.E.; Khan, S.J. *Water Reuse for Drinking Water Augmentation Chapter 16*; Water Quality and Treatment, 6th Edition. 16.1-16.48; Edzwald, J., Ed.; American Water Works Association: Denver, CO, USA, 2011.

38. Lazarova, V.; Peregrina, C.; Dauthuille, P. *Water-Energy Interactions in Water Reuse*; IWA Publishing: London, UK, 2012.

39. CPHEEO. Manual on Sewerage and Sewage Treatment. Available online: https://www.sswm.info/sites/default/files/reference_attachments/CPHEEO-2012-Manual-on-sewerage-and-sewage-treatment.-Part-A-Engineering.pdf (accessed on 20 February 2018).

40. WSP; IWMI. *Recycling and Reuse of Treated Wastewater in Urban India: A Proposed Advisory and Guidance Document*; International Water Management Institute: Colombo, Sri Lanka, 2016.

41. Meda, A.; Lensch, D.; Schaum, C.P. Chapter 2: Energy and Water: Relations and Recovery Potential. In *Water-Energy Interactions in Water Reuse*; IWA Publishing: London, UK, 2012; pp. 21–35.

42. Hawley, C.; Fenner, R. The potential for thermal energy recovery from wastewater treatment works in Southern England. *J. Water Clim. Chang.* **2012**, *3*, 287–299. [CrossRef]

43. Schmid, F. Sewage Water: Interesting Heat Source for Heat Pumps and Chillers. Available online: http://www.bfe.admin.ch/php/modules/publikationen/stream.php?extlang=en&name=en_508290240.pdf (accessed on 20 February 2018).

44. Mortensen, J.G.; Gonzaíez-Pinzo, R.; Dahm, C.N.; Wang, J.; Zeglin, L.H.; Van Horn, D.J. Advancing the Food-Energy–Water Nexus: Closing Nutrient Loops in Arid River Corridors. *Environ. Sci. Technol.* **2016**. [CrossRef] [PubMed]

45. Childers, D.L.; Corman, J.; Edwards, M.; Elser, J.J. Sustainability Challenges of Phosphorus and Food: Solutions from Closing the Human Phosphorus Cycle. *BioScience* **2011**, *61*, 117–124. [CrossRef]

46. WERF. *Nutrient Recovery State of the Knowledge as of 2011*; Water Research Foundation: Denver, CO, USA, 2011.

47. Zhang, W.; Dou, Z.-X.; He, P.; Ju, X.-T.; Powlson, D.; Chadwick, D.; Sayer, J. New technologies reduce greenhouse gas emissions from nitrogenous fertilizer in China. *Proc. Natl. Acad. Sci. USA* **2013**, *110*, 8375–8380. [CrossRef] [PubMed]

48. Mo, W.; Zhang, Q. Energy-nutrients-water nexus: Integrated resource recovery in municipal wastewater treatment plants. *J. Environ. Manag.* **2013**, *127*, 255–267. [CrossRef]

49. Jönsson, H. Source separation of human urine separation efficiency and effects on water emissions, crop yield, energy usage and reliability. In Proceedings of the First International Conference on Ecological Sanitation, Nanning, China, 5–8 November 2001.

50. Verstraete, W.; Van de Caveye, P.; Diamantis, V. Maximum use of resources present in domestic "used water". *Bioresour. Technol.* **2009**, *100*, 5537–5545. [CrossRef] [PubMed]

51. Muñoz, I.; Milà-i-Canals, L.; Fernández-Alba, A.R. Life Cycle Assessment of Water Supply Plans in Mediterranean Spain. *J. Ind. Ecol.* **2010**, *14*, 902–918. [CrossRef]

52. Parkinson, J.; Tayler, K. Decentralized wastewater management in peri-urban areas in low-income countries. *Environ. Urb.* **2003**, *15*, 75–90. [CrossRef]

53. Engin, G.O.; Demir, I. Cost analysis of alternative methods for wastewater handling in small communities. *J. Environ. Manag.* **2006**, *79*, 357–363. [CrossRef] [PubMed]

54. US EPA. *Response to Congress on Use of Decentralized Wastewater Treatment Systems*; United States Environmental Protection Agency: Washington, DC, USA, 1997.

55. Crites, R.; Tchobanoglous, G. *Small and Decentralized Wastewater Management Systems*; WCB/McGraw-Hill: New York, NY, USA, 1998.

56. Nelson, K. Water Conservation, Reuse, and Recycling. In *An Iranian-American Workshop*; National Academies Press: Washington, DC, USA, 2005.

57. Gutterer, B.; Sasse, L.; Panzerbieter, T.; Reckerzügel, T. Decentralised Wastewater Treatment System (DEWATS) and Sanitation in Developing Countries: A Practical Guide. Available online: http://www.cabdirect.org/abstracts/20103140120.html;jsessionid=9D0874D07B97A0501DC93F272AD3E4B2 (accessed on 20 February 2018).

58. Vymazal, J.; Brix, H.; Cooper, P.F.; Haberl, R.; Perfler, R.; Laber, J. Removal Mechanisms and Types of Constructed Wetlands. Available online: http://mit.biology.au.dk/~{}biohbn/cv/pdf_files/Con_Wet_Was_Treat_Eur(1998)17-66.pdf (accessed on 20 February 2018).

59. Mitsch, W.J.; Gosselink, J.G. *Wetlands*; Wiley: New York, NY, USA, 2007.

60. Kadlec, R.H.; Knight, R.L. *Treatment Wetlands*; CRC Press/Lewis Publishers: Boca Raton, FL, USA, 1996.

61. Albold, A.; Wendland, C.; Mihaylova, B.; Ergünsel, A.; Galt, H. Constructed Wetlands Sustainable Wastewater Treatment for Rural and Peri-Urban Communities in Bulgaria. Available online: https://www.gwp.org/globalassets/global/toolbox/case-studies/europe/bulgaria.-constructed-wetlands-for-rural-and-peri-urban-waste-water-treatment-431.pdf (accessed on 20 February 2018).

62. Hemkend-Reis, B.; Henseler, M.; Güdel, K.; Lehnhard, Y. Faecal Sludge Management (FSM). Available online: https://www.sswm.info/sites/default/files/reference_attachments/EAWAGSANDEC2008Module5FSMLecture.pdf (accessed on 20 February 2018).

63. WHO; UNICEF. Global Water Supply and Sanitation Assessment 2000 Report. Available online: http://www.who.int/water_sanitation_health/monitoring/jmp2000.pdf (accessed on 20 February 2018).

64. Dolma, K.; Rishi, M.S.; Lata, R. Evaluation of Groundwater Quality and its Suitability for Drinking Purposes—A Case of Leh Town, Ladakh (J&K), India. *Int. J. Sci. Eng. Res.* **2015**, *6*, 576–590.

65. Dolma, K.; Rishi, M.S.; Lata, R. An Appraisal of Centralized Waste Water Treatment Plant with Respect to Leh Town. *Int. J. Sci. Res.* **2015**, *4*, 2011–2016.

66. Kasturirangan, K. *Natural Resource Management Ladakh Region. A Remote Sensing Based Study*; Regional Remote Sensing Service Centre: Dehradun, India, 2003.

67. Chevuturi, A.; Dimri, A.P.; Thayyen, R.J. Climate change over Leh (Ladakh), India. *Theor. Appl. Climatol.* **2018**, *131*, 531–545. [CrossRef]

68. Sherratt, K. Social and Economic Characteristics of Ladakh, India. Available online: https://www.geolsoc.org.uk/~{}/media/shared/documents/Events/PastMeetingResources/Himalaya14CulturalBackgroundtoLadakh.pdf (accessed on 20 February 2018).

69. Pelliciardi, V. Tourism Traffic Volumes in Leh District: An Overview. *Ladakh Stud.* **2010**, *11/10*, 14–23.

70. Leh Tourism Department. *Statistics of Number of National and International Tourists in Leh Town*; Ladakh Autonomous Hill Development Council: Leh, Jammu & Kashmir, India, 2017.

71. Gondhalekar, D.; Nussbaum, S.; Akhtar, A.; Kebschull, J. *Planning Under Uncertainty: Climate Change, Water Scarcity and Health Issues in Leh Town, Ladakh, India*; Springer: Cambridge, UK, 2015; pp. 293–312.

72. Santra, P. Scope of Solar Energy in Cold Arid Region of India at Leh Ladakh. *Ann. Arid Zone* **2015**, *54*, 109–117.

73. Daultrey, S.; Gergan, R. Living with Change: Adaptation and Innovation in Ladakh. Climate Adaptation Series. Available online: http://www.intrepidexplorers.co.uk/download/i/mark_dl/u/4011885869/4607751139/DaultreyandGergan2011-Livingwithchange.pdf (accessed on 20 February 2018).

74. Harinarayana, T.; Abdul Azeez, K.K.; Murthy, D.N.; Veeraswamy, K.; Eknath Rao, S.P.; Manoj, C.; Naganjaneyulu, K. Exploration of geothermal structure in Puga geothermal field, Ladakh Himalayas, India by magnetotelluric studies. *J. Appl. Geophys.* **2006**, *58*, 280–295. [CrossRef]

75. Farooq, M.; Shah, I.K.; Mushtaq, S.M.; Khaki, B.A.; Marazi, A.A.; Shah, N.; Mushtaq, M. *Emission Inventory of CO_2 in Jammu and Kashmir—A Sectoral Analysis*; Government of Jammu & Kashmir: Srinagar, Jammu & Kashmir, India, 2016.

76. SuSanA. Towards More Sustainable Sanitation Solutions. Available online: http://www.susana.org/_resources/documents/default/3-267-7-1452594644.pdf (accessed on 20 February 2018).

77. WHO; UNICEF. Progress on Drinking Water, Sanitation And Hygiene: 2017 Update and SDG Baselines. Available online: http://www.wipo.int/amc/en/mediation/rules (accessed on 20 February 2018).

78. WHO; UNICEF. Countries WASH Data. Available online: https://washdata.org/data (accessed on 20 February 2018).

79. India Census. India Census 2011. Available online: http://www.censusindia.gov.in/2011-prov-results/indiaatglance.html (accessed on 20 February 2018).

80. Government of Jammu & Kashmir. Faecal Sludge Septage Management. Available online: http://jkhudd.gov.in/pdfs/FSMPolicy_J&K.pdf (accessed on 20 February 2018).

81. CRWQCB. A Review of the Nitrate Problems in the Ground Waters of the Santa Ana Region and Their Relationship to High Density Developments on Septic Tank-Subsurface Disposal Systems. Available online: http://www.waterboards.ca.gov/santaana/water_issues/programs/septic_tanks/docs/nit%0Arate_study.pdf (accessed on 20 February 2018).

82. BORDA. Fact Sheet: Faecal Sludge Treatment Plant (FSTP). Available online: https://smartnet.niua.org/content/ded33e3d-535f-4946-af0a-556244a79537 (accessed on 20 February 2018).

83. Tetra Tech. *Detailed Project Report for Sewerage System of Leh Town*; Tetra Tech: New Delhi, India, 2009.

84. Gao, T.; Liu, Q.; Wang, J. A comparative study of carbon footprint and assessment standards. *Int. J. Low Carb. Technol.* **2014**, *9*, 237–243. [CrossRef]

85. Wiedmann, T.; Minx, J. A Definition of "Carbon Footprint." Ecological Economics Research Trends. Available online: http://citeseerx.ist.psu.edu/viewdoc/download?doi=10.1.1.467.6821&rep=rep1&type=pdf (accessed on 20 February 2018).

86. IPCC. *Climate Change 2013: The Physical Science Basis*; Cambridge University Press: Cambridge, UK, 2013.

87. Bonanno, G. Water Resources Sustainability Assessment in Leh Town, Ladakh, India, and Alternative Solutions Using the Water-Energy-Food Nexus. Master's Thesis, Technical University of Munich, Munich, Germany, 2018.

88. Bunday, B.D. Methods for population estimation by random sampling. *Powder Technol.* **1975**, *12*, 283–286. [CrossRef]

89. Akhtar, A.; Gondhalekar, D. Impacts of tourism on water resources in Leh town. *Ladakh Stud.* **2013**, *30*, 25–37.

90. WBCSD; WRI. *The Greenhouse Gas Protocol. A Corporate Accounting and Reporting Standard*; World Business Council for Sustainable Development: Geneva, Switzerland; World Resources Institute: Washington, DC, USA, 2001.

91. Mander, U.; Kuusemets, V.; Lõhmus, K.; Mauring, T.; Teiter, S.; Augustin, J. Nitrous oxide, dinitrogen and methane emission in a subsurface flow constructed wetland. *Water Sci. Technol.* **2003**, *48*, 135–142. [CrossRef] [PubMed]

92. Mander, U.; Dotro, G.; Ebie, Y.; Towprayoon, S.; Chiemchaisri, C.; Nogueira, S.F.; Mitsch, W.J. Greenhouse gas emission in constructed wetlands for wastewater treatment: A review. *Ecol. Eng.* **2014**, *66*, 19–35. [CrossRef]

93. Wernet, G.; Bauer, C.; Steubing, B.; Reinhard, J.; Moreno-Ruiz, E.; Weidema, B. The ecoinvent database version 3 (part I): Overview and methodology. *Int. J. Life Cycle Assess.* **2016**, 1218–1230. [CrossRef]

94. GIZ; IWA. Energy Performance and Carbon Emissions Assessment and Monitoring (ECAM) Tool. Available online: http://wacclim.org/ecam/index.php (accessed on 20 February 2018).

95. Dos Santos, I.F.S.; Barros, R.M.; Tiago Filho, G.L. Electricity generation from biogas of anaerobic wastewater treatment plants in Brazil: An assessment of feasibility and potential. *J. Clean. Prod.* **2016**, *126*, 504–514. [CrossRef]

96. Ravindranath, N.H.; Ramakrishna, J. Energy options for cooking in India. *Energy Policy* **1997**, *25*, 63–75. [CrossRef]

97. Vu, T.K.V.; Vu, D.Q.; Jensen, L.S.; Sommer, S.G.; Bruun, S. Life Cycle Assessment of Biogas Production in Small-scale Household Digesters in Vietnam. *Asian Australas. J. Anim. Sci.* **2015**, *28*, 716–729. [CrossRef]

98. DEFRA. *Future Water: The government's Water Strategy for England*; District Statistics & Evaluation Office: Norwich, UK, 2008.

99. Hutson, S.S.; Barber, N.L.; Kenny, J.F.; Linsey, K.S.; Lumia, D.S.; Maupin, M.A. Estimated Use of Water in the United States in 2000. Available online: https://pubs.usgs.gov/circ/1344/pdf/c1344.pdf (accessed on 20 February 2018).

100. Friedrich, E.; Pillay, S.; Buckley, C. Environmental life cycle assessments for water treatment processes—A South African case study of an urban water cycle. *Water Sa* **2009**, *35*, 2731–2760. [CrossRef]

101. Venkatesh, G.; Brattebø, H. Energy consumption, costs and environmental impacts for urban water cycle services: Case study of Oslo (Norway). *Energy* **2011**, *36*, 792–800. [CrossRef]

102. Ball, A.; Ahmad, S.; Mccluskey, C.; Pham, P.; Ahn, I.; Dawson, L.; Nowakowski, D. Australian Energy Update 2016. Available online: https://www.industry.gov.au/Office-of-the-Chief-Economist/Publications/Documents/aes/2016-australian-energy-statistics.pdf (accessed on 20 February 2018).

103. Miller, L.A.; Ramaswami, A.; Asce, M.; Ranjan, R. Contribution of Water and Wastewater Infrastructures to Urban Energy Metabolism and Greenhouse Gas Emissions in Cities in India. *J. Environ. Eng.* **2013**, *139*, 738–745. [CrossRef]

104. Gikas, P. Towards energy positive wastewater treatment plants. *J. Environ. Manag.* **2017**, *203*, 621–629. [CrossRef] [PubMed]

105. De Haas, D.; Hartley, K. Greenhouse gas emissions from BNR plant-do we have the right focus? In Proceedings of the Sewage Management-Risk Assessment and Triple Bottom Line, Queensland, Australia, 7 June 2004.

106. Pescod, M.B. Wastewater Treatment and Use in Agriculture. Food and Agriculture Organization of the United Nations. Available online: http://www.fao.org/docrep/t0551e/t0551e00.htm#Contents (accessed on 20 February 2018).

107. US EPA. *Process Design Manual: Land application of Sewage Sludge and Domestic Septage*; US Environmental Protection Agency, Office of Research and Development: Washington, DC, USA, 1995.

108. Sophie, B. The State of the Promotion of Biogas from Wastewater Plants in France and Europe. Available online: http://www2.agroparistech.fr/IMG/pdf/syn08-eng-Bonnier.pdf (accessed on 20 February 2018).

109. Suzuki, A.B.P.; Fernandes, D.M.; Pereira Faria, R.A.; Vidal, T.C.M. Use of biogas in internal combustion engines. *Brazil. J. Appl. Technol. Agric. Sci.* **2011**, *41*, 221–237. [CrossRef]

110. Mohr, M.; Iden, J.; Beckett, M. Guideline: Vacuum Sewer Systems. Available online: http://www.unescap.org/sites/default/files/Guideline_VacuumSewerSystems_FraunhoferIGB_2016_0.pdf (accessed on 20 February 2018).

111. Cooper, P.F. The performance of vertical flow constructed wetland systems with special reference to the significance of oxygen transfer and hydraulic loading rates. *Water Sci. Technol.* **2005**, *51*, 81–90. [CrossRef] [PubMed]

112. Leverenz, H.L.; Tchobanoglous, P.E.G.; Jeannie, P.E.; Darby, L. Evaluation of greenhouse gas emissions from septic tanks. Available online: www.werf.org (accessed on 20 February 2018).

113. Hoffmann, H.; Platzer, C.; Winker, M.; Münch, E.V. *Technology Review of Constructed Wetlands*; von Muench, E., Ed.; GIZ GmbH: Eschborn, Germany, 2011.

114. Genzowsky, K.; Rohn, A.; Bolle, F.-W.; Merkel, W. *Methodenentwicklung zur Bewertung von Siedlungswasserwirtschaftlichen und Wasserwirtschaftlichen Anlagen Hinsichtlich ihres Ökologischen Fußabdrucks*; IWW Zentrum Wasser: Aachen, Germany, 2011.

115. Tharani, K.; Dahiya, R. Comparative analysis of DG and solar PV water pumping systems. *AIP Conf. Proc.* **2016**, 20019. [CrossRef]

116. Odeh, I. Introducing new design and performance points for photovoltaic water pumping systems based on long-term field data analysis. *J. Renew. Sustain. Energy* **2013**, *5*, 23135. [CrossRef]

117. Posorski, R. Photovoltaic water pumps, an attractive tool for rural drinking water supply. *Sol. Energy* **1996**, *58*, 155–163. [CrossRef]

118. CPHEEO. *Manual on Water Supply and Treatment*; Ministry of Drinking Water and Sanitation, Government of India: New Delhi, India, 1999.

119. Lohan, S.K.; Dixit, J.; Kumar, R.; Pandey, Y.; Khan, J.; Ishaq, M.; Kumar, D. Biogas: A boon for sustainable energy development in India's cold climate. *Renew. Sustain. Energy Rev.* **2015**, *43*, 95–101. [CrossRef]

120. Balasubramaniyam, U.; Zisengwe, L.S.; Meriggi, N.; Buysman, E. Biogas Production in Climates with Long Cold Winters. Available online: http://www.build-a-biogas-plant.com/PDF/BALASUBRAMANIYAM-2008-Biogas-Production-in-Climates-with-long-cold-Winters.pdf (accessed on 20 February 2018).

121. Bansal, N.K. A Technoeconomic Assessment of Solar-Assisted Biogas Systems. *Energy Sour.* **1988**, *10*, 213–229. [CrossRef]

122. Stranddorf, H.K.; Hoffmann, L.; Schmidt, A.; FORCE Technology Denmark. Impact Categories, Normalisation and Weighting in LCA. Available online: https://www2.mst.dk/udgiv/publications/2005/87-7614-574-3/pdf/87-7614-575-1.pdf (accessed on 20 February 2018).

![water logo] *water*

MDPI

Article

Storm Water Management and Flood Control in Sponge City Construction of Beijing

Shuhan Zhang [1,*], Yongkun Li [1], Meihong Ma [2], Ting Song [1] and Ruining Song [1]

[1]   Beijing Water Science and Technology Institute, Beijing 100048, China; dxj521@yeah.net (Y.L.);
      song4560@126.com (T.S.); songruining@163.com (R.S.)
[2]   School of Geographic and Environmental Sciences, Tianjin Normal University, Tianjin 300387, China;
      mmhkl2007@163.com
*   Correspondence: bjzhangshuhan@126.com; Tel.: +86-010-6873-1921

Received: 25 May 2018; Accepted: 3 August 2018; Published: 6 August 2018

Abstract: To solve the problems of increasing local flooding, water shortage, and water pollution caused by the traditional model of urban development, the Chinese government proposed a new model of urban development—the Sponge City. In Beijing, the capital of China, research on storm water management in urban areas has been carried out since 1989 and has put forward the concept of urban storm water harvesting and flood control. The further research and demonstration application started in 2000. So far, a series of policies and technology standards on storm water management have been formulated, which promote the application of technologies on comprehensive urban storm water harvesting and flood control. A significant number of storm water harvesting and flood control projects have been built in Beijing, which are now playing important roles in runoff reduction, local flood control, non-point source pollution reduction, and storm water utilization. However, it does not solve the above problem completely. Storm water management and flood control needs to be further strengthened. The "Sponge City" is based on natural and ecological laws, which allows storm water to be managed with natural infiltration, natural retention and detention, and natural cleaning facilities. Through in-depth analysis of the connotation, characteristics, and construction path of "Sponge City", this paper summarizes the status quo of urban rainwater flooding, flood control technology development and application, and Beijing policy and engineering to introduce the overall ideas and methods of Sponge City construction. All the above will provide a reference for cities with similar problems in the construction of sponge cities.

Keywords: Sponge City; water ecology; storm water management; flood control; resilience

1. Introduction

With global warming, rainstorms and other extreme weather events are occurring frequently, leading to flood and non-point source pollution [1]. In order to solve this problem, in the late 1990s, the state of Maryland in the United States developed Low Impact Development (LID) technology to achieve runoff and pollution control caused by heavy rain, mainly through decentralized, small-scale source control. After nearly 20 years of development, it has become the urban green rainwater infrastructure (GSI) technology most commonly used in the United States and many developed countries [2]. Similarly, Australia's Water Sensitive Urban Design (WSUD) [3], New Zealand's Low Impact Urban Design and Development (LIUDD) [4], and the UK's Sustainable Urban Drainage System (SUDS) [5], are like-minded technologies. The concepts and related measures of storm water harvesting and storm water management in Japan and rainwater storage infiltration in Japan are also offer important precedents for different countries to deal with urban flooding and runoff pollution. Since 1949, China's management of urban rainwater runoff has generally experienced three stages, namely the direct rain stage (1949~2000), the combined use stage (2000~2013), and the system management stage (2013~present). In 2013, General Secretary Xi Jinping proposed the construction of "Sponge City", which indicates that China's urban rainwater management has entered the stage of system's management. Beijing has conducted research and applied rainwater utilization since 1989. It is the earliest city in China and has achieved good results. It has played an important role in reducing and controlling urban rainfall runoff, reducing non-point source pollution and preventing urban infighting. Therefore, studying the current situation of sponge cities in Beijing, especially in the management of storms and floods, will help to provide experience and reference for the international defense construction of cities affected by floods [6].

2. The Connotation of Sponge City

2.1. Concept

General Secretary, Xi Jinping put forward the construction of a "sponge city" with natural accumulation, natural infiltration, and natural purification at the Central Working Conference on Urbanization in December 2013. At an important speech on the protection of water safety in March 2014, General Secretary Xi reiterated that urban planning and construction must consciously reduce the intensity of development, retain, and restore an appropriate proportion of ecological space, and build a "sponge home" and a "sponge city". In November 2014, the Ministry of Housing and Urban-Rural Development issued a technical guide for the construction of sponge cities—the construction of rainwater systems with low impact development (Trial), pointing out that sponge cities, like sponges, had good "elasticity" in adaptation to environmental changes and response to natural disasters [7]. When it rains, the city will absorb, store, infiltrate, and purify water, if necessary, the storage water will be released and utilized. As the Guidance of the Ministry of Water Resources on promoting the water conservancy work in Sponge City construction defines it, the sponge city is based on the low-impact development and construction model, and is supported by the flood control system, giving full play to the natural accumulation, penetration, purification and relaxation of green space, soil, rivers, and lakes. The release of runoff effect has achieved source reduction, decentralized retention, sustained release, and rational utilization. Urban storm water runoff enables cities to mitigate or reduce the impact of natural disasters and environmental changes, and to protect and improve the ecological environment as sponges. As the Guidance of General Office of the State Council on promoting the water conservancy work in Sponge City construction defines (No. 75 [2015] issued by General Office of the State Council), Sponge City is a city development model with natural accumulation, infiltration, and purification. By strengthening urban planning and construction management, it effectively controls rainwater runoff and gives full play to the construction, roads, green spaces, water systems and other ecosystems on absorption, storage-infiltration and slow release effect to rain. In this paper, we take the concept of Sponge City proposed by the State Council of China's highest government agency as the research object, and further put forward our own understanding.

Therefore, sponge city should have at least the following three aspects of meaning: first and foremost, from the perspective of water resources utilization, urban planning and construction must conform to the laws of nature. By constructing water-saving urban underlying surfaces, the urban rainfall can be accumulated, purified, reused, or recharged to groundwater. Second, from the perspective of flood control and disaster reduction, the city can live in harmony with rain and flood. Measures will be taken to minimize the risk of floods, reduce disaster losses, and recover production and life rapidly. Third, from the perspective of ecological environment, the city should be placed in nature, establish scientific development, and make residents "clear and lush". In other words, "Sponge City" should be able to cope with different return periods of rainfall to prevent flood disasters, rationally use rainwater, and maintain a good hydrological and ecological environment. Sponge city construction should adopt corresponding engineering and non-engineering measures according to the rainfall return periods at residential, regional and basin levels, so as to realize scientific management and use of different scales of rainwater. Since sponge city mainly deals with rainfall in different return periods, the technical framework is shown in Figure 1.

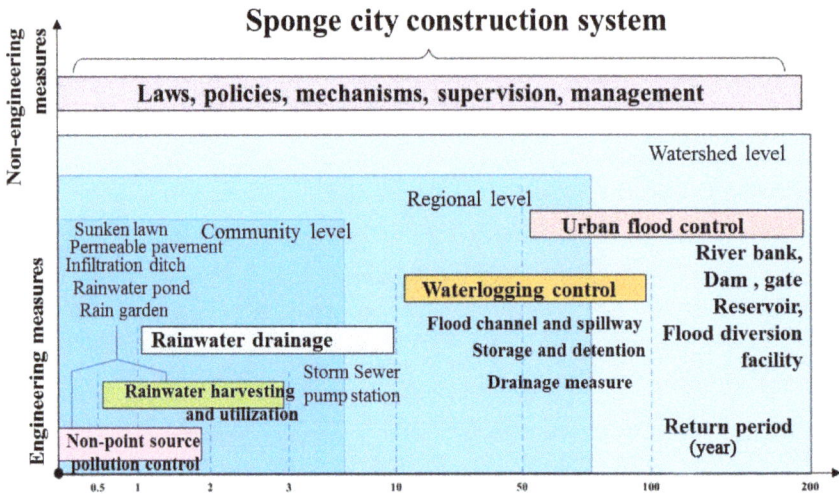

Figure 1. Diagram of sponge city construction system.

2.2. Characteristics of Sponge City Construction

In sponge city construction, water cycle regulation is the foundation and the framework gives full play to the function of storage, diversion and regulating of urban rainwater pipe network and river system. The work should be carried out according to local hydrological, geological, geographic, economic, and social conditions. Therefore, the sponge city construction has the following characteristics:

(1)   Multi-scales. The sponge city construction needs to be carried out on community scale, drainage scale, city scale and watershed scale so that urban rainfall and flood can be dealt with from multi-scales, and urban water cycle is regulated well.

(2)   Wide rainfall reappearing period. Sponge city construction needs to deal with the rainfall and flood in rainfall return period from small to large, including the rainfall and flood events that exceed the standard of the project [8].

(3)   Versatility. Sponge city construction has the function of preventing urban floods and relieving urban flood disaster, and also has the function of reducing rainfall and runoff pollution and improving water ecological environment. It also has the function of increasing the available water resources by direct or indirect use [9].

(4)   Systematisms. Sponge city construction is a systematic project involving water conservancy, architecture, gardens, landscapes, municipalities, and planning, etc. The project requires professional and technical personnel to plan, design, construct, operate, and maintain the engineering facilities, as well as to supervise, organize and coordinate the various departments.

(5)   Long-term nature. Sponge city is a beautiful vision not built overnight. It needs more than ten years or even decades of continuous construction and management.

2.3. Path of Sponge City Construction

As a systematic project, sponge city construction has the following procedures:

First, protect the city's original ecosystem. In the process of urban construction and development, maximize the protection of the original rivers, lakes, wetlands, ponds, ditches and other "sponges".

Leave enough water and maintain the natural hydrological characteristics of forest, grassland, lakes and wetlands in the process of an urban development faced with intensive precipitation [10].

Second, do ecological restoration and repair work. For damaged waters and other natural environments, use ecological means to recover and repair, maintain a certain proportion of ecological space, and then build a water ecological infrastructure through systematic scientific methods.

Third, implement the concept of low impact development. According to the development and construction concept of the minimum impact on the urban ecological environment, control the intensity of development reasonably to minimize destruction of the original urban water ecological environment, such as retaining sufficient ecological land and controlling the proportion of urban impermeable surface areas. According to the needs, we should properly excavate rivers, lakes, and ditches, increase the water area, promote the accumulation of rainwater, infiltrate and purify, and improve the size and quality of the urban "sponge".

Fourth, strengthen the comprehensive utilization and management of urban rain and flood resources. Regulate comprehensively and manage the surface runoff generated by urban precipitation and rain flood in pipe networks and rivers. Under the premise of ensuring the safety, utilize rainwater and control pollution, establish a flood disaster risk management system as well as personal hedging and disaster prevention and mitigation for buildings and other adaptive measures system.

Since the core issue of sponge city construction is the management of urban rainwater and the prevention of floods, this article will specifically introduce the current situation of rainwater management and flood control in Beijing.

3. Rainwater Management and Flood Control in Beijing

3.1. Progress of Rainwater Management Research and Practice

The technical research and practice of urban rainwater in Beijing has experienced three stages, namely the direct discharge of rainwater, the combination of discharge and harvesting, and system's management (Figure 2) [11]. Before 2000, the urban rainwater in Beijing was directly discharged, and a rainwater drainage system was built, including community rainwater pipe networks, municipal rainwater pipe networks, and drainage channels. In the early 1990s, due to the severe situation of water shortage, Beijing launched the National Natural Science Fund project: "one of the key issues of water resources development and utilization in Beijing is rain flood utilization research". At this time the concept of urban rain and flood utilization was put forward. Research is now complete, but due to various conditions, there were no exemplary applications. At the beginning of the year 2000, with the support of the Sino-German International Cooperation Project and the Beijing Major Science and Technology Project, the project "Beijing Rainwater Flood Control and Utilization Research and Demonstration" was launched, and the first batch of urban rainwater control and utilization demonstration projects in China completed. Now, Beijing urban flood management has entered the stage of "combination of discharge and harvesting". As the first city in China to carry out research and application of urban rain flood utilization, the "combination of discharge and harvesting" of urban rainwater in Beijing has gone through four stages: exploration (1989–2000), research and demonstration (2000–2005), integration and initial promotion (2006–2012), and comprehensive promotion (2012–present). At the present stage of development, the main task is to improve the relevant policies and measures to further strengthen the mandatory use of urban rainwater. In the whole city's perspective, urban rainwater utilization will be comprehensively and profoundly promoted.

With the continuous improvement of the urban sewage treatment rate in Beijing, the problem of urban rainfall runoff pollution has become increasingly prominent. The effect of reducing and controlling the non-point source pollution of rain flood control and utilization measures is recognized and valued. In recent years, frequent urban rainstorms have led to waterlogging disasters, which have become the focus of attention of governments at all levels and the whole society in general. As an important measure for resource utilization and flood control, the effective use of urban rainwater is critical. Therefore, in order to solve the water resource problems faced by urban development, such as rainwater runoff pollution, urban floods, and water shortages, Beijing began to explore how to effectively reduce water pollution, floods etc., as well as the pattern in which the upstream and downstream and hydrological process was systematically managed. In other words, Beijing's rainwater management began to enter the "system management" phase, which is in line with the concept of "sponge city" construction.

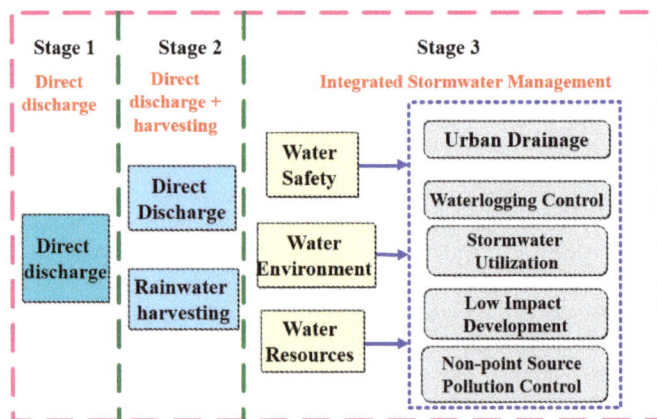

Figure 2. Diagram of Beijing urban rainwater management development.

The total amount of rainwater utilization in Beijing has been increasing steadily and the number of new projects each year increased from 129 in 2010 to about 169 in 2016. Comprehensive rainwater utilization capacity increased from 3300 million m^3 in 2010 to 4962 million m^3 in 2016. And the collection and utilization provided an important source of water for the lake and river environment, green land irrigation, car washing, and groundwater recharging etc. [12].

3.2. Status of Rainwater Management Technology

In addition to traditional drainage technology, rainwater management technology in Beijing is mainly embodied in rain flood control and utilization. After more than ten years of research, demonstration and application, Beijing has initially built a five-in-one roof green land-drainage-network and urban rainwater control system, which has been initially applied in residential, regional, and many other levels to reduce non-point source pollution and increase water infiltration [13]. Some technical measures have also been incorporated into the International Low Impact Development (LID) and have gradually become ecological and easy to use. Rain flood control and utilization at the residential level is organized into different blocks divided by urban roads. It includes mainly the basic measures for rainwater infiltration, collection and reuse, and discharge regulation [14]. Rainwater infiltration is a method allowing rainwater to penetrate underground through green spaces, permeable surfaces, and special infiltration facilities. Collect and reuse rainwater from roofs, roads, courtyards, plazas, and other undersides, irrigating green areas, flushing, car washing, landscape water supply, and road spraying, etc. Discharge regulations use storage facilities such as storage tanks, control facilities like flow control wells, and overflow weirs to keep rainwater in place in the pipeline and storage facilities in the appropriate position before rain is discharged from the area, then discharging rainwater to the downstream according to the flow controlled. In further research and application of urban flood control and utilization technologies, some new international ideas and technologies are gradually introduced and applied, such as best management practices (BMPs), low impact development (LID), and water-sensitive urban design (WSUD) and so on. A number of technologies that blend with the environment are formed, such as rain gardens, rain raised roof greening, road biological detention tank, dry pond, and wet pond, etc.

Rain flood control and utilization on a regional level is in municipal rainwater pipelines and river systems at all levels outside the city blocks. Since rainwater on the ground is discharged into municipal rainwater pipelines directly, runoff reduction and pollution control measures at the source of the residential rain flood control and utilization are often preferred, such as rainwater clean drainage and utilization technology in urban motor vehicle lanes. For rainwater in municipal pipelines and river systems, the following techniques can be adopted: separation and disposal of pollutants in municipal rainwater pipelines, rainwater storage technology in municipal pipelines, wetland regulation and purification technology before rainwater is discharged into river, regulation and utilization technology of rainwater in the public green lands along the river, rainwater infiltration increasing technology, utilization of sand and gravel pit to store and utilize rainwater in rivers, or real-time optimal operation and utilization of rainwater in urban rivers and lakes.

3.3. Status of Rainwater Management Policy

There are no special regulations on rainwater management in Beijing at present. Local regulations set by the operation of law can provide some guidance for urban rainwater management policies, such as "Water Resources Management Ordinance in Beijing", "Approach to Implement the Water Law of the People's Republic of China in Beijing", "Approach to Implement the Flood Prevention Law of the People's Republic of China in Beijing", "Approach to Save Water in Beijing", "Regulation of Drainage and Recycled Water Management in Beijing", and "Regulation of Soil and Water Conservation in Beijing". Particularly, "Regulation of Soil and Water Conservation in Beijing" issued by Beijing in 2015 put forward specific demands of rain control and utilization for all types of production and construction projects.

The Beijing Municipal Government has issued a series of official documents on the use of rainwater, such as the "Interim Provisions on Strengthening the Use of Rainwater Resources for Construction Lands" in March 2003; the "Beijing Water Conservation Measures" in October 2005; the "Key Points for Rainwater Control and Utilization of New and Reconstructed Projects (Temporary)" in August 2012; the "Notice on Printing and Distributing the Opinions on Further Strengthening Urban Rain and Flood Control and Utilization Opinions in August 2013, etc., which put forward the requirements for rainwater utilization or rain flood control and utilization, and promote urban rain flood management vigorously. In particular, a water impact assessment system of construction projects implemented in December 2013 made rainwater control and utilization indexes, which included layout and typical designs that corresponded to measures intended to uphold water and soil conservation schemes such as "Permeability rate of hardened ground", "Sunken rate of green lands", "storage modulus", "Permeability rate of hardened ground in pavements", and "permeable rate of side ditches". "The analysis and calculation of the influence of rainwater exclusion", "the analysis and calculation of waterlogging", "the evaluation of the rainwater exclusion influence" and "the evaluation of waterlogging influence" were listed in contents of reviewing flood impact assessments. Each contribute to the feasibility of sponge city construction.

From "The Comprehensive Planning of Water Resources Utilization in Capital in the Early 21st Century" approved by the State Council, Beijing has formulated the following documents: "Eleventh Five-Year Special Planning for Rain Flood Utilization of Beijing Water Affair Development Planning", "12th Five-Year Rainwater and Recycled Water Use Planning in Beijing" and four other municipal plans on urban rainwater utilization. There are also thirteen rainwater utilization planning documents for the district, county or regional level such as "Special Planning of Rain Flood Utilization on Beijing Economic and Technological Development Zone", "The Planning of Rainwater Utilization in the Central Area of the Olympic Park", "11th Five-Year Special Plan of Rainwater Utilization in Fengtai District", a series of local standards such as "Standard of Roof Greening (DB11/T281-2005)", "Content and Depth of Landscape Design Documents (DB11/T 335-2006)", "Technical Standard of

Urban Rainwater Utilization Engineering (DB11/T685-2009)", "Rules for Construction and Acceptance of Grounds with Permeable Bricks (DB11/T686-2009)", "Standard for Calculation of Storm Runoff on Urban Rainwater System Planning and Design (DB11/T969-2013)", "Design Code of Rainwater Control and Utilization Engineering (DB11/T685-2013)", and ten enterprise, industry or association standards such as "Sand Permeable Brick", and "Blocks with Silica Sand Well", and "Technical Rules for Silica Sand Rainwater Utilization Engineering".

3.4. Flood Preventive Measures

Beijing still lacks specialized prevention and control projects for urban waterlogging caused by the inadequate drainage capacity of storm sewers. However, the urban lakes in Beijing can retard and store the rain, which plays an important role in controlling waterlogging. Before liberation, there are many depressions and lakes that retard and store the rain in the urban water system of Beijing downtown, most of which are buried in the process of urban construction. According to statistics, the area of rivers and lakes buried reached 71.84 km^2. According to "Beijing Basic Data Handbook for Safe Flood Season" (2009), the area of urban rivers and lakes at present is 621.59 km^2. The total maximum capacity is 114.501 million m^3, and the total normal water storage capacity is 9,861,800 m^3, of which the effective water storage capacity is 1,677,300 m^3. At the same time, there are 355.8 kilometers of drainage rivers in the urban area. However, due to demolition and land acquisition, some rivers cannot be managed as planned.

3.4.1. Facilities for Drainage and Preventing Waterlogging

Drainage facilities includes sewage treatment plants (reclaimed water plants), network facilities, and pumping stations, etc. According to the results of the first water affair census in Beijing "Water Drainage Survey Results", there are 1108 sewage treatment plants in Beijing, of which 58 sewage treatment plants in the central urban area (including the new city) have a designed sewage treatment capacity of 3.735 million m^3/d. The total length of the public drainage network in Beijing is 11,619.88 kilometers. The rainwater pipe network is 4230.20 kilometers long, including 683.53 kilometers of integrated sewer network. Moreover, there are 167 drainage pump stations, including 133 public rain water pump stations.

3.4.2. Flood Prevention Projects

First of all, the city must have a system of flood prevention engineering for security. Beijing has a good system of flood prevention projects, including reservoirs, rivers, dikes, and flood detention areas, etc. There are 84 reservoirs in Beijing with a total capacity of 9.35 billion m^3, 210 locks on the river with the total water storage capacity of 80~100 million m^3, 256 small reservoirs with the total capacity of 8.3 million m^3, and 65 inflatable rubber dams with the total storage capacity of 70 million m^3. In the right bank of the Yongding River, a flood detention area—the Xiaoqing river flood diversion area—was built. Meanwhile, there are dikes of main rivers with the length of 429 km, the flood prevention standard is 20 to 50 years. The flood control standard of the Yongding River to the left bank of the Lugou Bridge is the maximum possible flood, which usually occurs during a perennial period.

3.4.3. Flood Hazard Mapping and Risk Management

The flood risk map is an important basis for "flood risk management". It is of great significance in regulating land use, formulating watershed or regional flood control plans, deploying non-engineering measures and engineering measures for flood control and disaster mitigation, and raising awareness of flood control and disaster mitigation among citizens. Therefore, the mapping of risk maps from "flood prevention" to "flood management" is an important basis and support for establishing risk management systems and developing flood risk management. At present, there are many studies on flood risk maps of basins in Beijing, but the flood hazard map system for an urban flood has not been fully established, especially for the waterlogging risk.

3.4.4. Non-Engineering Measures Based on Resilience

The non-engineering measures for flood control and disaster mitigation mainly include daily management and emergency response capabilities. For daily management, key measures mainly include strengthening public emergency management education, establishing an emergency response system that is led by the government, involving the whole society, public welfare and market-oriented enterprises; establishing a concept of flood coexistence and strengthening floods risk management concept; improving flood risk management processes, reducing the vulnerability of social and natural ecosystems, bearing risks, and standardizing and restricting the development, utilization and protection of flood risk zones based on flood disaster maps. Furthermore, it is also necessary to improve the ability of society to resist disasters and reduce disasters, strengthen social mobilization capacity, build with complementary strengths and positive interactions with social functions, give full play to the role of the market, society and citizens in flood control and disaster mitigation, adopt distributed data atlas, network communication, training and other measures to improve people's awareness of disaster prevention and mitigation, self-rescue ability, mutual assistance and crisis rational behavior.

In terms of emergency capacity building, it is essential to improve the real-time flood monitoring system, information sharing system, and flood disaster investigation and correction system, as well as dynamic early warning and forecasting, information release and dispatching systems to strengthen the collaboration and response speed between departments. In short, the timely release of waterlogging warning and emergency response information, in-depth and sustained emergency training and drills, attention to the particularity and operability of emergency drills, and specifying corresponding operational mechanisms before, during and after the event play effective guiding role in the emergency plan.

3.4.5. Emergency Response Mechanism

In 2013, Beijing established the "1 + 7 + 5 + 16" flood control system, namely the Beijing Municipal People's Government Flood Control and Drought Relief Headquarters, which established 7 special branches of propaganda, housing and construction, road traffic, urban underground pipelines, geological disasters, scenic spots and comprehensive support, 5 basin and 16 district-level flood control headquarters. During flood season, the command mechanism plays an important role and each sub-headquarters can perform respective tasks to effectively reduce the influence of rainstorm on social order and the production and life of residents [15].

The meteorological department monitors severe weather such as heavy rain, hail, strong winds, and thunder. The land department monitors areas threatened by geological disasters. The water department monitors the safe operation of flood control projects such as reservoirs, rivers, dams, and lakes, as well as precipitation data and hydrological information. Through the monitoring information, the city's flood control and drought relief headquarters, all branches, districts, counties, and units will conduct timely consultations to analyze the situation of flood, determine the possibility, scope, damage and impact of flood disasters, and then implement prevention measures and inform the relevant units of the results in a timely manner. According to the "Beijing Flood Control Emergency Plan (2014 Revision)", the Beijing flood warning response can be divided into three categories: rainstorm early warning response, geological hazard meteorological risk early warning, and flood early warning response. According to the severity and scope, it can be divided into IV, III, II, and I from low to high. After consultation with the municipal meteorological department, the land department, the hydrological department and the urban flood control office, the above-mentioned water level warnings are issued, changed, and deleted in blue, yellow, orange and red, respectively.

Above all, the core of sponge city construction is the scientific management of urban rainwater. When it rains, it can absorb water, store water, seepage water and clean water. When water is needed, the stored water can be "released" and used. At the same time, to ensure the safety of flood control in cities with large return period rainfall, and to minimize disasters when encountering super-standard storm floods. This part analyzes the construction of China's sponge city from the aspects of rainwater management research and practice progress, technology and policy status, and flood control measures. Although China has developed in areas such as policy, law, technology, human resources, industry, and investment mechanisms, it is still in the stage of germination and promotion, facing greater challenges. For example, the national norms and standards system is still not perfect; multiple departments fail to achieve unified and coordinated management, multi-disciplinary integration is difficult, and the talent team and industrial system gap is large.

4. The Vision of Sponge City Construction in Beijing

As the first city to adopt storm rain control and utilization measures, Beijing should innovate and lead the construction of sponge cities, breaking through the original ideas of urban rain flood control and utilization. The construction of sponge cities from multi-scale and multi-return period to manage rainfall runoff and prevent disasters has made Beijing a truly livable ecological sponge city with no fear of the storm, no flood, clear water, and green embankments. Therefore, the goal of the sponge city should be diversified and the measures should be more comprehensive and feasible.

There is a need to coordinate the relationship between sponge facilities, roads, green spaces, and water systems in the construction area, and formulate specific and feasible management measures. Therefore, in order to ensure the comprehensive benefits of the sponge city facilities, the sponge city construction requirements of the construction area, the new district, the park and the film district should be fully met. When the construction area of the newly built area exceeds 2000 square meters, the rainwater storage capacity in the hardened area shall not be less than 30 cubic meters per thousand square meters. There should be at least 50% sunken green space for stranding rain of green lands. The permeable pavement rate should be no less than 70% in buildings and residential areas, public parking lots, sidewalks, pedestrian streets, bike lanes and leisure squares, and outdoor courtyards. The old city should be combined with shanty towns, dilapidated buildings, and old districts to promote the city's overall management. For roads and squares, use permeable pavements in non-motorized vehicles roads and sidewalks of the newly built and converted roads should be incorporated. Permeable pavements are gradually promoted to use in squares and parking lots. Roads can make use of green belts to strand rainwater. Rain gardens, sunken green space, constructed wetlands or other ways can be used in parks and green lands to enhance the function of the sponge in the park and green land system, store rain in this and the surrounding region and purify rainwater quality. For the rivers and lakes system, strengthen the protection and restoration of natural forms such as ponds, rivers and lakes, and wetlands, maintain the connectivity of rivers and lakes natural water system, build a good urban water circulation system, and gradually improve the water quality and water environment. It is also necessary to recover deep pools, shoals and flood plains with ecological restoration to create a diverse living environment.

It is important to reflect upon and implement the concept of sponge city construction fully, and regard the volume capture ratio of annual rainfall as a rigid control index in compiling an overall plan, a regulatory detailed plan, and a municipal plan about roads, green lands, and water etc.

It is necessary to comprehensively consider the layout and space requirements of sponge facilities and scientifically delineate the urban blue line and urban watershed; sort out relevant urban construction standards and norms; propose and highlight the key content and technical requirements of sponge city construction; accelerate the preparation of Beijing sponge city construction plan, technical guide, and standard drawings.

One must establish a cooperation mechanism between government and social capital risk sharing and revenue sharing; encourage social capital to participate in the construction and operation management of a sponge city through business income rights, government procurement services, and financial subsidies, etc.; increase the investment in research and development of sponge city technology; support the innovation and application of new materials, new processes and new methods, the rational construction of sponge cities, and then provide scientific and technological support for the industrialization of sponge city construction and economic growth. At the same time, strengthen the construction of sponge city construction, create a good atmosphere for the construction of sponge cities, and then promote the broad support and active participation of the whole society.

5. Conclusions

Although in the construction of the sponge city Beijing made use of good theory and technology, engineering practices, management policies, technical standards and an industrial basis and has made some achievements, Beijing still has a long way to go to achieve its goal of building a harmonious and livable sponge capital. The main problems are as follows: Firstly, the urban flood hazard and water supply security situation are still grim; secondly, the contradiction between supply and demand of water resources is still very serious, and there is no fundamental solution to the problem of water shortage; thirdly, improving the water environment is more urgent; fourthly, the function of water ecological services is increasingly important. Therefore, we should establish a priority water-saving, space-balanced governance system, and comprehensively develop the construction of sponge cities. Under the background of the coordinated development of Beijing, Tianjin and Hebei, Beijing will become a world-class city with harmonious Chinese characteristics, ecology, and livability.

Author Contributions: This study was carried out through collaboration among all authors. S.Z. and Y.L. designed the study, wrote the paper, and gave comments and helped to revise the paper. M.M. helped improve its flow, progression, and clarity. Y.L., T.S., and R.S. support for development of the manuscript. All authors provided corrections and suggestions for improving the text.

Funding: This research was funded by [National Key R&D Program of China] grant number [2016YFC0401405]; [Study on monitoring and evaluation of the effect of sponge city construction in Beijing] grant number [8161002]; [Research and demonstration on key technologies and management mechanism of sponge city construction in Beijing] grant number [2017ZX07103-002].

Acknowledgments: The authors wish to thank the anonymous reviewers for their comments and suggestions.

Conflicts of Interest: The authors declare no conflict of interest.

References

1. Trenberth, K.E.; Dai, A.; van der Schrier, G.; Jones, P.D.; Barichivich, J.; Briffa, K.R.; Sheffield, J. Global warming and changes in drought. *Nat. Clim. Chang.* **2014**, *4*, 17–22. [CrossRef]
2. Dagenais, D.; Thomas, I.; Paquette, S. Siting green stormwater infrastructure in a neighbourhood to maximise secondary benefits: Lessons learned from a pilot project. *Landsc. Res.* **2017**, *42*, 1–16. [CrossRef]
3. Wong, T.H.F. *Australian Runoff Quality: A Guide to Water Sensitive Urban Design*; Engineers Media: Crows Nest, NSW, Australia, 2006.
4. Ignatieva, M.; Stewart, G.; Meurk, C.D. Low impact urban design and development (LIUDD): Matching urban design and urban ecology. *Landsc. Rev.* **2008**, *12*, 61–73.
5. Mitchell, G. Mapping hazard from urban non-point pollution: A screening model to support sustainable urban drainage planning. *J. Environ. Manag.* **2005**, *74*, 1–9. [CrossRef] [PubMed]
6. Ma, M.; Wang, H.; Jia, P.; Liu, R.; Hong, Z.; Labriola, L.G.; Hong, Y.; Miao, L. Investigation of inducements and defenses of flash floods and urban waterlogging in Fuzhou, China, from 1950 to 2010. *Nat. Hazards* **2018**, *91*, 1–16. [CrossRef]
7. The Ministry of Housing and Urban-Rural Development. *Technical Guide for Sponge City Construction—Construction of Rainwater System with Low Impact Development (Trial Implementation)*; Architecture& Building Press: Beijing, China, 2014; pp. 4–6.
8. Yu, K.; Li, D.; Yuan, H.; Fu, W.; Qiao, Q.; Wang, S. Sponge City: Theory and Practice. *City Plan.* **2015**, *39*, 26–36.
9. Wu, D.; Zhan, S. New Trends and Practical Research on the Sponge Cities with Chinese Characteristics China Soft Science. *Chin. Soft Sci.* **2016**, *1*, 79–97.
10. Wang, W.; Li, J.; Wang, E.; Zhang, L.; Cao, Y.; Xu, H. The key points of sponge city construction. *Constr. Sci. Technol.* **2015**, *1*, 19–21.
11. Zhang, S. Stormwater management in Beijing. *Constr. Technol.* **2013**, *2*, 39–42.

12. Zhao, Z.; Li, H. Research of stormwater utilization projection management in Beijing. *Beijing Water* **2015**, *4*, 29–31.

13. Ding, Y.; Chen, J.; Zhang, S. Discussion on the current situation and development of urban stormwater and flood utilization in Beijing. In Proceedings of the International Conference on Sustainable Utilization of Water Resources in Beijing; Beijing Water Conservancy Bureau; Water&Power Press: Beijing, China, 2007; pp. 201–204.

14. Zhang, W.; Luo, Q.; An, G.; Zhang, R. Analysis of the sponge City construction factors of the city's New District. *Chin. Rural Water Hydropower* **2017**, *58*, 62–67.

15. Xie, Y. Discussion on emergency response mechanism of flood control in Beijing. *Beijing Water* **2014**, *4*, 46–48.

water

MDPI

Article
Stormwater Reservoir Sizing in Respect of Uncertainty

Bartosz Szeląg [1],*, Adam Kiczko [2] and Lidia Dąbek [3]

[1] Department of Geotechnics and Water Engineering, Faculty of Environmental, Geomatic and Energy Engineering, Kielce University of Technology, 25-314 Kielce, Poland
[2] Department of Hydraulics, Faculty of Civil and Environmental Engineering, Warsaw University of Life Sciences (WULS-SGGW), 02-787 Warsaw, Poland; adam_kiczko@sggw.pl
[3] Department of Water and Wastewater Technology, Faculty of Environmental, Geomatic and Energy Engineering, Kielce University of Technology, 25-314 Kielce, Poland; ldabek@tu.kielce.pl
* Correspondence: bartoszszelag@op.pl

Received: 29 December 2018; Accepted: 11 February 2019; Published: 14 February 2019

Abstract: The sizing of the stormwater reservoir, as the design of its properties, usually requires simulations of a basin runoff for a long rainfall series using a hydrodynamic model. In the case of insufficient observations, the rainfall series can be reproduced using empirical approaches. One of the crucial elements in the sizing of the stormwater reservoir is determination of duration time and intensity of rainfall (design rainfall event), for which the maximum reservoir capacity is being obtained. The outcome is, however, affected by significant uncertainty of runoff modeling. The aim of the study is to analyze the effect of the uncertainty of a rainfall-runoff model on calculated capacities of stormwater reservoirs, along with estimated duration times of the design rainfall. The characteristics of the rainfall events—intensity, duration, and frequency—were reproduced using an empirical approach of IDF (Intensity–Duration–Frequency). The basin response to the precipitation was modeled using the SWMM (Storm Water Management Model) and its uncertainty was estimated on the basis of the GLUE (Generalized Likelihood Uncertainty Estimation) method. The obtained probabilistic solution was compared with the deterministic one, neglecting the uncertainty. Duration times of the design rainfall determined in respect of the reservoir outflow using the probabilistic model were longer than those found with a deterministic approach. This has an effect on the desired capacities of the stormwater reservoir, which were overestimated when uncertainty was neglected.

Keywords: stormwater reservoir; Generalized Likelihood Uncertainty Estimation (GLUE); design rainfall event; Storm Water Management Model (SWMM)

1. Introduction

Intensive precipitation might result in inflows exceeding the capacity of stormwater drainage, causing flooding and releases of stormwater to the recipient (i.e., river). One of the possible solutions is attenuating stormwater discharges with storage reservoirs, built within sewer systems. The problem of designing these structures have attracted many researchers [1,2]. Their main focus are methods for finding the required reservoir capacity, ensuring proper reduction of flood flows. The problem can be also stated in respect of desired reduction of pollutant load in stormwaters [3].

The storage reservoirs are usually designed to operate as overflow reservoirs. Their main advantage is maintaining a constant outflow [4–6]. Many theoretical studies were devoted to methodology of designing this type of the reservoir [7–10]. Its capacity is usually determined for so-called design rainfall. Its duration is found to maximize required reservoir capacity, that allows for maintenance of the desired outflow [11]. The determination of the design rainfall duration is usually done on the basis of many simplifications to the runoff model and description of the stochasticity of the rain events. In practice, it is done using nomograms of the rainfall duration and required reservoir capacity, elaborated for the given physio-geographic properties of the basin, drainage standards, and outflow devices [4,6,11]. Capacities obtained in such a graphical approach are considered to be affected with high uncertainty, because of very rough parametrization of the basin. A more detailed representation of the basin response to the rainfall can be obtained using hydrodynamic modeling [7,12–14]. In such an approach, reservoir capacities are calculated on the basis of so-called continuous simulations [2,7,12], where the input consists of long precipitation time series (around 30 years long). In the case where an observation period is significantly shorter than 30 years, available time series can be used to develop a synthetic precipitation generator that allows for extension of the series length artificially [15,16]. Recently, multidimensional probability density distributions parametrized with copula functions [17,18] are gaining popularity for precipitation generators, allowing high resolution of simulated time series to be obtained (below 5 min). In such an approach, in a precipitation time series, rainfall events are identified and then parametrized in a sense of height and duration, assuming usually the rectangular shape of the precipitation graph. On the basis of empirical distributions of these precipitation parameters, two dimensional functionals are being developed, which can be used in precipitation simulations. In other approaches, the dependency between the precipitation and duration is explained using the correlation coefficient, and then the rainfall events can be generated using the Iman-Conover method [19]. In many successful studies, duration and the rainfall height were considered as independent variables. One of the most sophisticated methods for a rainfall model includes canonical microsaccades, based on the fractal theory [12].

In practical studies, usually there is a lack of observed time series. The design rainfall can be determined using regional precipitation models, which on the basis of empirical relationships allows for estimation of its height for the given duration and occurrence frequency [20,21]. The approach takes a form of so-called Intensity–Duration–Frequency (IDF) curves and can be also used for a rainfall generator. Because the observation series of precipitation are rare in engineering applications, IDF approach is often used for the design of hydraulic structures in sewer systems [8,22].

Runoff models used in the design of the storage reservoirs are affected with strong uncertainty [23–26]. Surprisingly, its effect on the design of sewer system devices is, however, rarely analyzed. Kiczko et al. [27] showed that it significantly affects the determined capacities of the storage reservoir. This study included the design of the stormwater reservoir for a single rainfall event, using SWMM model. The model uncertainty was estimated using the well-known hydrology method of Generalized Likelihood Uncertainty Estimation (GLUE) [28]. However, the study focused on a single event, and did not address the problem of dependency of the design uncertainty on precipitation height and duration (in a result, the rainfall probability).

The aim of the present study was to develop a general method for the design of a stormwater reservoir in respect to the runoff model uncertainty, using the IDF approach for the design rainfall. It analyzes the effect of the model uncertainty for different probabilities of the design rain and shows its significance on the determination of duration times for the rainfall event that maximizes the reservoir capacity. As in the previous study, model uncertainty was assessed using the GLUE approach. Reservoir capacities obtained in respect of the model uncertainty were compared with those calculated neglecting it. The study utilizes IDF curves developed for Poland.

2. Materials and Methods

2.1. Object of Study

The analyzed catchment is located in the central-eastern part of Kielce and its area is 62 ha (Figure 1). The total area of the City Kielce is 109 km^2, with the population reaching 200,000 (density 21.4 people/ha). The highest point of the catchment is 271.20 m above sea level and the lowest one is 260 m above sea level. The total length of the sewer network is 5583 m, of which the main canal is 1569 m long and its diameter from top to outlet varies between 600–1250 mm. The main channel receives rainwater from 17 side channels, whose diameters vary in the range of 300–1000 mm. The total volume of pipes with stormwater wells is 2032 m^3. The drop of the collector in individual sections varies from 0.04% to 3.90%, while the drop of side channels is 2.61% maximum. The area is covered with housing estates, public utility buildings, as well as main and side streets. The catchment contains 6 types of runoff areas: roofs (14.3%), pavements (8.4%), roads (17.7%), parking lots (11.2%), greenery (47.2%), and school pitches (1.3%). Road density in the basin is about 108 m/ha. More detailed data concerning the characteristics of the catchment are described by Dąbkowski et al. [29].

According to the DWA A-117 (2006) method, the annual number of rainfall events in the observation period (2008–2016) was around 36–58. The total rainfall height and its duration were varying in ranges of 3.0–45.2 mm and 20–2366 min, respectively. The duration of rainless periods was from 0.16 to 60 days. Annual precipitation heights in the observation period were changing between 537–757 mm, with rainfall days 155–266. The annual air temperature was 8.1–9.6 °C and the number of days with snowfall 36–84.

A stormwater treatment plant (STP) is located on the outflow from the catchment. At a distance of approximately 4.0 m from the distribution chamber (DC), an ultrasonic flow meter MES1 is installed in channel S1. The MES1 ultrasonic flow meter measures the filling and flow of stormwater in a 1-min step. Obtained time series of the discharge (MES1 point in Figure 1) were used for the identification of the runoff model.

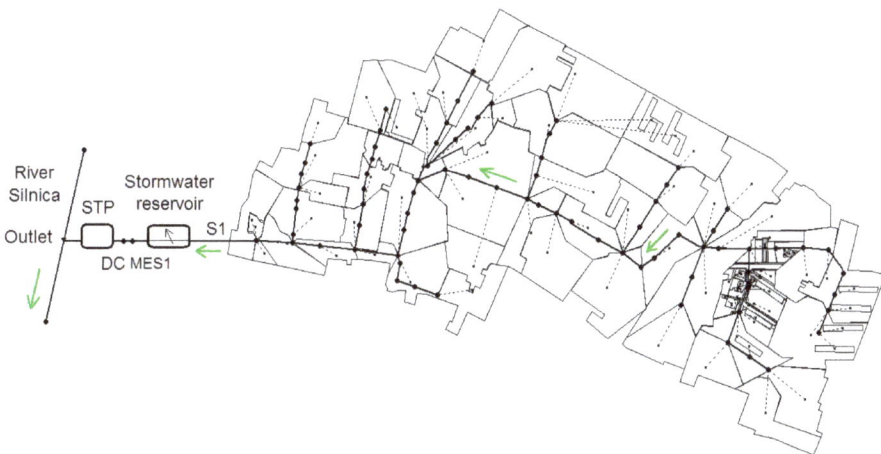

Figure 1. Location of the analyzed catchment and model diagram in Storm Water Management Model (SWMM) software.

The catchment, considered in operation, has already been the subject of numerous studies, in which surface runoff, stormwater quality, storm overflow, and stormwater treatment plant operation were modelled [3,16]. As part of these analyses, the model was calibrated using the trial and error method, where a set of parameters determining satisfactory results of calculations was searched for.

2.2. Methodology

In the absence of continuous, long-term rainfall data, the storage reservoir capacity is determined on the basis of the calculated rainfall frequency (C = 1, 2, 5, 10), using the category of drainage standard [11,30]. The calculation diagram of the developed method is presented in Figure 2.

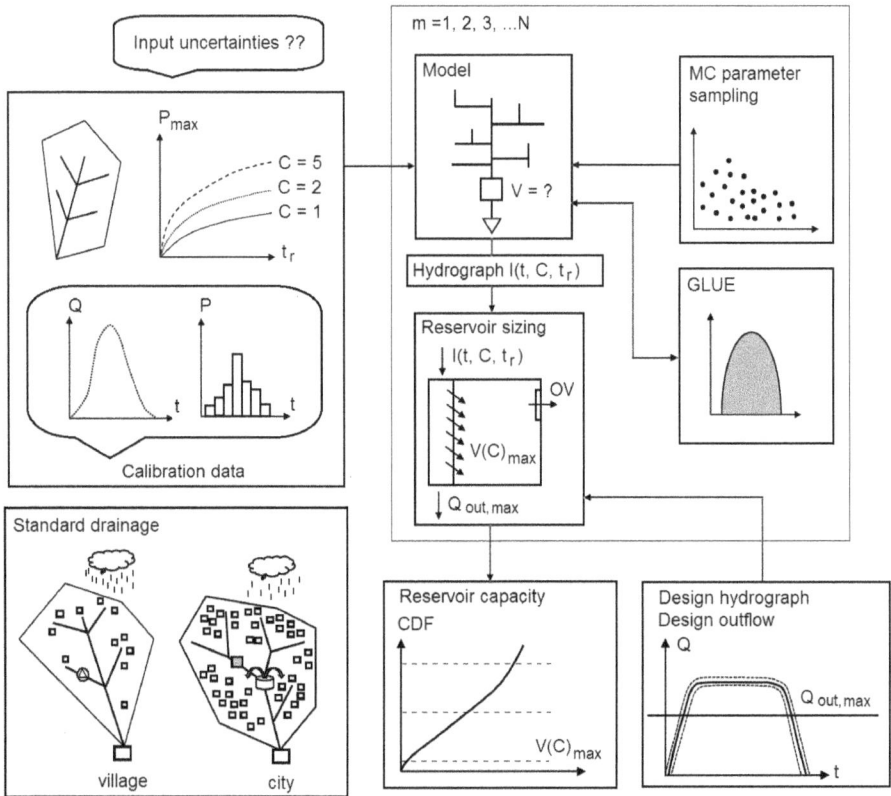

Figure 2. Calculation diagram of the method of dimensioning the retention capacity (where $I(t, C, t_r)$—reservoir inflow hydrograph for given rain duration time (t_r) and occurrence probability C, m = 1, 2, 3, ..., N—subsequent Monte Carlo simulations with varying parameters [α,d_{imp}, d_{perv}, n_{imp}, n_{perv}, n_n, γ, β], for which reservoir inflow hydrograph is compute, $Q_{out,max}$—maximal allowed outflow form the reservoir, $V(C)_{max}$—maximal reservoir retention capacity for given values of C and $Q_{out,max}$).

The computational data adopted in the study include the physical and geographical characteristics of the catchment and rainfall and flow measurements used to calibrate the hydrodynamic model. In order to determine the storage capacity of the reservoir, it is necessary to transform the calculated rainfall depth values with Equation (3) in the outflow values. In order to take into account the uncertainty of the catchment model, the Monte Carlo method is used to simulate the model parameters for the assumed ranges of their variability. Then, on the basis of measurements, the identification of distributions of analyzed parameters is performed with the GLUE method [31]. In this way the basin outflow hydrograph is determined, and using the balance of Equation (7), the reservoir capacity is calculated. In order to determine the maximum capacity of the reservoir for the assumed standard of drainage (C) and assumed outflow ($Q_{out,max}$), it is necessary to perform calculations for different (m) rainfall duration (t_r) in order to determine the extremum of variability of $V = f(C, t_r, Q_{out,max})$ and to determine the duration of the determining rain, where $V \rightarrow max\{(C, t_{rd}, Q_{out,max})\}$ (where: t_{rd}—design rainfall event). The values of the outflow from the retention reservoir ($Q_{out,max}$) and the determined accumulation capacities $V(C)_{max}$ were normalized by reference of the above variables to the impermeable area of the catchment (F_{imp}) and determination of the unit capacity index (V_q) and unit outflow (q).

2.3. Dimensioning the Retention Reservoir

A review of the literature [5–7,27] shows that the most common task of the retention reservoir in the stormwater system is to reduce the maximum flow at the outflow from the catchment and to improve the quality of the stormwater. At the reservoir design stage it is necessary to meet the condition that for the assumed return period at a given time period of rainfall (assumed repeatability of rainfall over a given period of time) and at its maximum filling $H = H_{max}$, the outflow value $Q_{out} \leq Q_{out,max}$ (design flow) does not lead to overloading of the stormwater treatment plant. Exceeding the H_{max} value leads to an increase in outflow from the reservoir ($Q_{out} \geq Q_{out,max}$) and discharges with emergency overflow (OV, Figure 2). Therefore, the determination of the reservoir capacity is limited to the determination of the combination of P_{max} and t_r values, for which the desired capacity reaches the maximum value.

2.4. Uncertainty Analysis by the GLUE Method

Uncertainty analysis was performed using a method popular in hydrology: Generalized Likelihood Uncertainty Estimation (GLUE) [28,31]. It is based on the probabilistic formulation of the parameter identification problem, where instead of a single parameter set, as in the deterministic approach, a distribution of parameters is estimated. The uncertainty is modeled by the parameter variability. Because the output distribution is conditioned with the observations, a total model uncertainty is obtained. The method uses the Bayes formula:

$$P(Q/\Theta) = \frac{L(Q/\Theta) \cdot P(\Theta)}{\int L(Q/\Theta) \cdot P(\Theta) d(\Theta)} \tag{1}$$

where $P(\Theta)$ denotes *a priori* parameter distribution, $L(Q/\Theta)$ likelihood function, and $P(Q/\Theta)$ the *a posteriori* distribution, as the result of the parameter identification. In practical cases, the assumption on the *a priori* distribution is weak and usually takes a form of the uniform distribution limited to the parameter physical variability. Therefore, the choice of the likelihood function is crucial. In the present study the following function was used [32]:

$$L(Q/\Theta) = exp\left[\frac{-\sum_{i=1}^{N}(Q_i - \hat{Q}_i)^2}{\kappa \cdot \sigma^2}\right] \tag{2}$$

with Q_i and $\hat{Q}_i$ standing for modeled and observed discharge in the time step i, σ^2 the variance of the model residua, κ the factor used to control the variance of the *a posteriori* distribution. The value of the κ factor was estimated in order to maximize the likelihood of the observations in the output discharge distribution, ensuring that 95% of observation points were enclosed by 95% confidence intervals. For the $\hat{Q}_i$ flow rates measured with the ultrasonic flow meter, MES1 (at S1, Figure 1) were used.

2.5. Rainfall Depth

In engineering practice, IDF curves are widely used in the dimensioning of sewer systems and design of objects located on it (overflows, reservoirs, etc.) [4,8,14,22]. These curves describe the relation between the duration of rainfall (t_r) and the frequency of its occurrence (C), which can be generally written with the relation:

$$J = f(t_r, C, \xi) \tag{3}$$

in which ξ—regional parameter differentiating the variability of the obtained curves, J—mean rain intensity in $dm^3 \cdot (ha \cdot s)^{-1}$, expressed as $J = 166.7 \cdot P_{max}$.

Due to a number of conditions, such as the genesis of rainfall, climatic conditions, land use, the length of rainfall measurements, etc., the relationships described by Equation (1) may take different forms [20,33,34]. Taking into account the above remarks and the climate conditions in Poland, which was reflected in a series of works [20,35] aimed at developing empirical rainfall models, the formula of Bogdanowicz and Stachy [32] was used in the analyses to determine the maximum rainfall depth (P_{max}) for the assumed precipitation frequency:

$$P_{max} = 1.42 \cdot t_r^{0.33} + \xi(R, t_r) \cdot (-lnp)^{0.584} \tag{4}$$

where t_r—duration of rain (min), p—probability of rainfall exceeding ($p = 1/C$), $\xi(R, t_r)$—regional parameter determined depending on the location and duration of rainfall.

The above relationship was developed on the basis of 60 years of precipitation measurements in Poland and is now used in engineering practice [4,5]. For the region covered by the research, i.e., the Świętokrzyskie Voivodship and the city of Kielce, the values of parameter ξ are determined on the basis of equations:

$$\xi(R.t_r < 120 \ min) = 4.693 \cdot \ln(t_r + 1) - 1.249 \tag{5}$$

$$\xi(R.t_r < 1080 \ min) = 2.223 \cdot \ln(t_r + 1) + 10.639 \tag{6}$$

In further considerations at the stage of reservoir dimensioning for $C = 2, 5, 10$, the rainfall duration was considered in the range of $t_r = 15$–240 min [9]. The results for $C = 1$ were omitted in the analyses, as they lead to an underestimation of reservoir sizes, which was reported in various studies [4,5,9].

2.6. Surface Runoff Modelling

One of the factors determining the capacity of the reservoir and the type of drains designed is the outflow from the catchment. This outflow is the result of the runoff on the surface of the catchment area, and then of the flow of stormwater through the sewer system. In this paper, the SWMM model was used to simulate the drain from the catchment. This model is used commonly to simulate the quantity and quality of stormwater and the phenomenon of rainfall–runoff is modelled in it, taking into account non-linear reservoirs, where infiltration and surface runoff are simulated [36].

Due to the complex process of wastewater accumulation in the overflow reservoir, requiring large calculation outlays in the SWMM model, a simplified model for the reservoir was used in the conducted analyses [5,6,9]. In this model, the overflow reservoir capacity can be determined on the basis of the differential equation of the wastewater volume balance of the form:

$$\frac{dV(t)}{dt} = S \cdot \frac{dH(t)}{dt} = I(t) - Q(t)_{out} \tag{7}$$

where S—surface area of the reservoir in the projection, $H(t)$—reservoir depth, $I(t)$—inflow to the reservoir determined on the basis of SWMM simulations, $Q(t)_{out}$—outflow from the reservoir, with negligible error that can be assumed to be constant [36].

3. Results

On the basis of SWMM model calculation results and literature data, the ranges of variability of individual model parameters were determined (Table 1). Due to strong interactions between the calibrated parameters and their significant influence on the results of surface runoff calculations, the following modification coefficients were introduced at the calculation stage: coefficient (α) for the flow path of width (W_j), coefficients (d_{imp}; d_{perv}) for the retention depth of impervious and pervious areas, coefficients (n_{imp}; n_{perv}) of roughness of impervious and pervious areas, coefficients (n_n) of roughness of sewer, coefficient (γ) for the share of impervious areas (Imp_j) in the whole catchment area, coefficient (β) for the average slope of a partial catchment (I_j). In order to reduce the multidimensionality of the solution and limit the outlays of calculations, it was assumed that the flow path width is determined as $W = \alpha \cdot A^{0.5}$. The values of coefficients (β, γ) are used to correct the value of the sealed area (Imp_j) and the slope of the catchment ($I = \beta \cdot I_j$), where j—the number of partial catchments.

Table 1. Ranges of variability of model parameters for the *a priori* uniform distribution [16].

Parameters	Unit	Range
Coefficient for flow path width (α)	-	2.7–4.7
Retention of impervious areas (d_{imp})	mm	0.8–4.8
Retention of pervious areas (d_{perv})	mm	0.8–6.8
Roughness coefficient for impervious areas (n_{imp})	$m^{-1/3} \cdot s$	0.01–0.022
Roughness coefficient for pervious areas (n_{perv})	$m^{-1/3} \cdot s$	0.16–0.20
Roughness coefficient of sewer channels (n_n)	$m^{-1/3} \cdot s$	0.01–0.048
Correction coefficient for percentage of impervious areas (γ)	-	0.7–1.275
Correction coefficient for sub-catchments slope (β)	-	0.8–1.375

The Monte Carlo sample was conditioned using a likelihood function (Equation (7)) on two sets of observations: for 15 September 2010 (time and total rainfall depth equal to $t_r = 107$ min and $P_{tot} = 9.2$ mm, volume of runoff $V_{tot} = 2221$ m^3), and 7 August 2011 ($t_r = 60$ min; $P_{tot} = 8.6$ mm, volume of the runoff $V_{tot} = 1733$ m^3), events [37]. The scaling factor κ was adjusted in order to ensure that 95% of observations in calibration sets are enclosed in 95% confidence intervals. In Figure 3, resulting parameter distributions in respect of the likelihood measure (Equation (8)) are showed as box-plots. The total variability of the likelihood function is presented with dashed lines extents. A red center line denotes the median value and 25th and 75th quantiles are indicated by box edges. Box-plots are used instead of dot-plots, found in Kiczko et al. [27], as they provide a more precise characterization of the *a posteriori* distribution. For α, n_{perv}, n_{imp}, d_{perv}, and γ, the model response is almost uniform—similar high and low likelihood values were found in the whole parameter span (Figure 3a–c,e,g). It is different in the case of remaining parameters, d_{imp}, β, and n_n (Figure 3d,f,h). This suggests that the model is more sensitive to the second parameter set, as it is possible to spot the dependency of the likelihood function on parameter values. For example, it is clear that the highest model likelihood was obtained for β values close to 0.9 (Figure 3f). Figure 4 presents verification of the probabilistic solution for an independent data set. On the basis of the simulations performed (Figure 4), it can be stated that the measured and modelled outflow hydrographs for the calibration set are characterized by a high adjustment for high intensity rainfall (time and total rainfall depth equal to $t_r = 270$ min and $P_{tot} = 16.5$ mm, which results in rainfall intensity $q = 10.19$ dm$^3 \cdot$(ha·s)$^{-1}$ at the volume of $V_{tot} = 3415$ m^3 hydrograph).

Using the simulation results obtained with the Monte Carlo method and *a posteriori* distributions of the model parameters (Table 1), hydrograph calculations of outflow from the catchment were performed for precipitation $t_r = 10$–240 min and $C = 1, 2, 5, 10$, determined on the basis of IDF curves described with Equations (4)–(6). Next, on the basis of Equation (7) for the assumed design flows ($Q_{out,max} = 0.1$–1.0 m$^3 \cdot$s^{-1}), capacities of retention reservoirs were determined, thus obtaining a probabilistic solution and determining for individual values of $V(t_r, C, q)$ the mean value and 95% confidence interval, respectively.

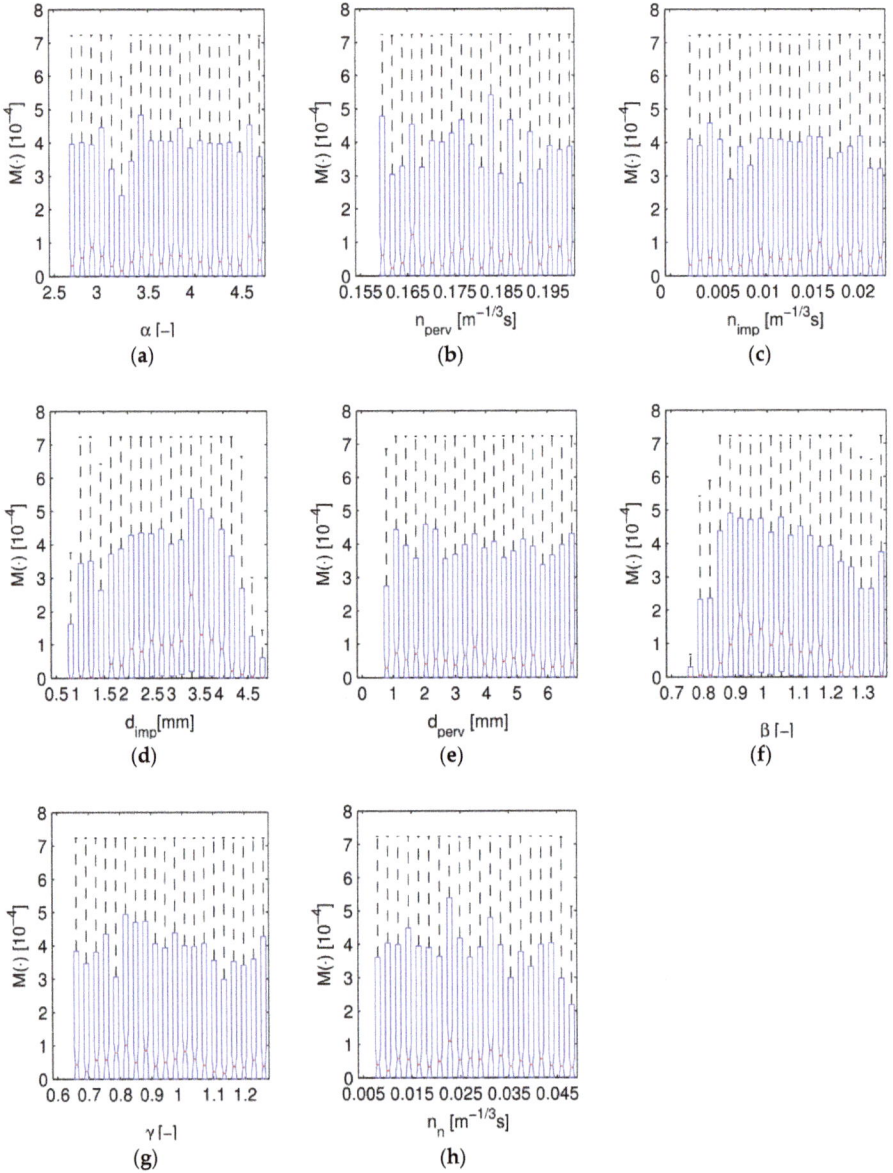

Figure 3. Box-plots for likelihood function values in the respect of parameter values (parameters symbols explained in Table 1).

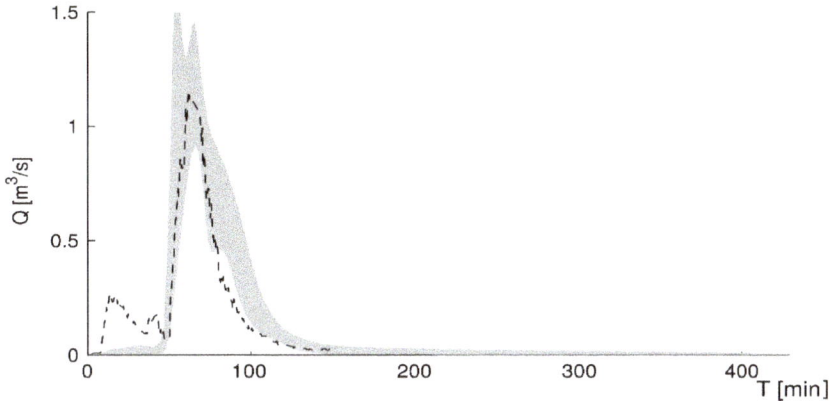

Figure 4. Exemplary results of calculations and measurements for the confidence interval of 95% using the Generalized Likelihood Uncertainty Estimation (GLUE) method for the hydrograph of outflow from the catchment (30 July 2010).

The results were compared with the reservoir capacities obtained neglecting the uncertainty, for a single parameter set, ensuring satisfactory simulation results. This approach is commonly used in engineering practice and is called the deterministic solution. The results of calculations obtained in this way allowed to assess the influence of uncertainty of the calibrated parameters of the SWMM model on the determined reservoir capacities.

Using the above remarks and based on simulations performed with the SWMM model, the influence of the uncertainty of the hydrodynamic model of the catchment and unit outflow on the design rainfall duration used to dimension the reservoir and the unit reservoir capacity was analyzed. On the basis of performed calculations, the dependence $V_q = f(t_r)$ for selected values of q was first determined; an example of the dependence $V_q = f(t_r)$ for $C = 5$ and $q = 17$ dm³·(ha·s)⁻¹ and $q = 6$ dm³·(ha·s)⁻¹ is shown in Figure 5.

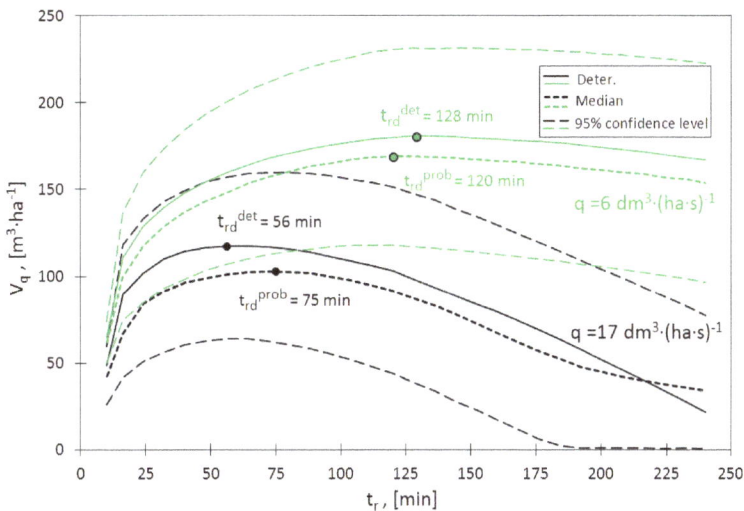

Figure 5. Influence of the unit outflow from the reservoir (q) and the time duration rainfall (t_r) on the unit accumulation capacity (V_q), taking into account the model uncertainty.

From the results obtained it can be concluded that the uncertainty of the hydrodynamic model of the catchment influences the unit accumulation capacity of the reservoir, which is confirmed by the range of variability of the confidence interval of 95% for individual values of $V_q = f(t_r)$ and by the dependence of $V_q = f(q, C)$ shown in Figure 8. Moreover, it was found that the reservoir capacity determined taking into account the model uncertainty (median) is smaller than the value obtained in the deterministic solution, which is also indicated by the variability of V_q values illustrated in Figure 5. Due to the fact that curves in Figure 5 were prepared for selected values of q, it is impossible to generalize them for the remaining unit values of reservoir outflows. For this purpose, it was necessary to analyze the relations between unit reservoir capacities obtained for t_{rd}^{prob} and t_{rd}^{det} and the assumed outflows (q) and the assumed rainfall frequency (C), which is discussed in the further part of the paper.

On the basis of the curves shown in Figure 5 and determined for $q = 6 \text{ dm}^3 \cdot (\text{ha·s})^{-1}$ and $q = 17 \text{ dm}^3 \cdot (\text{ha·s})^{-1}$, it is possible to determine the influence of the SWMM model uncertainty on the duration of the design rainfall used to dimension the reservoir in the deterministic and probabilistic solutions. On the basis of the presented curves, it can be stated that the value of the design rainfall duration in the deterministic solution (t_{rd}^{det}) is shorter than the value obtained in the probabilistic one (median)—t_{rd}^{prob}. In the first solution for $q = 17 \text{ dm}^3 \cdot (\text{ha·s})^{-1}$, the t_{rd} value is 56 min and is 19 min shorter than in the second one, and the $\frac{t_{rd}^{prob}}{t_{rd}^{det}}$ ratio is 1.33, while for $q = 6 \text{ dm}^3 \cdot (\text{ha·s})^{-1}$ the shift between rain duration in the considered solutions is shortened to 8 min and the $\frac{t_{rd}^{prob}}{t_{rd}^{det}}$ of 1.09 decreases accordingly. The obtained result indicates that the smaller the unit outflow from the reservoir (q), the value of the shift between the duration of rainfall obtained in deterministic and probabilistic solutions decreases, which translates into the size of the required storage capacity of the reservoir. Bearing in mind the need to carry out a detailed evaluation of the model uncertainty influence on individual aspects of reservoir capacity selection and its operation, on the basis of the obtained results the $\frac{t_{rd}^{prob}}{t_{rd}^{det}} = f(q, C)$ and $\frac{V_q^{prob}}{V_q^{det}} = f(\frac{t_{rd}^{prob}}{t_{rd}^{det}}, C)$ curves were prepared and shown in Figures 6 and 7.

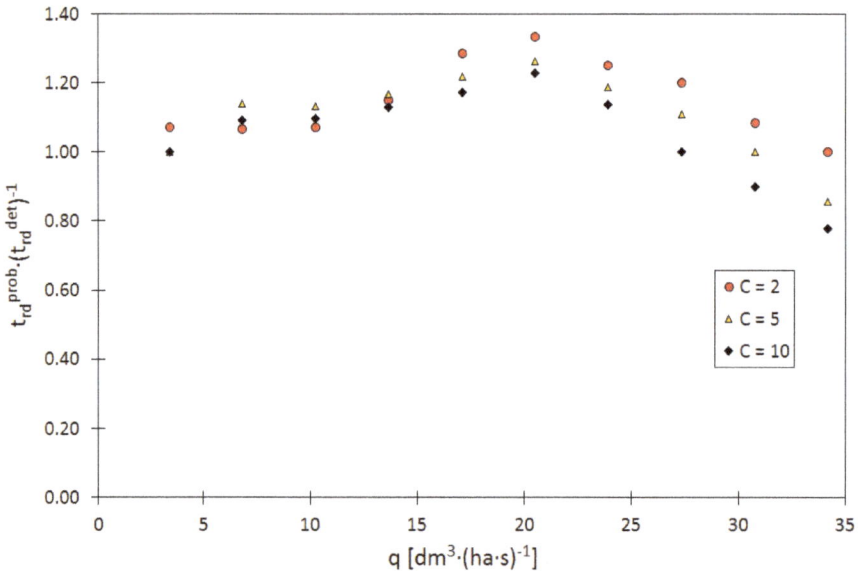

Figure 6. Influence of unit outflow from the reservoir (q) and rainfall frequency (C) on the difference between the design rainfall duration obtained for deterministic and probabilistic solutions.

On the basis of the data in Figure 6, it can be stated that the difference in the shift between the rain duration determined in the probabilistic and deterministic solutions for designing the retention reservoir is influenced by the unit outflow (q) and the rainfall frequency (C). The results of calculations shown in Figure 6 showed that the highest relative difference between the t_{rd} time obtained in the deterministic and probabilistic solution was obtained for $q = 21$ dm^3·(ha·s)$^{-1}$ for $C = 2$–10, and it is the highest for $C = 2$ and exceeds 35%. However, for the value of $C = 10$, it is smaller than for $C = 5$ and $C = 2$, and is 1.15. A further increase in the unit value of the outflow from the reservoir (q) leads to a decrease in the time lag between the duration of rainfall obtained in the deterministic and probabilistic solutions. The issue discussed above is important from the point of view of the reservoir design, because identification of the most unfavorable distribution of t_{rd}^{prob} values in relation to the deterministic solution defines the area of operation of the reservoir, in which it is most exposed to potential underestimation.

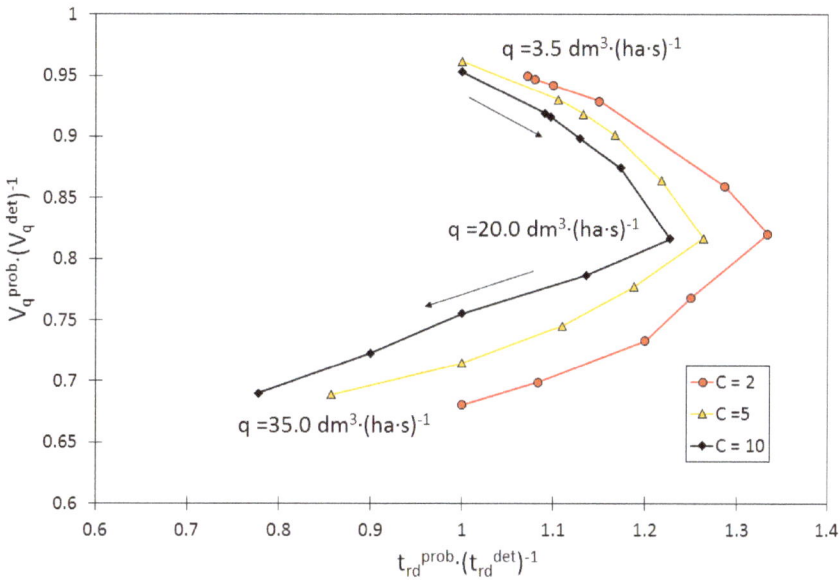

Figure 7. Influence of relative shift between probabilistic and deterministic solution and unit reservoir capacities (V_q).

Taking into account the above considerations, the variability of $\frac{V_q^{prob}}{V_q^{det}} = f(\frac{t_{rd}^{prob}}{t_{rd}^{det}}, C)$, allowing for determination of the effect of the relative shift between the t_{rd} value determined for deterministic and probabilistic solutions on the unit storage capacity, is presented in Figure 7. Analyzing the shape of the curves, it can be stated that the increase in the relative shift between the design rainfall duration obtained in the deterministic and probabilistic solution (median) relative to the maximum value of $\frac{t_{rd}^{prob}}{t_{rd}^{det}}$ corresponding to $q = 15$ dm^3·(ha·s)$^{-1}$ ($C = 2$–10) leads to a decrease in the relative difference in the capacity of reservoirs obtained in the SWMM model and taking into account the uncertainty, and reaches the value equal to $\frac{V_q^{prob}}{V_q^{det}} = 0.8$. As a result of a decrease in the $\frac{t_{rd}^{prob}}{t_{rd}^{det}}$ value, which is related to the increase in the outflow from the reservoir (Figure 5), the $\frac{V_q^{prob}}{V_q^{det}}$ quotient decreases further, the value of which indicates an underestimation of the storage capacity in the probabilistic solution (the median is a search solution) in relation to the value obtained only by the calibrated SWMM model.

In the next stage of analyses, on the basis of calculations carried out with the SWMM program, curves were determined (Figure 8) showing the influence of the unit outflow from the reservoir (q) on the unit maximum accumulation capacity (V_q) at preset values C for the design rainfall duration (t_{rd}). The calculations done showed that the size of the unit storage capacity of the reservoir (V_q) determined in the deterministic solution is larger than the median obtained in the probabilistic solution, and its value decreases with the increase of the unit outflow from the reservoir, which is also confirmed by the variability of $V_q = f(t_r, C)$ in Figure 8. This result indicates that the unit storage capacity of the reservoir (V_q) obtained in the deterministic solution is overestimated in relation to the capacity obtained in the probabilistic solution (median), which means that the reservoir may be oversized. From the point of view of the reliability of the reservoir operation, this is a beneficial solution, however, taking into account the 95% confidence interval, it is difficult to draw far-reaching generalizations, as it may turn out that the capacity of the reservoir designed on the basis of a deterministic solution may still be insufficient.

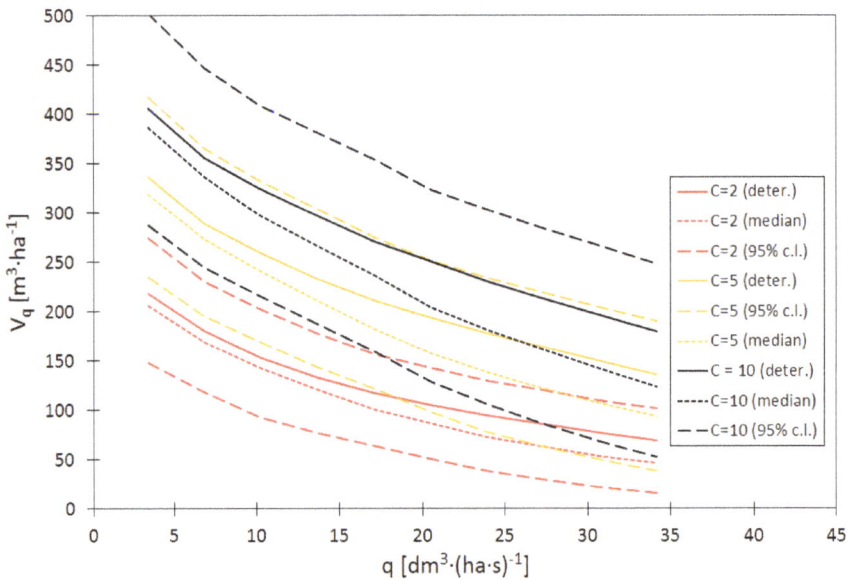

Figure 8. Influence of the unit outflow from the reservoir (q) and the rainfall frequency (C) on the unit accumulation capacity (V_q), taking into account the model uncertainty (where: c.l.—confidence level).

Therefore, further analyses are necessary in this respect, all the more so as retention reservoirs are important objects on the stormwater networks and the appropriate selection of their dimensions is of key importance for the high reliability of the drainage systems operation.

The analysis of the obtained curves $V_q = f(q, C)$ indicates an increase in the relative unit difference of reservoir capacities obtained in deterministic and probabilistic solutions, depending on the value of q. For example, for $C = 2$ an increase in the value from $q = 6.8$ dm³·(ha·s)$^{-1}$ to $q = 13.7$ dm³·(ha·s)$^{-1}$ leads to an increase in the relative difference in reservoir capacity from 6% to 19%. Referring the obtained result to the above analyses, it can be stated that with the increase in the amount of outflow from the reservoir, the degree of its oversizing (taking the probabilistic solution as the starting point) increases, which from the point of view of economy and operation of the stormwater system below, the reservoir is unfavorable.

Based on Figure 8, it can be concluded that an increase in the unit outflow from the reservoir (q) leads to a reduction in the required unit accumulation capacity (V_q). This means that the uncertainty of the model is important for the selection of discount devices. Comparing the values of unit outflows from the reservoir for the assumed unit accumulation capacity, it can be stated that for the deterministic solution the outflows, *q* are greater than taking into account the uncertainty. This is important when designing the reservoir, as omitting the uncertainty leads to an overstatement of the outflow with the drain, which may lead to unfavorable phenomena (system overload) in the stormwater system located below the outlet from the reservoir.

4. Summary and Conclusions

The paper presents a methodology of reservoir dimensioning taking into account the uncertainty of identification of parameters calibrated in the hydrodynamic model. The calculations made in the paper showed that the uncertainty of the model has a significant impact on the design of the retention reservoir. The simulations showed that due to the increase of the unit outflow from the reservoir to q = 20 dm$^3 \cdot$(ha$\cdot$s)$^{-1}$, the design rainfall duration taken into account when dimensioning the reservoir (probabilistic solution) is longer than in the deterministic solution. Simultaneously, with the increase of q, the difference in the values of the design rainfall duration determined in the probabilistic and deterministic solutions increases. On the other hand, after exceeding a certain limit, a further increase in q leads to a decrease in the difference between precipitation duration obtained in the probabilistic and deterministic solutions.

On the basis of the analyses carried out, it was found that in the case under consideration, the unit storage capacity of the reservoir (median) obtained taking into account the uncertainty is smaller than that determined in the probabilistic solution. At the same time, it was shown that with the increase in unit outflow (q), the absolute difference in reservoir capacity (median) obtained in deterministic and probabilistic solutions increases. Therefore, the reservoir capacity (median value) determined on the basis of the hydrodynamic model simulation, where the uncertainty of the model was omitted, is underestimated compared to the deterministic solution. From the point of view of reliability of reservoir operation, this is a positive result, as the reservoir will not be overfilled. However, due to the range of variability of the established 95% confidence interval and the fact that these analyses were performed for a single catchment, it is difficult to draw far-reaching guidelines and generalizations. Therefore, further analyses for urban catchments with diversified physico-geographical characteristics are advisable.

Author Contributions: Conceptualization, B.S.; Methodology, B.S. and A.K.; Software, A.K. and B.S.; Writing—Original Draft, B.S., A.K., L.D.

Funding: The work was founded by the Polish Ministry of Science and Higher Education, the RID project, according to the agreement: 025/RID/2018/19 of 28/12/2018 with total budget of 12,000,000 PLN.

Conflicts of Interest: The authors declare no conflict of interest.

Abbreviations

A	subcatchment area in the runoff model, ha;
C	rainfall frequency;
d_{imp}	retention depth of impervious areas in the runoff model, mm;
d_{perv}	retention depth of pervious areas in the runoff model, mm;
F_{imp}	impermeable of the catchment, ha;
$H(t)$	reservoir depth, m;
H_{max}	maximum reservoir depth, m;
$I(t)$	inflow to reservoir, m$^3 \cdot$s^{-1};
I	subcatchments slope in the runoff model;
Imp	percentage impervious areas in the runoff model;
$L(Q/\Theta)$	likelihood function;
n_{imp}	roughness coefficient for impervious areas in the runoff model, m$^{-1/3} \cdot$s;

n_n	roughness coefficient for sewer channels in the runoff model, $m^{-1/3} \cdot s$;
n_{perv}	roughness coefficient for pervious areas in the runoff model, $m^{-1/3} \cdot s$;
$P(\Theta)$	denotes *a priori* parameter distribution;
$P(Q/\Theta)$	the *a posteriori* distribution;
P_{max}	maximum rainfall depth, mm;
P_{tot}	rainfall depth in an episode, mm;
p	probability of rainfall exceeding;
Q_{out}	outflow from stormwater reservoir, $m^3 \cdot s^{-1}$;
$Q_{out,max}$	maximum outflow from stormwater reservoir, $m^3 \cdot s^{-1}$
q	unit outflow from reservoir; $dm^3 \cdot (ha \cdot s)^{-1}$
S	surface area of the reservoir in the projection, m^2;
t_r	rainfall duration, min;
t_{rd}^{det}	design rainfall event—deterministic solution, min;
t_{rd}^{prob}	design rainfall event—probabilistic solution, min;
V_q	unit capacity index, $m^3 \cdot ha^{-1}$;
V_q^{det}	unit capacity index—deterministic solution, $m^3 \cdot ha^{-1}$;
V_q^{prob}	unit capacity index—probabilistic solution, $m^3 \cdot ha^{-1}$;
V_{tot}	volume of runoff, m^3;
W	flow path width in the runoff model, m;
α	coefficient for flow path width in the runoff model;
β	correction coefficient for the percentage of impervious areas;
γ	correction coefficient for subcatchments slope in the runoff model;
κ	the factor used to control the variance of the *a posteriori* distribution.

References

1. Paola, D.P.; Martino, D.F. Stormwater tank performance: Design and management criteria for capture tanks using a continuous simulation and a semi—probabilistic analytical approach. *Water* **2013**, *5*, 1699–1711. [CrossRef]

2. Andrés-Doménech, I.; Montanari, A.; Marco, J.B. Stochastic rainfall analysis for storm tank performance evaluation. *Hydrol. Earth Syst. Sci.* **2010**, *14*, 1221–1232. [CrossRef]

3. Water Framework Directive, European Union (WFD EU). *Establishing a Framework for Community Action in the Field of Water Policy*; Direcive 2000/60/EC of the European Parliament and the Council of 23 October 2000; EU: Brussels, Belgium, 2000.

4. Kotowski, A.; Wartalska, K.; Nowakowska, M. Uogólniona metoda analityczna wymiarowania przelewowych zbiorników retencyjnych ścieków deszczowych [Generalized analytical methods of overfall storm water retention reservoir sizing]. *Ochr. Srod.* **2016**, *38*, 45–52. (In Polish)

5. Dziopak, J.; Słyś, D. *Modelowanie Zbiorników Klasycznych i Grawitacyjno—Pompowych w Kanalizacji [Classical and Pump—Gravitational Storage Reservoirs Modeling in Sewage Systems]*; Oficyna Wydawnicza Politechniki Rzeszowskiej: Rzeszów, Poland, 2007. (In Polish)

6. Hong, Y.M. Graphical estimation of detention pond volume for rainfall of short duration. *J. Hydrol. Environ. Res.* **2008**, *2*, 109–117. [CrossRef]

7. DWA A-117. *Bemessung von Regenrückhalteräumen*; Deutsche Vereinigug für Wässerwirschaft, Abwasser und Abfalle e.V.: Hennef, Germany, 2006. (In German)

8. Guo, J.C.Y. Hydrology-based approach to storm water detention design using new routing schemes. *J. Hydrol. Eng.* **2004**, *9*, 1–13. [CrossRef]

9. Mrowiec, M. *Efektywne Wymiarowanie i Dynamiczna Regulacja Kanalizacyjnych Zbiorników Retencyjnych [The Effective and Dimensioning and Dynamic Regulation Sewage Reservoirs]*; Wydawnictwo Politechniki Częstochowskiej: Czętochowa, Poland, 2008; ISBN 978-83-7193-424-7. (In Polish)

10. Wang, J.; Guo, J.C.Y. An analytical stochastic approach for evaluating the performance of combined sewer overflow tanks. *Water Resour. Res.* **2018**, *54*, 3357–3375. [CrossRef]

11. DWA A-118E. *Hydraulic Dimensioning and Verification of Drain and Sewer Systems*; DWA German Association for Water, Wastewater and Waste: Hennef, Germany, 2006. (In German)

12. Licznar, P. Wymiarowanie zbiorników retencyjnych ścieków deszczowych na podstawie syntetycznych szeregów czasowych opadów deszczów [Stormwater Reservoir Dimensioning Based on Synthetic Rainfall Time Series]. *Ochr. Srod.* **2013**, *35*, 27–32. (In Polish)

13. Xie, J.; Wu, H.; Li, H.; Chen, G.A. Study on storm-water management of grassed swales and permeable pavement based on SWMM. *Water* **2017**, *9*, 840. [CrossRef]

14. Strzec, M.; Dziopak, J.; Słyś, D.; Pochwat, K.; Kordana, S. Dimensioning of required volumes of interconnected detention tanks taking into account the direcion and speed of rain movement. *Water* **2018**, *10*, 1826. [CrossRef]

15. Fu, G.; Butler, D. Copula-based frequency analysis of overflow and flooding in urban drainage systems. *J. Hydrol.* **2014**, *510*, 49–58. [CrossRef]

16. Osorio, F.; Muhaisen, O.; García, P.A. Copula based simulation for the estimation of optimal volume for detention basin. *J. Hydrol. Eng.* **2009**, *14*, 1378–1382. [CrossRef]

17. Vandenberghe, S.; Verhoest, N.; Buyse, E.; Baets, B. A stochastic design rainfall generator based on copulas and mass curves. *Hydrol. Earth Syst. Sci.* **2010**, *14*, 2429–2442. [CrossRef]

18. Vernieuwe, H.; Vandenberghe, S.; Baets, B.; Verhoest, N. A continuous rainfall model based on vine copulas. *Hydrol. Earth Syst. Sci.* **2015**, *19*, 2685–2699. [CrossRef]

19. Szeląg, B.; Kiczko, A.; Studziński, J.; Dąbek, L. Hydrodynamic and probabilistic modeling of storm overflow discharges. *J. Hydroinform.* **2018**, *20*, 1100–1110. [CrossRef]

20. Burszta-Adamiak, E.; Licznar, P.; Zalewski, J. Criteria for identifying maximum rainfall rainfalls determined by peak-over-threshold (POT) method under the Polish Atlas of Rainfalls Intensities (PANDa) project. *Meteorol. Hydrol. Water Manag. Res. Oper. Appl.* **2018**, *7*, 3–13. [CrossRef]

21. Tfwala, W.M.; Rensburg, L.D.; Schall, R.; Mosia, S.M.; Dlamini, P. Precipitation intensity-duration-fequency curves and their uncertainties for Ghaap plateau. *Clim. Risk Manag.* **2017**, *16*, 1–9. [CrossRef]

22. Goncalves, M.L.R.; Zischg, J.; Rau, S.; Sitzmann, M.; Rauch, W.; Kleidorfer, M. Modelling the efect itroducing low impact development in a tropical city: A case study from Joinville, Brazil. *Sustainability* **2018**, *10*, 728. [CrossRef]

23. Knighton, J.; White, E.; Lennon, E.; Rajan, R. Development of probability distributions for urban hydrologic model parameters and a Monte Carlo analysis of model sensitivity. *Hydrol. Process.* **2014**, *28*, 5131–5139. [CrossRef]

24. Knighton, J.; Lennon, E.; Bastidas, L.; White, E. Stormwater detention system parameter sensitivity and uncertainty analysis using SWMM. *J. Hydrol. Eng.* **2016**, *21*, 05016014. [CrossRef]

25. Sun, N.; Hong, B.; Hall, M. Assessment of the SWMM model uncertainties with generalized likelihood uncertainty estimation (GLUE) framework for a high-resolution urban sewershed. *Hydrol. Process.* **2014**, *28*, 3018–3034. [CrossRef]

26. Zheng, Y.; Keller, A.A. Understaning parameter sensitivity and its management implications in watershed-scale water quality modeling. *Water Resour. Res.* **2006**, *42*, 1–14. [CrossRef]

27. Kiczko, A.; Szeląg, B.; Kozioł, A.; Krukowski, M.; Kubrak, E.; Kubrak, J.; Romanowicz, J. Optimal capacity of a stormwater reservoir for flood peak reduction. *J. Hydrol. Eng.* **2018**, *23*, 1–9. [CrossRef]

28. Beven, K.; Binley, A. The future distributed models: Model calibration and uncertainty prediction. *Hydrol. Process.* **1992**, *6*, 279–298. [CrossRef]

29. Dąbkowski, S.L.; Górska, K.; Górski, J.; Szeląg, B. Wstępne wyniki badań ścieków deszczowych w jednym z kanałów w Kielcach [Introductory results of examining precipitation sewage in one Kielce channels]. *Gaz Woda i Technika Sanitarna* **2010**, *34*, 20–24. (In Polish)

30. EN-752. *Drain and Sewer Systems Outside Building—Part 4: Hydraulics Design and Environmental Cosideration*; British Standards Institution (BSI): London, UK, 2017.

31. Beven, K.; Binley, A. GLUE: 20 year on. *Hydrol. Process.* **2014**, *28*, 5897–5911. [CrossRef]

32. Romanowicz, J.; Beven, K. Comments on generalized likelihood uncertainty estimation. *Reliab. Eng. Syst. Safe* **2006**, *91*, 1315–1321. [CrossRef]

33. Bogdanowicz, E.; Stachy, J. *Maksymalne Opady w Polsce—Charakterystyki Projektowe [Maximum Rainfall in Poland—Design Characteristics]*; Oficyna Wydawnicza Instytu Meteorologii i Gospodarki Wodnej: Warszawa, Poland, 1998. (In Polish)

34. Licznar, P.; Łomotowski, J. Analiza chwilowych natężeń deszczów miarodajnych we Wrocławiu [Analysis of the time frequncy distribution on maximal intensities of storms in the city of Wrocław]. *Ochr. Srod.* **2005**, *27*, 25–28. (In Polish)

35. Licznar, P.; Siekanowicz-Grochowina, K.; Oktawiec, M.; Kotowski, A.; Burszta-Adamiak, E. Empiryczna weryfikacja formuły Błaszczyka do obliczania wartości natężenia deszczu miarodajnego [Empirical verfication of Błaszczyk's formula for design rainfall intensity calculation]. *Ochr. Srod.* **2018**, *40*, 17–22. (In Polish)

36. Rossmann, L.A. *Storm Water Management Model User's Manual*; Version 5.0; National Risk Management Research Laboratory, Office of Research and Development, U.S. Environmental Protection Agency: Cincinatti, OH, USA, 2010.

37. Szeląg, B.; Kiczko, A.; Dąbek, L. Analiza wrażliwości i niepewności modelu hydrodynamicznego (SWMM) do prognozowania odpływu wód opadowych ze zlewni zurbanizowanej—Studium przypadku [Sensitivity and uncertainty analysis of hydrodynamic model (SWMM) for storm water runoff forecating in an urban basin—A case study]. *Ochr. Srod.* **2016**, *38*, 15–22. (In Polish)

water

MDPI

Article

Conceptualization and Schematization of Mesoscale Sustainable Drainage Systems: A Full-Scale Study

Salar Haghighatafshar [1,*], Jes la Cour Jansen [1], Henrik Aspegren [1,2] and Karin Jönsson [1]

[1] Water and Environmental Engineering, Department of Chemical Engineering, Lund University, P.O. Box 124, SE-221 00 Lund, Sweden; jes.la_cour_jansen@chemeng.lth.se (J.l.C.J.); henrik.aspegren@chemeng.lth.se (H.A.); karin.jonsson@chemeng.lth.se (K.J.)

[2] VA SYD, P.O. Box 191, SE-201 21 Malmö, Sweden

* Correspondence: salar.haghighatafshar@chemeng.lth.se; Tel.: +46-46-222-8998

Received: 18 June 2018; Accepted: 3 August 2018; Published: 6 August 2018

Abstract: Sustainable Drainage Systems (SuDS) can be considered the joint product of water engineering and urban planning and design since these systems must comply with hydraulic, hydrologic, and social-ecological functions. To enhance this joint collaboration, a conceptual model of mesoscale SuDS is introduced based on the observed rainfall-runoff responses from two catchments with SuDS and a pipe-bound catchment. The model shows that in contrast to pipe systems, SuDS disaggregates the catchment into a group of discrete mini catchments that have no instant connection to the outlet. These mini catchments start to connect to each other (and perhaps to the outlet) as the rainfall depth increases. It is shown that the sequence of stormwater control measures (SCMs as individual components of SuDS) affects the system's overall performance depending on the volumetric magnitude of the rainfall. The concept is useful in the design and implementation of mesoscale SuDS retrofits, which include several SCMs with different retention and detention capacities within a system.

Keywords: rainfall-runoff; storm water control measure; SuDS; urban drainage; urban landscape; urban planning

---

1. Introduction

Sustainable Drainage Systems (SuDS) within the context of green infrastructure are becoming more accepted and popular in urban landscapes. Numerous studies indicate that these systems, besides delivering multiple ecosystem services and promoting public health [1–3], have large retention capacities for the management of rainfall events up to the design magnitude [4]. It has also been pointed out that SuDS have positive effects on flood mitigation [5,6]. Therefore, SuDS are occupying more space in urban landscapes either as an alternative solution or as a complement to the existing combined or separate wastewater collecting infrastructure. The large retention capacity associated with SuDS is achieved by introducing extended pervious areas, which allows increased infiltration along with larger retention and retention volumes as well as slow transport of runoff towards the outlet point [7]. In other words, the management of storm water with SuDS utilizes urban spaces and, therefore, affects their functionality. This means that the urban surfaces occupied by SuDS have to comply with social-ecological qualities besides fulfilling their hydraulic role in an urban drainage perspective. Therefore, the planning and designing of SuDS has to be brought about collaboratively by water engineers and urban planners [8–10].

SuDS in urban areas can be implemented at three different levels, i.e., microscale, mesoscale, and macroscale, which was proposed by Haghighatafshar et al. [11]. A graphical illustration of these three levels is presented in Figure 1. A microscale implementation of SuDS (Figure 1a) consists of scattered individual stormwater control measures (SCMs) from which the excess discharge is directly connected to the urban drainage pipe-network (either separated or combined sewer networks). The procedure for designing an individual SCM is already established and widely practiced based on applying existing methods such as the Rational Method or the Time-Area Method. Details of the design process for individual SCMs can be found in e.g., Water Sensitive Urban Design [12].

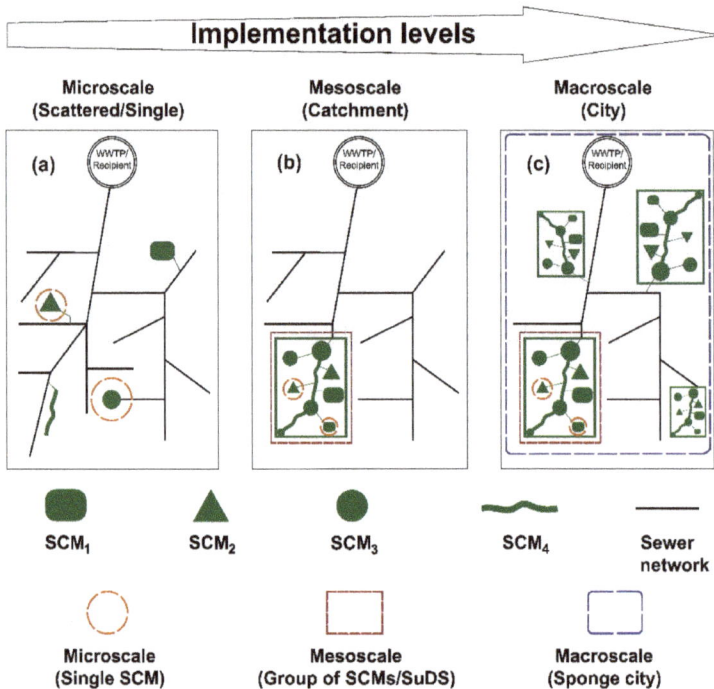

Figure 1. Different levels of implementation for SuDS/SCMs, (**a**) Microscale, (**b**) Mesoscale, and (**c**) Macroscale.

Mesoscale SuDS (Figure 1b) is implemented at the catchment level. This means that a group of interconnected SCMs are integrated in an urban catchment. In this type of implementation, SCMs are connected to each other so that the collected stormwater could flow from an upstream SCM to a downstream SCM. Mesoscale, in this context, has been referred to as "SuDS management train" by [13]. The extensive implementation of SuDS over the entire city catchments could be considered a macroscale approach (Figure 1c) through which the city could be transformed to a sponge city [14]. In contrast to microscale, studies regarding the hydraulic performance of SuDS at meso-scales and macroscales, as the train of several individual SCMs, are comparatively rare in the literature, e.g., [13,15,16].

In order to facilitate the implementation of SuDS, it is necessary to provide tools and models to enhance the communication between the urban water engineers and urban planners [8]. This can be done by characterizing SCMs as well as understanding their cumulated affect in a larger system, which is reported to be challenging and empirically less attended [17].

One of the early standard frameworks for implementation of SCMs was introduced by Stahre [7] in Sweden in the city of Malmö. Peter Stahre developed administrative procedures where it was outlined how different SCMs could be implemented on private and public land, respectively. As a result, several SCMs where introduced in the late 1990s in Malmö as part of the drainage system. A list of the implemented SCMs/SuDS in Malmö was presented by Haghighatafshar et al. [18]. The most prominent of these implementations is the neighborhood Augustenborg in Malmö, which in contrast to others, is located in the densely constructed and populated part of the city. The implemented SuDS were, to some extent, demonstration facilities showing the potential and the benefits of a new game-changing type of planning process where aesthetically designed open drainage systems were part of the urban landscape in accordance with the motto "make space for the water". While some potentially suitable measures were tentatively suggested at each level from upstream to downstream, the hydraulic and hydrologic performance of the suggested SCMs and of the entire system were not addressed. With the more intense rainfall events that have been experienced in many parts of the world as well as an elevated densification of our cities, the interest in SCMs especially in already built areas has grown [6,11].

The aim of this study is to introduce a new conceptual function-oriented description of the SuDS at a mesoscale level. The suggested model is based on observed rainfall-runoff data from the perspective of connectedness of surfaces and, to what extent, they contribute to the observed runoff. Consequently, the concept is applied to schematize the existing SuDS in Augustenborg as a demonstration. This approach aims to bridge an engineering design to urban planning and design by providing a simple hydraulic scheme for mesoscale SuDS.

2. Methodology

This study is based on rainfall-runoff measurements in an urban catchment of about 20 ha in which the runoff from most surfaces is managed through combinations of SCMs. All the implemented SCMs in the study area are surface-based (open) stormwater solutions. The study area known as Augustenborg was originally drained through the underground pipe-system of the city. For two years, flow measurements were carried out at the most downstream of catchments where the excess runoff is diverted into the major wastewater collecting system of the city. The following subsections present the adopted parametrical assessment method, a brief description of the study area, and the employed measurement instrument.

2.1. The Study Area—Augustenborg

Augustenborg is located in the inner city of Malmö, Sweden and is one of the most renowned SuDS retrofits. The area is often regarded as a unique example of an integrated collaboration of urban planners and urban water engineers. In this scenario, an area about 20 ha, which was originally drained with a combined sewer network, is managed via interconnected combinations of SCMs (i.e., mesoscale). The area was retrofitted in the late 1990s and has been in operation for about 20 years now. Augustenborg has been associated with many tentatively positive effects over the years such as, among other benefits, mitigation of basement flooding [19]. However, the effect and the in-depth understanding of the function of the SCMs have never been described in detail and the ideas have not been reproduced elsewhere despite the very positive verdict. One prerequisite for the reproduction is understanding the functionality so that the results, rather than the layout, can be transferred to other places. There is, therefore, a need to develop concepts that discuss SuDS and their functionality in a city-wide perspective to help urban planners and water engineers systematically design and reshape the urban landscapes through a shared perspective. An enhanced communication between engineers and planners can help alleviate some of the institutional shortcomings [20] for the widespread adoption of SCMs.

The Augustenborg area, which is shown in Figure 2, handles the storm water runoff through three different systems with each serving its unique catchment; i.e., the pipe-system (3.5 ha), the Northern SuDS (6.3 ha), and the Southern SuDS (9.5 ha). Figure 2 also shows the location of the flow and rainfall monitoring points as well as the systems' connection points to the municipal wastewater collection network of Malmö.

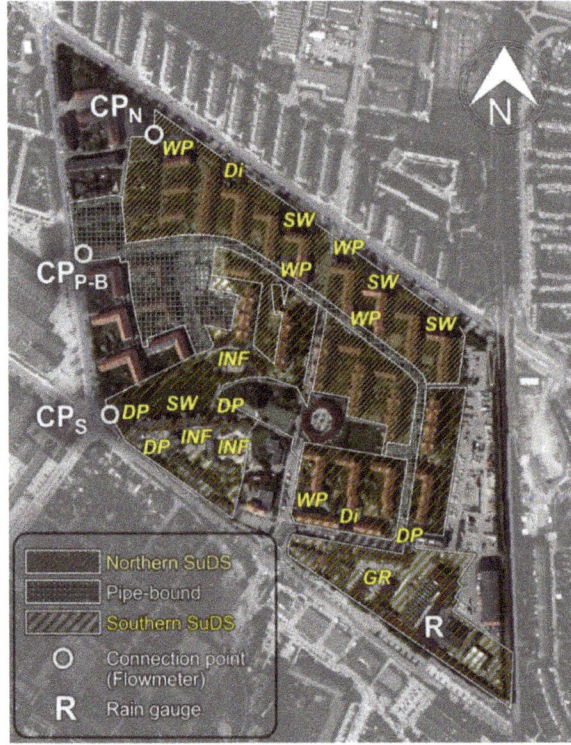

Figure 2. The locations of rainfall and runoff monitoring points as well as the catchment delineation in Augustenborg. The unmarked areas within the borders of the Augustenborg area are directly drained into the existing municipal pipe-bound combined sewer network. Connection points marked as CP_N, CP_{P-B}, and CP_S are the discharge points for the Northern retrofit, the pipe-bound catchment, and the Southern retrofit, respectively. Note that the flow is in the north-west direction, i.e., towards the connection points [21]. For SCM types, see Figure A1. Background picture: GSD-Orthophoto, courtesy of The Swedish Mapping, Cadastral and Land Registration Authority, ©Lantmäteriet (2015). [This figure—slightly modified—is adopted from Haghighatafshar et al. [11] with permission].

Catchments in all three stormwater subsystems consist of various types of surfaces such as tile roofs, green roofs, asphalt surfaces, concrete surfaces, grass, and sand covered areas. A Geographical Information System (GIS) analysis of different land uses in each of the subsystems in Augustenborg, using orthophotos of the area, shows that about 50% of the catchment in both Northern and Southern SuDS is occupied by surfaces assumed to be impervious from an engineering point of view (i.e., tile roofs, asphalt, and concrete) while the corresponding value in the pipe-bound catchment is above 70%. Green roofs make up a considerable part in the Southern SuDS (about 11%) while it is almost negligible in the pipe-bound catchment as well as the Northern SuDS. A schematic representation of different land uses in the area are presented in Table 1. The numbers are based on a GIS-analysis of the land use and the digital elevation model (DEM) of Augustenborg by Nordlöf [21].

Table 1. Distribution of different types of surfaces in the catchments in Augustenborg extracted from Reference [21].

Surface Type	Pipe-Bound		Northern SuDS		Southern SuDS	
	ha	%	ha	%	ha	%
Tile roof	0.5	15	1.7	27	1.7	17
Asphalt/Concrete	2.0	56	1.5	24	3.0	32
Grass area	1.0	28	2.9	46	3.0	31
Green roof	0.0	0	0.0	0	1.0	11
Sand	0.0	1	0.1	1	0.8	8
Gravel	0.0	0	0.1	2	0.1	1
Total	3.5	100	6.3	100	9.5	100

Different types of the implemented SCMs in the Northern and Southern SuDS are shown in Figure 2. The Northern SuDS consists of a major flow-path of swales and a stormwater ditch to which some stormwater ponds are also connected. Outflow from the Northern systems occurs in the form of overflow from the final pond (Figure 2 (CP$_N$)). In contrast, the Southern SuDS includes several relatively large retention ponds (with larger areas/freeboards) with a considerable area of green roofs at the most upstream parts of the catchment (Figure 2 (GR)). Outflow from the Southern SuDS is the result of overflow from the final pond in the system (Figure 2 (CP$_S$)). Some photos of the SCMs in Augustenborg are presented in Appendix A.

2.2. On-Site Measurements

Discharges from the sub-catchments were monitored and logged at connection points (marked as CP in Figure 2). The flow was measured using Mainstream Portable AV-Flowmeters with velocity and level sensors. Flow-monitoring was carried out for a period of over two years, which is shown in Figure 3. A total of 10 rainfall events (denoted A–J) with reliable corresponding flow measurements were selected. The selected rainfalls were all volumetrically considered, which means that they led to a discharge from at least one of the SuDS in Augustenborg. As seen in Figure 3, all selected rainfalls belong to the period of May–August during which most intense rainfalls were observed. Details of the selected rainfall events are shown in Table 2. The rainfall was monitored and logged by a Casella CEL tipping bucket rain gauge with 0.2 mm resolution, which was installed at the south-east part of the area.

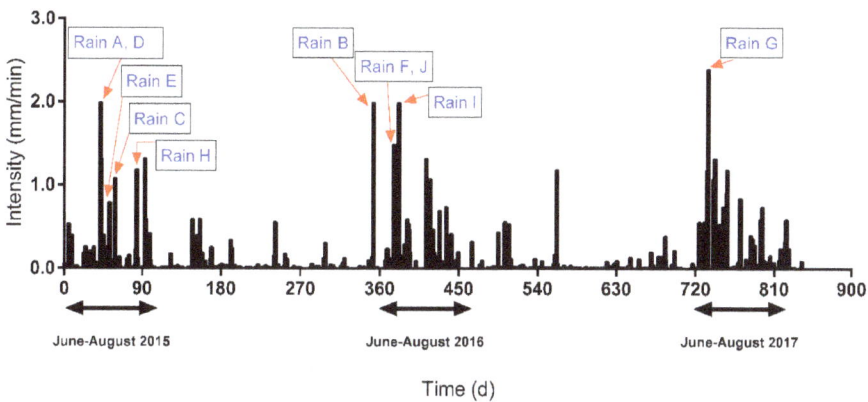

Figure 3. Hyetograph over the period of the study.

Table 2. Calculated REIAs and their corresponding contribution coefficients for 10 rain events.

Rainfall Event ID	Rainfall Depth (mm)	Rainfall Duration (h)	V_{out} (m³)			Contribution Coefficient—Equation (2) (-)		
			Pipe-System (3.5 ha)	Northern SuDS (6.3 ha)	Southern SuDS (9.5 ha)	Pipe-System (3.5 ha)	Northern SuDS (6.3 ha)	Southern SuDS (9.5 ha)
A	7.8	0.45	107.47	24.28	0.00	0.55	0.10	0
B	10.6	3.37	129.56	n/a	5.40	0.49	n/a	0.01
C	13.4	2.18	189.55	107.78	14.00	0.57	0.25	0.02
D	13.8	9.25	197.15	86.22	12.76	0.57	0.19	0.02
E	15.6	4.78	174.31	96.43	11.14	0.45	0.19	0.02
F	17.4	3.90	165.17	142.27	132.66	0.38	0.25	0.16
G	17.8	15.5	154.73	117.23	101.4	0.35	0.20	0.12
H	19.0	22.7	268.9	164.49	93.75	0.57	0.27	0.11
I	22.6	4.15	273.23	258.04	215.89	0.48	0.35	0.20
J	28.4	9.14	352.19	288.53	293.02	0.50	0.32	0.22

2.3. Runoff-Equivalent Impervious Area

The parts of the impervious surfaces in a catchment that are hydraulically connected to the drainage network within the context of pipe-systems are known as directly connected to an impervious area (DCIA) [22]. DCIA is often regarded as an effective impervious area (EIA) in an interchangeable manner [22–24], which implies that the effectiveness of the surfaces from a runoff contribution point of view is reflected in DCIA. DCIA has widely been employed to understand the rainfall-runoff patterns in urban basins. Lee and Heaney [25] report that connectedness of the impervious area has the most noticeable effect on urban hydrology. It has also been shown that mild changes of imperviousness are reflected as amplified runoff responses. For instance, grass areas contribute to runoff as soon as rain intensity exceeds the infiltration rate [22]. It is also important to consider that the routed runoff from ineffective impervious areas onto the pervious surfaces would lead to rapid consumption of percolation capacity, which makes the previous surface react as impervious [26]. The generated runoff under such scenarios is then not only contributed by DCIA, but also other types of impervious and pervious surfaces start to contribute.

Using the same indicators for functionality of various types of stormwater handling systems makes it easy to compare and understand the role of these systems in urban runoff management. While DCIA can be quantified through GIS maps of high spatial resolution as well as intensive in-situ assessment of the catchment connected to the pipe network [27], it is not convenient to apply the same method to SuDS since the boundaries between the "catchment" and the "system" cannot be clearly drawn in case of SuDS. Therefore, a lumped parameter representing the *runoff-equivalent impervious area* (REIA) is introduced in this paper to explain the activeness of the surfaces. This parameter, REIA, is the equivalent surface area with 100% contribution to runoff, which is calculated based on the observed accumulated outflow from systems. It should be noted that REIA and DCIA could be identical parameters in case of pipe-bound conventional drainage systems. The difference between these two parameters lies in their conceptual definitions through which REIA could be used for evaluating the efficiency of SuDS as alternative solutions for urban runoff management and is estimated using Equation (1).

$$REIA = \frac{V_{out}}{R} \tag{1}$$

in which *REIA* is expressed in m², V_{out} is the total volume of the observed runoff outflow at the most downstream point (m³), and R is the rainfall depth (m). Total runoff volume was measured until the discharge was either zero or reached a minimum before the subsequent rainfall. The ratio between the observed *REIA* and the GIS-based quantified total impervious area (TIA) is then considered as the contribution coefficient of the system (Equation (2)).

$$Contribution\ coefficient = \frac{REIA}{TIA} \tag{2}$$

3. Results and Discussion

3.1. Development of the Conceptual Model

All 10 rainfall-runoff datasets (hyetographs and hydrographs) included in this study are provided as supplementary material. Figure 4 shows two examples of the observed rainfall-runoff events. Rainfall **I** (Figure 4(I)) is the most intensive rainfall event with a recurrence interval of about two years according to Dahlström (2010) [28]. It has one peak with a large depth that leads to discharges from both the Northern and the Southern SuDS. Rainfall **D** (Figure 4(D)), however, consists of two peaks while the discharge occurs only from the Northern SuDS and only in connection to the second peak.

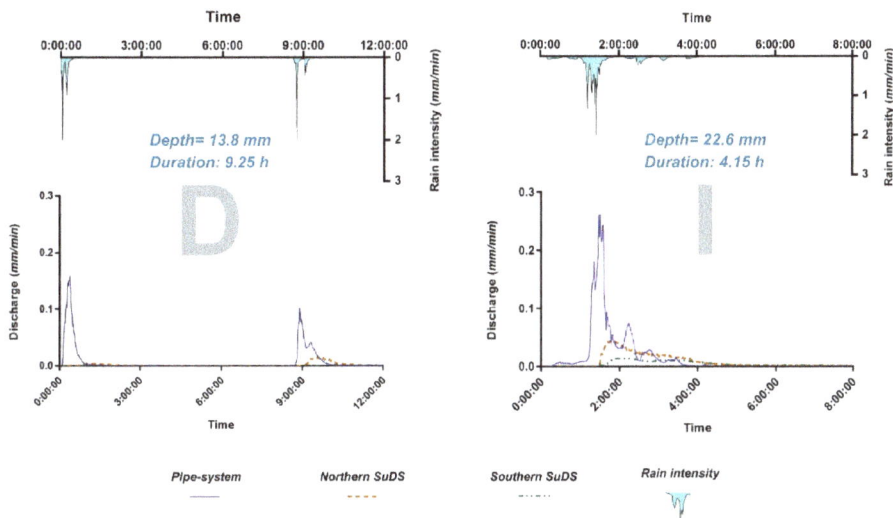

Figure 4. Two examples from the monitored hyetographs and hydrographs during this study. See rainfalls **D** and **I** in Table 2. The details of all rainfall-runoff observations are provided as supplementary material (available online).

Analysis of the hydrographs, which are the normalized outflows against the total catchment area, shows that the pipe system is very sensitive to rain peaks even in smaller magnitudes. In other words, there is always an observed peak in the hydrograph, which corresponds to a certain peak monitored in the rain pattern (see Figure 4(D) and Figure 4(I)). The correlation between the rain intensity and the outflow from the pipe-system indicates that pipe-systems are flow oriented and should be designed in accordance with flow capacity. In contrast to the pipe-system, the outflow from SuDS is observed to be a function of the rainfall depth rather than rainfall intensity. For instance, as seen in Figure 4(D), outflow from the Northern SuDS occurs in connection with the second peak observed at about 8 h 45 min after the start of the event while there is no outflow from the system at the first peak in the rain (at about 30 min after the start of the event). This means that the first part of the rainfall (60 min from the start, depth = 8 mm) is retained in the SuDS and to some extent fills the existing capacity while the second peak, although lower (depth = 6 mm), exceeds the threshold and initiates an outflow from the Northern SuDS.

The monitored hydrographs indicate an almost negligible delay in the flow initiation in the pipe network followed by relatively shorter lag times, i.e., 5–20 min depending on the rainfall pattern. The observed lag time for Northern SuDS and Southern SuDS was found to be about 20–100 min and 90–190 min, respectively.

The observed range of accumulated rainfall required for the initiation of runoff in the pipe system is found to be 0.8 mm to 2.2 mm, which aligns with findings of Albrecht [29] who reported a runoff initiation threshold of 0.8 mm to 2.3 mm for cool and hot weather, respectively. The almost immediate initiation of flow in the pipe system can be an indication that the major part of the flow is contributed by the DCIA, which lies close to the measurement point. In contrast to the pipe system, the small contribution coefficient in the catchments with SuDS (see Table 2) agrees well with the observed long periods of delay in the flow initiation, which is an indication of larger retention capacity of the SuDS catchments.

The larger retention capacity in the SuDS can be accredited to storage, evapotranspiration, and direct and indirect infiltration. The term indirect infiltration is assigned to the infiltration that takes place when the runoff from impervious surfaces is diverted to pervious surfaces for infiltration. In contrast, direct infiltration is when the rain falls on a pervious surface and is infiltrated directly.

Table 2 summarizes the total discharge volumes (V_{out}) and their corresponding contribution coefficients for the catchments in Augustenborg for 10 rainfall events were monitored for two years. Figure 5 presents the graphical illustration of the calculated REIA values for the subsystems in Augustenborg. As observed in Table 2, the contribution coefficient for the catchment with pipe-system is about 50% (i.e., 0.48 ± 0.08%) of the TIA. This is in agreement with the published literature in which the proportion of contributing surfaces (also regarded as EIA) is reported to vary from 14% to 60% depending on the physical characteristics of the catchment such as slopes, gutters, curbs, and more [30–33].

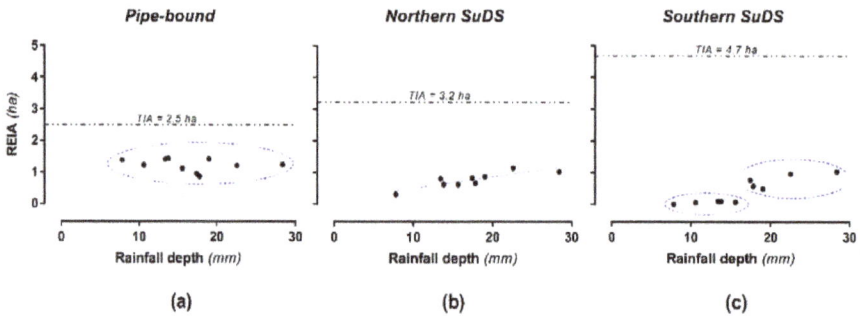

Figure 5. Calculated REIA in case of the observed rainfalls for all three catchments in Augustenborg compared to the TIA (based on field surveys and GIS maps as given in Nordlöf [21]): (**a**) Pipe-bound; (**b**) Northern SuDS; (**c**) Southern SuDS.

The contribution coefficient is considerably lower in the catchments with SuDS implementations (varying values) but note that the two SuDS (Northern and Southern) differ with respect to how they react under different rain depths. The REIA in the Northern SuDS tends to increase gradually when the rainfall depth increases (Figure 5b) while, in the Southern SuDS, the REIA is generated first when the rainfall depth exceeds a larger threshold of about 17 mm (Figure 5c). It was also observed that the outflow from the SuDS in Augustenborg is not only levelled out and flat (no intensive peaks as seen in Figure 4) but is also much smaller in accumulative volume when compared to the pipe system (compare the contribution coefficients given in Table 2). This implies that the retention capacity including surface storage, infiltration and evapotranspiration in catchments with SuDS in Augustenborg is higher than the pipe-bound catchment.

The observed gradual increase in the REIA for the Northern SuDS (REIA from 0.3 ha to 1.1 ha corresponds to rainfall depths of 8 mm to 23 mm, which is shown in Figure 5b) means that the contributing proportion of the catchment grows as the rainfall depth increases. The corresponding projection of this observation in the field could be considered if the system is constituted of a network of several small disaggregated (discrete) individual mini catchments with each having a certain retention volume. These discrete mini catchments are filled gradually as the rainfall depth grows. Wventually, when their threshold is exceeded, overflow to the corresponding immediate downstream mini catchment. Accordingly, if the rainfall depth is large enough, the number of connected mini catchments increases and the accumulated overflows might finally contribute to the final discharge from the system.

The same conceptual model is also valid for the REIA trend observed for the Southern SuDS. As seen in Figure 5c and Table 2, the calculated REIA is almost negligible for rainfall depths up to about 16 mm while a dramatic increase is observed in the case of 17 mm of rainfall. The same concept presented above (discrete mini catchments) explains the observed phenomenon. A possible explanation for this very abrupt alteration in behavior (sudden jump in REIA from 16 mm to 17 mm of rainfall) could be associated with the relatively large retention volume at the most downstream part of the Southern SuDS (see Figure A1(DP 4)). Another possible explanation could be that some other mini catchments further upstream join the rest of the system when a threshold is exceeded. Both hypotheses could generate a relatively large outflow volume considering the possible connectedness of the catchment at that stage after the initial 16 mm has filled up the capacity up to the system's threshold. However, application of the concept to the systems in Augustenborg could possibly reveal what hypothesis is a valid explanation for the observed phenomenon.

A schematic illustration of the conceptual model is presented in Figure 6. In this illustration, five retention cells (SCMs) with each having a connected mini catchment area = A are presented. The constant connected area, A, for each SCM is assumed to promote the comprehensibility of the conceptual model. Each of these SCMs has a certain retention capacity as different multiples of an assumed unit capacity (i.e., V [mm]). The retention capacity of each SCM is reflected in the size of the schematic circles in Figure 6. As evident in Figure 6, both illustrated models have identical total retention capacity (=25 × V), but the circumstances under which a discharge is initiated from the systems depend on the spatial distribution of the mini catchments with respect to their retention capacity. It is important to note that retention capacity, V, in this context is considered the sum of surface storage, retention, and losses in the form of infiltration and evapotranspiration.

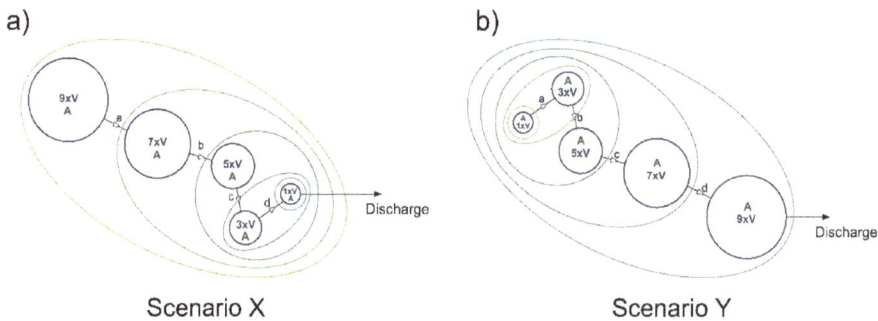

Figure 6. Conceptual illustration of two extreme setups for construction of SuDS with different components. Please note that all the shown mini catchments have the same area, i.e., A, while the size of the circles represents the retention capacity of SCMs: (**a**) Scenario X; (**b**) Scenario Y.

Figure 6a (Scenario X) is comparable to Northern SuDS in Augustenborg. The mini catchments with smaller retention capacities are placed close to the discharge point. A consequence of this configuration is that a discharge from the system will be observed as soon as the most downstream mini catchment (1 × V) is saturated in capacity, which is when the rain depth exceeds 1 × V. As the rainfall depth continues to increase, more mini catchments are connected to each other and contribute more to the final discharge. The growth direction of the contributing catchments in this case is downstream-to-upstream, i.e., links are activated from **d** towards **a** (Table 3).

Table 3. Response matrix of the conceptual model in the case of scenarios X and Y (see Figure 6).

Rain Depth (mm)	Scenario X			Scenario Y		
	Active Links	Discharge (mm)	Contributing Area	Active Links	Discharge (mm)	Contributing Area
<1 × V	-	-		-	-	-
2 × V	-	1 × V	A	a	-	-
3 × V	-	2 × V	A	a, b	-	-
4 × V	d	4 × V	2 × A	a, b, c	-	-
5 × V	d	6 × V	2 × A	a, b, c, d	-	-
6 × V	d, c	9 × V	3 × A	a, b, c, d	5 × V	5 × A
9 × V	d, c, b	20 × V	4 × A	a, b, c, d	20 × V	5 × A
10 × V	d, c, b, a	25 × V	5 × A	a, b, c, d	25 × V	5 × A

In Figure 6b (Scenario Y), comparable to the situation in Southern SuDS, the final discharge would not flow out unless a certain rain depth is obtained. In the specific example, the outflow from the model presented as Scenario Y is initiated when a rainfall larger than 5 × V mm is applied on the system while all rainfalls up to 5 × V mm would result in higher connectedness of the system without any downstream discharges. In this type of setup, the connectedness of the system propagates from the upstream towards the downstream, i.e., links are activated from **a** towards **d** (Table 3).

In the presented conceptual model, if a longer lag time is desired for the system, it is more beneficial that the SCMs with higher retention or retention capacity are placed downstream.

Additionally, relatively smaller volumes of discharge can also be expected for rainfalls up to a certain magnitude (see the data for rainfall 6 × V in Table 3). These advantages become especially important and effective when the final recipient for the SuDS is the municipal sewer system, which is the case in Augustenborg. In the municipal sewer system, the bought time in terms of longer lag times may be enough for the receiving pipe-bound stormwater network to maintain some pressure relief.

Figure 6 along with Table 3 illustrate the basic concept behind the functionality and behavior of SuDS in a full-scale urban catchment by demonstrating two straightforward examples under simplified circumstances in which connectedness grows along a single pathway. Basically, the SCMs can be visualized as a flow train of interconnected bowls with physical properties that, at least from theoretical point of view, should be quantifiable. Once the properties have been determined, the functionality of the flow train is set.

However, in contrast with the unique setups of the concept (Figure 6), each SCM in a real implementation of mesoscale SuDS is assigned to a specific mini catchment varying in area and characteristics. In addition to the local retention depth available (= storage depth in the freeboard, S_{fb}^i+ storage depth in the infiltration layer, S_{inf}^i) in the SCM, the area of the connected catchment is also important in the overall retention performance of the SCM. It is also assumed that the effect of evapotranspiration is negligible in case of short term individual rainfall events. Therefore, it is excluded from the model. Consequently, in order to be able to compare the retention capacity of each SCM, the effective retention of each SCM is calculated, according to Equation (3).

$$R_e^i = \frac{(S_{fb}^i + S_{inf}^i) \times A_{SCM}^i}{DCIA^i} \tag{3}$$

in which R_e^i is the effective retention capacity of the SCM i (mm), S_{fb}^i is the storage depth in the freeboard of the SCM (mm), S_{inf}^i is the storage depth in the infiltration layer, A_{SCM}^i is the area occupied by the SCM (m²), and $DCIA^i$ is the directly connected impervious area to the SCM (m²). Please notice that DCIA (mainly tile roofs and some asphalt in the Northern SuDS [21]) is used to simplify the model since it is anticipated that DCIA is the dominant parameter for runoff volume, which is reported by Shuster et al. [19].

3.2. Remarks on the Schematized Augustenborg

The developed conceptual model is used to characterize the processes in the Northern and Southern SuDS in Augustenborg. Figure 7 shows the conceptualized representations of Northern (top) and Southern (bottom) SuDS in Augustenborg based on the effective retention depth (Equation (3)) of SCMs.

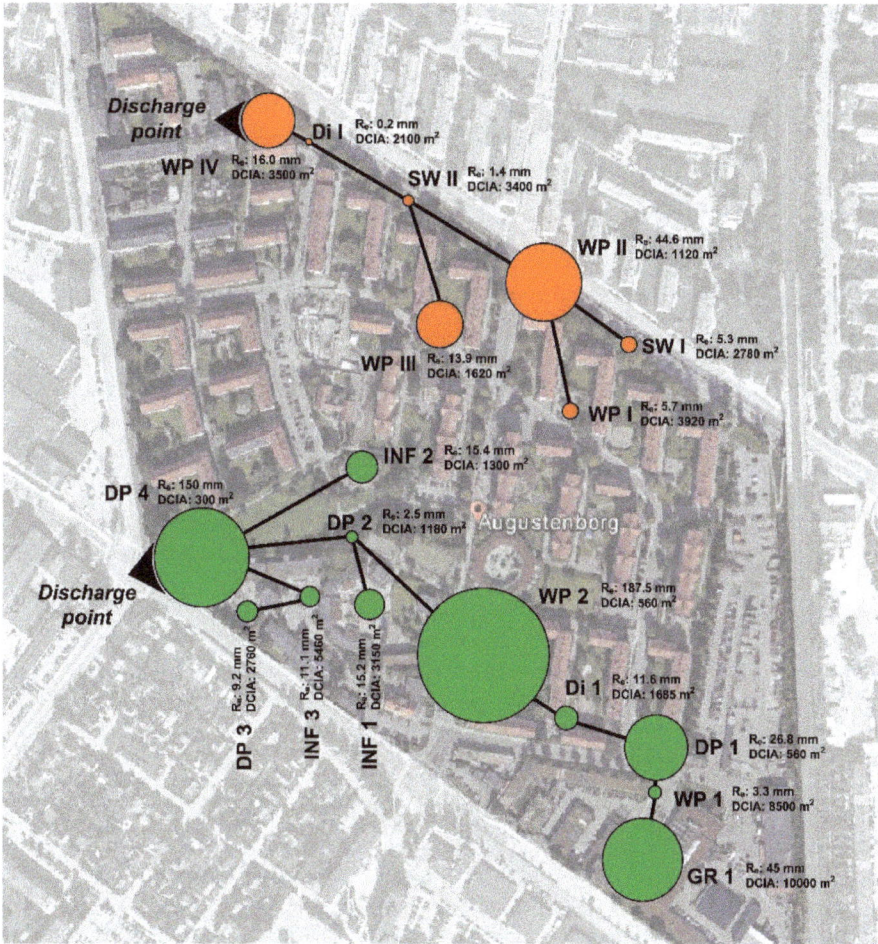

Figure 7. Conceptual illustration of Northern SuDS (top) and Southern SuDS (bottom) in Augustenborg. Each SCM is represented as a circle. The size of the circle corresponds to the effective retention capacity of the SCM, i.e., R_e^i. Please notice that the SCMs belonging to the Northern SuDS are indexed with Roman numerals while SCMs in the Southern SuDS are numbered with Arabic numerals. The size of the circle representing SCM is an indicator of its effective detention depth (sizes are not proportionally correct). The background picture is acquired from Google Earth.

In Table 4, all the different SCMs have been systematically assigned names and properties accordingly. The details of the characterized SCMs shown in Figure 7 are also presented in Table 4. Equation (3) and Equation (4) were subsequently employed to build a model to estimate the discharges from the SCMs along all the flow paths in the Northern and Southern SuDS. The model was built in an Excel spreadsheet.

Table 4. Characteristics of the SCMs in the Northern and the Southern SuDS in Augustenborg. The sequence of the SCMs with respect to flow path (upstream-downstream) is illustrated in Figure 7.

System	SCM ID	Storage, S_{fb}^i (mm)	Infiltration S_{inf}^i (mm)	SCM Area, A_{SCM} (m^2)	DCIA (m^2)	Effective Retention, R_e^i (mm)
Northern SuDS	SW I	5	15	740	2780	5.3
	WP I	250	0	90	3920	5.7
	WP II	250	0	200	1120	44.6
	WP III	250	0	90	1620	13.9
	SW II	5	15	240	3400	1.4
	Di I	5	0	80	2100	0.2
	WP IV	350	0	160	3500	16.0
Southern SuDS	GR 1	0	45	10,000	10,000	45
	WP 1	200	0	140	8500	3.3
	DP 1	105	45	100	560	26.8
	Di 1	200	0	98	1685	11.6
	WP 2	150	0	700	560	187.5
	INF. 1	35	25	800	3150	15.2
	DP 3	105	45	170	2760	9.2
	INF. 2	0	25	800	1300	15.4
	DP 2	0	15	200	1180	2.5
	INF. 3	500	25	115	5460	11.1
	DP 4	25	25	900	300	150.0

More information about the type of the implemented SCMs are shown in Figure 2 (also in Appendix A). The information regarding the characteristics of the SCMs (S_{inf}^i, S_{fb}^i, and A_{SCM}^i) as well as their corresponding mini catchment ($DCIA^i$) is collected from the hydrodynamic model of the area, which was developed by Haghighatafshar et al. [11], the on-site measurements, and the GIS maps. These parameters are relatively easy to estimate and can be measured on site.

As seen in Table 4, it is obvious that water ponds, (Dry Ponds and Wet Ponds) are the backbone of the systems regarding effective retention, which account for a total effective retention depth of 80 mm and 380 mm in the Northern and Southern SuDS, respectively. The second most important feature in terms of effective retention are the infiltration areas that contribute with approximately 42 mm of effective retention in the Southern SuDS. In case of infiltration areas, effective retention capacity in most cases has two components, which include a storage volume that can be determined from the geometrical properties of the basin and measured on site, and the infiltration capacity (i.e., the function of the underlying soil properties).

The ditches and the swales, however, have a limited storage volume since they add up to 7 mm and 12 mm of effective retention depth to the Northern and Southern SuDS, respectively. For the observed rain events, these components act as connection nodes between the different ponds and also provide a connected and diverse blue-green landscape. Despite relatively large retention at the most downstream pond in the Northern SuDS (WP IV), a discharge is initiated as soon as rainfall depth reaches around 7 mm. This rapid fill-up of retention capacity is due to the two upstream SCMs, i.e., Di I and SW II, from which the discharged volume overrides the remaining free capacity in WP IV and leads to a discharge. By comparing the conceptual approach with the onsite SCMs in the Northern SuDS, it can be claimed that using swales with large DCIA as the backbone of SuDS for the conveyance of runoff from upstream SCMs to downstream SCMs without introducing substantial retention structures on the flow path leads to decreased overall effectiveness of SuDS in runoff reduction. This aligns with findings of Qin et al. [34] who found that the retention capacity of swales is very limited and, therefore, is saturated quickly. Generally, ponds and infiltration basins with an overflow threshold (freeboard) have a pronounced role in the overall runoff retention when compared to other SCMs.

At this point, it is important to make clear that, although different SCMs can relatively easily be translated to effective retention volumes, the concept of SuDS needs to be studied through a combination of SCMs to be understood and adopted by the city planners as well as the individual house-owners. This is the key to the success of this technique. In order to better understand the response pattern of the different combinations of SCMs, it is beneficial to expand our knowledge on how SuDS as the flow train of different SCMs can be introduced in the best way.

At this stage, the conceptual model provides a better understanding of hydraulics that prevail in mesoscale SuDS implementations, which can promote the required dialogue between different actors at the planning phase. However, in the prospect of future studies, the concept can be further developed to estimate the discharge hydrographs from mesoscale SuDS. A mathematically simple representation of the hydraulics in mesoscale SuDS could result in computationally faster models. Such fast models can then be used for large-scale simulations as an alternative to the computationally costly and time-consuming 2-dimensional distributed hydrodynamic models. Fast and cheap models are needed to study the upscaling effects of SuDS on the city-level.

4. Conclusions

Extensive rainfall-runoff measurements at two urban catchments with SuDS along with one pipe-bound catchment were used to investigate the systems' responses at different rainfalls. Runoff measurements at the most downstream point of each catchment showed that, in contrast with the SuDS, the flow from the pipe-system was directly affected by the rainfall intensity. However, the total runoff volume was still a function of the total rain depth. In order to describe this transformation that takes place in a SuDS, a conceptual model was introduced from the viewpoint of catchments' runoff-equivalent impervious area. In the model, implementation of SuDS disintegrates the catchment area into a group of discrete and disaggregated mini catchments. These mini catchments establish connections with each other depending on the volume of the rain event. The dynamics of the conceptual model demonstrated that the order and placement of different stormwater control measures within the framework of SuDS with different retention capacities affects the overall performance of the system. The conceptualization of SuDS establishes a new platform for further evaluation and discussion of these systems at mesoscale. The model promotes the communication between urban planners and water engineers. This, in turn, can lead to the design of SuDS in which hydraulic performance alongside aesthetical and architectural quality is taken into consideration.

Supplementary Materials: The following are available online at http://www.mdpi.com/2073-4441/10/8/1041/s1, Figure S1: Rainfall A, Figure S2: Rainfall B, Figure S3: Rainfall C, Figure S4: Rainfall D, Figure S5: Rainfall E, Figure S6: Rainfall F, Figure S7: Rainfall G, Figure S8: Rainfall H, Figure S9: Rainfall I, and Figure S10: Rainfall J.

Author Contributions: Conceptualization, J.J., H.A. and S.H. Methodology, S.H., K.J., and J.J. Investigation, J.J. and H.A. Resources, K.J., J.J., and H.A. Data Curation, S.H. and H.A. Writing-Original Draft Preparation, S.H. Writing-Review & Editing, S.H., J.J., H.A., and K.J. Supervision, K.J., J.J., and H.A. Visualization, S.H. Project Administration, S.H. Funding Acquisition, K.J., J.J., and H.A.

Funding: This research was financially supported by VA SYD, Sweden Water Research, J. Gustaf Richert Foundation (grant number 2015-00181) at SWECO as well as the Swedish Water and Wastewater Association (*Svenskt Vatten*) via VA-teknik Södra (grant number 15-108).

Acknowledgments: The authors thank the personnel at VA SYD for their practical support in the installation of flowmeters and the collection of data. Tomas Wolf and John Hägg, VA SYD, are especially acknowledged for their invaluable contribution to the setup and the maintenance of the measurement system. Professor Magnus Larson, Division of Water Resources Engineering at the Department of Building and Environmental Technology, Lund University is also sincerely acknowledged for his thoughtful input to the final manuscript.

Conflicts of Interest: The authors declare no conflict of interest.

Appendix A

Figure A1. The photos of some of the implemented SCMs in Augustenborg. Different types of SCMs are shown in the figure as follows: (**DP 4**) dry pond, (**Di I & 1**) stormwater ditch, (**SW I & II**) swale, (**WP 2, II & IV**) wet pond, (**INF 3**) infiltration basin, (**GR**) green roof. The major runoff directions in the Northern and Southern systems are SW I → WP IV and GR → DP 4, respectively. Pictures for INF 3, WP 2, Di 1, and SW I & II are taken by Henrik Thorén (Rambøll). Background picture: GSD-Orthophoto, courtesy of The Swedish Mapping, Cadastral and Land Registration Authority, ©Lantmäteriet (2015).

References

1. Meerow, S.; Newell, J.P. Spatial planning for multifunctional green infrastructure: Growing resilience in Detroit. *Landsc. Urban Plan.* **2017**, *159*, 62–75. [CrossRef]
2. Tzoulas, K.; Korpela, K.; Venn, S.; Yli-Pelkonen, V.; Kaźmierczak, A.; Niemela, J.; James, P. Promoting ecosystem and human health in urban areas using Green Infrastructure: A literature review. *Landsc. Urban Plan.* **2007**, *81*, 167–178. [CrossRef]

3. Wolch, J.R.; Byrne, J.; Newell, J.P. Urban green space, public health, and environmental justice: The challenge of making cities 'just green enough. *Landsc. Urban Plan.* **2014**, *125*, 234–244. [CrossRef]

4. Palla, A.; Gnecco, I. Hydrologic modeling of Low Impact Development systems at the urban catchment scale. *J. Hydrol.* **2015**, *528*, 361–368. [CrossRef]

5. Ahiablame, L.; Shakya, R. Modeling flood reduction effects of low impact development at a watershed scale. *J. Environ. Manag.* **2016**, *171*, 81–91. [CrossRef] [PubMed]

6. Jato-Espino, D.; Charlesworth, S.; Bayon, J.; Warwick, F. Rainfall–Runoff Simulations to Assess the Potential of SuDS for Mitigating Flooding in Highly Urbanized Catchments. *Int. J. Environ. Res. Public Health* **2016**, *13*, 149. [CrossRef] [PubMed]

7. Stahre, P. *Sustainability in Urban Storm Drainage—Planning and Examples*, 1st ed.; Svenskt Vatten: Malmö, Sweden, 2006.

8. Sörensen, J.; Persson, A.; Sternudd, C.; Aspegren, H.; Nilsson, J.; Nordström, J.; Jönsson, K.; Mottaghi, M.; Becker, P.; Pilesjö, P.; et al. Re-Thinking Urban Flood Management—Time for a Regime Shift. *Water* **2016**, *8*, 332. [CrossRef]

9. Mottaghi, M.; Aspegren, H.; Jönsson, K. Integrated urban design and open storm drainage in our urban environments: Merging drainage techniques into our city's urban spaces. *Water Pract. Technol.* **2016**, *11*, 118–126. [CrossRef]

10. Fletcher, T.D.; Shuster, W.; Hunt, W.F.; Ashley, R.; Butler, D.; Arthur, S.; Trowsdale, S.; Barraud, S.; Semadeni-Davies, A.; Bertrand-Krajewski, J.-L.; et al. SUDS, LID, BMPs, WSUD and more—The evolution and application of terminology surrounding urban drainage. *Urban Water J.* **2015**, *12*, 525–542. [CrossRef]

11. Haghighatafshar, S.; Nordlöf, B.; Roldin, M.; Gustafsson, L.-G.; la Cour Jansen, J.; Jönsson, K. Efficiency of blue-green stormwater retrofits for flood mitigation—Conclusions drawn from a case study in Malmö, Sweden. *J. Environ. Manag.* **2018**, *207*, 60–69. [CrossRef] [PubMed]

12. Healthy Land and Water. *WSUD Water Sensitive Urban Design—Technical Design Guidelines for South East Queensland*; Healthy Land and Water: Brisbane, Australia, 2006.

13. Charlesworth, S.; Warwick, F.; Lashford, C. Decision-Making and Sustainable Drainage: Design and Scale. *Sustainability* **2016**, *8*, 782. [CrossRef]

14. Liu, H.; Jia, Y.; Niu, C. "Sponge city" concept helps solve China's urban water problems. *Environ. Earth Sci.* **2017**, *76*, 473. [CrossRef]

15. Jarden, K.M.; Jefferson, A.J.; Grieser, J.M. Assessing the effects of catchment-scale urban green infrastructure retrofits on hydrograph characteristics. *Hydrol. Process.* **2016**, *30*, 1536–1550. [CrossRef]

16. Loperfido, J.V.; Noe, G.B.; Jarnagin, S.T.; Hogan, D.M. Effects of distributed and centralized stormwater best management practices and land cover on urban stream hydrology at the catchment scale. *J. Hydrol.* **2014**, *519*, 2584–2595. [CrossRef]

17. Jefferson, A.J.; Bhaskar, A.S.; Hopkins, K.G.; Fanelli, R.; Avellaneda, P.M.; McMillan, S.K. Stormwater management network effectiveness and implications for urban watershed function: A critical review. *Hydrol. Process.* **2017**, *31*, 4056–4080. [CrossRef]

18. Haghighatafshar, S.; la Cour Jansen, J.; Aspegren, H.; Lidström, V.; Mattsson, A.; Jönsson, K. Storm-water management in Malmö and Copenhagen with regard to Climate Change Scenarios. *J. Water Manag. Res.* **2014**, *70*, 159–168.

19. Sörensen, J.; Mobini, S. Pluvial, urban flood mechanisms and characteristics—Assessment based on insurance claims. *J. Hydrol.* **2017**, *555*, 51–67. [CrossRef]

20. Burns, M.J.; Fletcher, T.D.; Walsh, C.J.; Ladson, A.R.; Hatt, B.E. Hydrologic shortcomings of conventional urban stormwater management and opportunities for reform. *Landsc. Urban Plan.* **2012**, *105*, 230–240. [CrossRef]

21. Nordlöf, B. *1D/2D Modeling of the Open Stormwater System of Augustenborg Using MIKE FLOOD by DHI*; Project Report Available at Water and Environmental Engineering at the Department of Chemical Engineering; Lund University: Lund, Sweden, 2016.

22. Shuster, W.D.; Bonta, J.; Thurston, H.; Warnemuende, E.; Smith, D.R. Impacts of impervious surface on watershed hydrology: A review. *Urban Water J.* **2005**, *2*, 263–275. [CrossRef]

23. Yao, L.; Wei, W.; Chen, L. How does imperviousness impact the urban rainfall-runoff process under various storm cases? *Ecol. Indic.* **2016**, *60*, 893–905. [CrossRef]

24. Bell, C.D.; McMillan, S.K.; Clinton, S.M.; Jefferson, A.J. Hydrologic response to stormwater control measures in urban watersheds. *J. Hydrol.* **2016**, *541*, 1488–1500. [CrossRef]

25. Lee, J.G.; Heaney, J.P. Estimation of Urban Imperviousness and its Impacts on Storm Water Systems. *J. Water Resour. Plan. Manag.* **2003**, *129*, 419–426. [CrossRef]

26. Leopold, L.B. Lag times for small drainage basins. *CATENA* **1991**, *18*, 157–171. [CrossRef]

27. Ebrahimian, A.; Gulliver, J.S.; Wilson, B.N. Effective impervious area for runoff in urban watersheds. *Hydrol. Process.* **2016**, *30*, 3717–3729. [CrossRef]

28. Dahlström, B. *Regnintensitet—En Molnfysikalisk Betraktelse (in Swedish) [English: Rain Intensity—A Cloud-Physical Contemplation]*; Svenskt Vatten AB: Stockholm, Sweden, 2010.

29. Albrecht, J.C. *Alterations in the Hydrologic Cycle Induced by Urbanization in Northern New Castle County, Delaware: Magnitudes and Projections*; University of Delaware: Newark, DE, USA, 1974.

30. Arnell, V. Estimating Runoff Volumes from Urban Areas. *J. Am. Water Resour. Assoc.* **1982**, *18*, 1–21. [CrossRef]

31. Miller, R.A. *Characteristics of Four Urbanized Basins in South Florida*; Open-File Report 79–694; United States Geological Survey: Reston, VA, USA, 1979.

32. Dinicola, R.S. *Characterization and Simulation of Rainfall-Runoff Relations for Headwater Basins in Western King and Snohomish Counties, Washington*; Water-Resources Investigations Report 89–4052; United States Geological Survey: Reston, VA, USA, 1990.

33. Alley, W.M.; Veenhuis, J.E. Effective impervious area in urban runoff modelling. *J. Hydraul. Eng.* **1983**, *109*, 313–319. [CrossRef]

34. Qin, H.; Li, Z.; Fu, G. The effects of low impact development on urban flooding under different rainfall characteristics. *J. Environ. Manag.* **2013**, *129*, 577–585. [CrossRef] [PubMed]

water

MDPI

Article

Rainwater Harvesting for Drinking Water Production: A Sustainable and Cost-Effective Solution in The Netherlands?

Roberta Hofman-Caris [1,*], Cheryl Bertelkamp [1], Luuk de Waal [1], Tessa van den Brand [1], Jan Hofman [2], René van der Aa [1] and Jan Peter van der Hoek [3,4]

[1] KWR Watercycle Research Institute, P.O. Box 1072, 3430 BB Nieuwegein, The Netherlands; cheryl.bertelkamp@kwrwater.nl (C.B.); luuk.de.waal@kwrwater.nl (L.d.W.); tessa.van.den.brand@kwrwater.nl (T.v.d.B.); rene.van.der.aa@waternet.nl (R.v.d.A.)

[2] Department of Chemical Engineering, Water Innovation and Research Centre, University of Bath, Claverton Down, Bath BA2 7AY, UK; j.a.h.hofman@bath.ac.uk

[3] Waternet (Public Water Utility of Amsterdam and Regional Water Authority Amstel, Gooi and Vecht), Postbus 94370, 1090 GJ Amsterdam, The Netherlands; j.p.vanderhoek@tudelft.nl

[4] Department of Water Management, Faculty of Civil Engineering and Geosciences, Delft University of Technology, P.O. Box 5, 2600 AA Delft, The Netherlands

* Correspondence: roberta.hofman-caris@kwrwater.nl

Received: 26 January 2019; Accepted: 4 March 2019; Published: 12 March 2019

Abstract: An increasing number of people want to reduce their environmental footprint by using harvested rainwater as a source for drinking water. Moreover, implementing rainwater harvesting (RWH) enables protection against damage caused by increasing precipitation frequency and intensity, which is predicted for Western Europe. In this study, literature data on rainwater quality were reviewed, and based on Dutch climatological data the usable quantity of rainwater in the Netherlands was calculated. For two specific cases, (1) a densely populated city district and (2) a single house in a rural area, the total costs of ownership (TCO) for decentralized drinking water supply from harvested rainwater was calculated, and a life cycle assessment (LCA) was made. For the single house it was found that costs were very high (€60–€110/m^3), and the environmental impact would not decrease. For the city district, costs would be comparable to the present costs of centralized drinking water production and supply, but the environmental benefit is negligible ($\leq$1‰). Furthermore, it was found that the amount of rainwater that can be harvested in the city district only covers about 50% of the demand. It was concluded that the application of rainwater harvesting for drinking water production in the Netherlands is not economically feasible.

Keywords: rainwater harvesting; footprint; lifecycle analysis; total cost of ownership; sustainability; urban water management; drinking water

1. Introduction

Water utilities in the Netherlands observe a societal trend of an increasing number of people adopting a more sustainable lifestyle and showing willingness to make personal efforts to reduce their ecological footprint. Some of them consider rainwater harvesting (RWH) as one of the measures that could significantly contribute to a more sustainable way of living. Rainwater is thought to be clean and many people have the impression that rainwater is amply available in the Netherlands. As a result, drinking water utilities are increasingly confronted with customers wishing to live "off-grid", and to use rainwater for the decentralized production of drinking water.

Climate change will result in an increasing frequency and intensity of precipitation in Western Europe, and particularly in the Netherlands. It is also likely that the balance between dry and wet periods will change [1,2]. Existing urban drainage systems are based on a centralized approach, with drainage networks that transport wastewater and storm water run-off away from the populated areas. Urban drainage systems in the Netherlands are designed for a peak capacity of 20 mm rain in 1 h with a repetition frequency of once per 2 years. It is expected that the present drainage capacity will not be sufficient for future climatic conditions [1]. This will result in more frequent water at the street level and associated nuisance, damage to property, and increasing health risks [3]. A potential solution to cope with the effects of climate change is increasing storage capacity for rainwater in tanks or aquifers. This is common practice in parts of Belgium for new buildings and after home improvement [4]. In a recent study in Portugal, Bellu et al. [5] developed a framework model for a flood mitigation system based on detention basins, which could retain river water during a flood. Terêncio et al. [6,7] studied rainwater harvesting systems in the rural areas of the Ave River and Sabor River Basin in Portugal, for controlling excess flows and floods on one hand, and using the water for agricultural purposes in rural areas.

Collecting and storing rainwater in the urban environment also opens up opportunities for re-using the rainwater as an alternative source for water applications within the city. Many cities in the world already suffer water stress, and harvested rainwater can be an interesting supplemental water resource in these areas. Rainwater harvesting is gaining much attention in the international scientific community and among urban planners as an alternative source in integrated water resources management (IWRM) programs. Important examples are the "sponge cities" in China [8,9] and the large RWH initiatives in South Korea [10]. In Portugal, the dimensioning of a rainwater harvesting system was optimized for low demanding applications, where water availability largely exceeds water demand [11].

Many studies in international literature focus on the application of harvested rainwater for non-potable applications such as toilet flushing, washing machines and garden watering. Studies into rainwater quality and harvesting systems are described for the USA [12–14], Australia [15], Malaysia [16,17], Spain [18–20], South Korea [21] and Mexico [22]. In general, the conclusions were that the quality of the harvested water strongly depended on the type of roof material, the length of the preceding dry period, the application of a first flush and general environmental conditions. For application of harvested rainwater for drinking water production, it was found that a robust disinfection treatment is required [12,15,17,18].

In the Netherlands, there is little experience with rainwater harvesting, as the availability of fresh water has not yet been a problem. However, water utilities are increasingly confronted with customers who want to decrease their environmental impact by preparing drinking water from rainwater. Individual households may collect rainwater and use it for toilet flushing, but already in the 1990s the safety of harvested rainwater was considered a point of attention [23,24]. In order to study the effects, some large scale pilot investigations were carried out. At the moment, the Dutch drinking water law does not allow the use of harvested rainwater for applications other than toilet flushing. This prohibition was set in 2003, after hundreds of people became ill after drinking low-grade household water, as a result of cross-connections between the drinking water and household water network in one of the pilots [25,26]. In the Netherlands, only one quality of water is distributed, which is used for all (potable) applications. The Dutch water sector has a proactive attitude towards societal trends and their effect on delivered water services, and therefore has initiated research with a focus on three specific questions: (1) what is known about rainwater quality in the Netherlands and/or Europe, (2) does the amount of rainwater that can be harvested cover the local drinking water demand, and (3) what are the costs, economic benefits and environmental impact in comparison with the centralized conventional drinking water supply?

In this paper a feasibility study was described for the production of drinking water from harvested rainwater in two Dutch situations: a densely populated city district area and a single house in the rural area. We describe the results of a literature study on the quality of harvested rainwater, as there are only limited data available for the Dutch situation. The water quality data found in literature were used to propose a robust water treatment that would be required to guarantee the production of safe drinking water. In case such a treatment process would actually be built, experimental data on water quality would have to be gathered to determine the optimum treatment process. In this study, the quantity of rainwater that can be harvested in the Netherlands was determined. For two specific cases, a city district and a single house in a rural area, the total cost of ownership (TCO) for decentralized drinking water supply from harvested rainwater was calculated. Furthermore, a life cycle assessment (LCA) was made for both situations. Other applications than drinking water were not taken into account within this study.

2. Methods

A literature study was carried out on rainwater quality, as only very limited data are available for the Dutch situation. With respect to quantitative aspects, data of the Royal Dutch Meteorological Institute (KNMI) for the Netherlands between 2006 and 2016 were studied [27] and combined with data on the average use of drinking water [28].

Six different scenarios were studied, two for the city district and four for the single house. Details of these scenarios are shown in Table 1. All scenarios were based on either reverse osmosis (RO) or advanced oxidation (UV/H_2O_2). In this way a double disinfection barrier was realized in order to be able to guarantee safe drinking water. For the city district it was assumed that water would be collected in an open pond, but as a large open pond near a single house may not be practical, scenarios with either an open pond or a closed tank were calculated.

The dimensions of the ponds and collection and storage tanks were estimated based on the assumptions that it would have to be possible to collect two heavy showers within 24 h, and that sufficient drinking water should be available to cover a period of 6 weeks of drought period.

The total costs of ownership (TCO) method calculates the total capital costs and operational costs for a chosen evaluation period (TCO = investment costs + operating costs + maintenance costs + residual value). In this case, a period of 20 years was taken into account with an interest rate of 1.5%. For the cost calculations, a handbook for the calculation of small treatment processes was used [29]. Detailed information of the process steps, investments, building, energy, chemical costs, and so on are shown in Table S1 in the supplementary information. According to the model, the uncertainty in cost calculations is about 30%. This was determined by validation of the model with real capital and operational costs of a large number of full scale installations for drinking water production in the Netherlands that actually have been built and are in operation. A life cycle analysis (LCA) can be used to determine the environmental impact of urban water systems [30] and water treatment processes [31–35]. An impact calculation was made by applying SimaPro 8 software, applying the ReCiPe endpoint (E) [36] and impact data from the EcoInvent 3.0 database for consumables (i.e., energy and chemicals). In this way, results were obtained covering a wide range of environmental impacts, including climate change effects on ecosystems and human health, fossil and metal depletion, human toxicity, terrestrial, marine and fresh water ecotoxicity, particulate and chemical oxidant formation, urban and agricultural land occupation and natural land transformation. All impacts have been weighed by a panel of experts resulting into one single score expressed in ecopoints per functional unit, with the total yearly impact of one western European person being about 1000 ecopoints [37]. In this study, the functional unit was 1 m^3 of produced drinking water. Effects were calculated for both a small scale and a larger scale installation, and for different types of processes. As such, the results can be regarded as a sensitivity analysis on both scale and type of process.

Table 1. Six rainwater harvesting scenarios. Scenarios 1 and 2 for city district, scenarios 3–6 for single house. RO = reversed osmosis, CT = contact time.

Process Step	Scenario 1	Scenario 2	Scenario 3	Scenario 4	Scenario 5	Scenario 6
1	Collection of rainwater from paved and built surfaces	Collection of rainwater from paved and built surfaces	Roof surface area 235 m²; grid [4]	Roof surface area 235 m²; grid [4]	Roof surface area 140 m²; grid [4]	Roof surface area 140 m²; grid [4]
2	Open pond, 14 × 10³ m³ concrete [1]	Open pond, 14 × 10³ m³ concrete [1]	Closed HDPE[5] tank, 12 m³	Closed HDPE tank, 12 m³	20 m³ open pond, concrete [1]	20 m³ open pond, concrete [1]
3	Pumping (Grundfos CR 15-05) 10 m³/h	Pumping (Grundfos CR 15-05) 10 m³/h	Pumping 1.25 m³/h (Grundfos CR1-7)	Pumping 1.25 m³/h (Grundfos CR1-7)	Pumping 1.25 m³/h (Grundfos CR1-7)	Pumping 1.25 m³/h (Grundfos CR1-7)
4	RO, membrane area 400 m², recovery 90%, 7.3 m³/h	Sand filter height 2 m, 1.3 m²; 3000 kg of sand	RO, Membrane area 40 m², recovery 90%, influent 1.11 m³/h,	Sand filter height 1.5 m, 0.167 m²; 650 kg of sand	Bag filter, pore size 25 µm	Bag filter (van Borselen X100), pore size 25 µm
5	Conditioning over calcite [2]	UV/H₂O₂ process (reactor with 4300 W LD UV lamps) 10 mg H₂O₂/L	Conditioning over calcite [2]	UV/H₂O₂ process (reactor with 120 W LD UV lamp Hereaus NNI 125-84-XL) 10 mg H₂O₂/L	RO, 5 µm sediment filter, two 5 µm AC filters; RO recovery 25%	UV/H₂O₂ Process (reactor with 120 W LD UV lamp Hereaus NNI 125-84-XL) 10 mg H₂O₂/L
6	UV disinfection (reactor with 120 W LD UV lamp Hereaus NNI 125-84-XL)	Activated carbon (CT = 20 min.); height 2 m, 1.65 m², 1320 kg of carbon	UV disinfection (reactor with 120 W LD UV lamp Hereaus NNI 125-84-XL)	Activated carbon (CT = 20 min.); height 1.5 m, 0.165 m², 99 kg of carbon	APEC in-line remineralization filter	Activated carbon filtration [6] (van Borselen VB06B005-090DP)
6	Storage, 2 vessels [3], 5000 m³ each, absolute filter.	Conditioning over calcite [2]	Storage, two 20 m³ HDPE vessels with absolute filter.	Conditioning over calcite [2]	UV disinfection (reactor with 120 W LD UV lamp Hereaus NNI 125-84-XL)	Addition of CaCO₃ to increase pH
7	Treatment and disposal of RO concentrate	UV disinfection	Treatment and disposal of RO concentrate	UV disinfection	Addition of CaCO₃ to increase pH	Storage in 2 m³ HDPE tank, absolute filter
8	Storage, 2 vessels [3], 5000 m³ each, absolute filter.	Storage, 2 vessels [3], 5000 m³ each, absolute filter.	Storage, two 40 m³ HDPE vessels with absolute filter.	Storage, 2 vessels [3], 5000 m³ each, absolute filter.	Storage in 2 m³ HDPE tank, absolute filter (van Borselen BorsoPTFE BPFI7SP002)	UV disinfection
9					Treatment of RO concentrate	

(1) Made of concrete, to prevent leakage of water in or out of the pond. (2) Using calcite from the softening process during centralized drinking water treatment. (3) Single coated steel buffer vessels, equipped with absolute filter. (4) Grid to remove branches, leaves, etc. (5) High Density Polyethylene (6) Removal of excess H₂O₂ and possibly formed byproducts and assimilable organic carbon (AOC).

The CO_2 footprint was calculated using the "single issue, greenhouse gas protocol" according to the international greenhouse gas protocol of the UNFCCC (United Nations Framework Convention on Climate Change) [38]. The CO_2 footprint (expressed as CO_2 equivalents = kg CO_2/m^3 drinking water) is calculated as the sum of fossil and biogenic CO_2 emissions and CO_2 from land transformation minus the CO_2 uptake. For the LCA, only consumables were taken into account, as it would be impossible to compare the impacts of buildings, installations and networks that differ in age by many decades and have been made from several materials. Consumables partly refer to the use of energy for pumps, membrane installations, and UV reactors. Here, we assumed that green energy (wind energy: electricity high voltage [NL] | wind, <1 MW turbine, onshore | Alloc Def S) was used, as all drinking water utilities in the Netherlands already use green energy. Furthermore, the use of chemicals and other compounds (activated carbon, sand, $CaCO_3$, H_2O_2, antifouling agents, etc.) were considered. The impact of all parameters was obtained from the EcoInvent database. The LCA included only consumables, such as energy, chemicals and materials (NaOH, HCl, CO_2, $CaCO_3$, H_2O_2, activated carbon, sand, etc.). Installations, buildings and networks were not taken into account. The impact of the centralized drinking water production was calculated from the following process [39]:

Water intake, coagulation by means of $FeCl_3$ and NaOH, sedimentation, filtration, infiltration, rapid sand filtration, ozonation, softening (by adding calcite and NaOH), pH correction by addition of HCl, filtration over activated carbon and aeration, addition of NaOH, and slow sand filtration. The total installed production capacity of the production site in Amsterdam was 12,000 m^3/h.

3. Results and Discussion

3.1. Rainwater Quality

In the Netherlands little data are available on rainwater quality, and therefore a literature search was carried out for quality data in other countries. In water, two types of contaminants can be distinguished: (a) chemical (either dissolved or suspended) and (b) microbiological. The uptake of contaminants occurs from the moment the raindrops leave the clouds. According to Grömping et al. [40], over 90% of atmospheric contaminants are removed by means of wet deposition. Although many ions present in rainwater are of natural origin (e.g., sodium, calcium and chloride) there also are anthropogenic contaminants like sulphate, nitrate, phosphate, ammonium, traces of iron, copper, cadmium, manganese, lead, zinc, nitrite, bromate and fluoride [41–48]. Concentrations are generally low, and in most cases below the Dutch standards for drinking water [49]. A comparison between some literature data [13] and both Malaysian and Dutch drinking water standards is shown in Table S2 in the Supplementary Information. In Tables S3–S6, analytical data from various countries in the world are summarized. The Dutch situation would be comparable to the situation in surrounding European countries like France (Paris city center and Île de France, department Ain, Seine Maritime in Normandy and a rural village), Ballinabrannagh in Ireland, Exeter in the UK and Bayreuth, Germany [50]. From the data it can be concluded that the variation in water-quality data in a certain area is similar in general to the variation between data from different areas all over the world. Furthermore, it can be concluded that for certain parameters, like heavy metals, treatment would be required, depending on local standards. The contents of iron, manganese and zinc in harvested rainwater may well be too high, but to which level concentrations would have to be decreased would be dependent on local legislation. In Malaysia, the standard for iron is 300 µg/L, whereas in the Netherlands this is 200 µg/L. On the other hand, the Malyasian standard for copper is 1000 µg/L, which is lower than the Dutch standard of 2000 µg/L. In comparing international treatment processes for harvested rainwater, it should be kept in mind that differences in local standards would affect the proposed treatment processes. In most investigations, inorganic compounds were measured, but Cindoruk and Ozturk [51] showed that organochlorine pesticides can be found in rainwater in several places in the world, and the presence of polycyclic hydrocarbons was demonstrated by Göbel et al. [52] and Angrill, Petit-Boix, Morales-Pinzón,

Josa, Rieradevall and Gabarrell [18]. An overview of the concentrations generally found in rainwater is shown in the Supplementary Information Tables S3–S8.

Problems with water quality mainly arise from contamination during the collection of water, when the rainwater is in contact with hard surfaces. These surfaces are often covered with contaminants from dry (e.g., dust) and wet precipitation (rain, fog, snow, etc.), animal urine and feces and plant debris, which end up in collected rainwater [14,17]. Also, because of the often acidic character of the water, metals and carbonate from roof material may dissolve [48,52,53]. As a result, the quality of the water collected from roofs is generally worse than that of the rainwater itself. Factors that affect the influence of the roof are the type of surface (a rough surface in general contains more contaminants than a smooth surface), and the angle and direction of the roof [50]. An overview of physico-chemical parameters and concentrations of ions, heavy metals and microbiological parameters in harvested rainwater is given in the Supplementary Information Tables S3–S8. In general, the pH ranges between 6 and 9, and TOC concentrations are low (mostly $\leq$ 10 mg/L, sometimes 10–20 mg/L), although sometimes, like in Ain in France, high values up to 8800 mg/L are reported (see Table S5 in the Supplementary Information). The inorganic content (Cl^-, Na^+, SO_4^{2-}, NO_2^- and NO_3^-) of harvested rainwater is also low, and all data are far below the standards for drinking water. Table S7 shows the concentrations of a number of heavy metals in harvested rainwater. These may occur from the settling of aerosols on the roof and dissolution of roofing and water collection materials. Most values in Table S7 are well below drinking water standards, except for the lead concentrations, which may exceed drinking water standards.

The microbial contamination of water is especially a problem if the water is to be used as drinking water, as shown in Tables S2 and S8. As this problem occurs in all rainwater harvesting systems, it can be assumed that also in the Netherlands adequate disinfection would be required. In many cases, the number of bacteria (strongly) exceeds the standards for drinking water. Health risks appear to be related to bad material selection and maintenance of the rainwater harvesting system. If the wrong material is selected for the roof and plumbing, either heavy metals may dissolve into the water, or microorganisms may be able to grow on it. Regular cleaning of the equipment would prevent the presence and growth of microorganisms. Two sources of contamination have to be distinguished: (1) direct contamination of the harvesting surface and system, and (2) regrowth of bacteria in the storage tank. A robust disinfection is a prerequisite for use as a source for drinking water [12,15,17,18]. According to an investigation by Boogaard and Lemmen [54], similar results were obtained for collected Dutch rainwater.

In order to improve the quality of harvested rainwater, a "first flush" could be applied, in which the first amount of rainwater is disposed of, as this contains the highest concentrations of contaminants [17,20,21,45,50,55,56]. How large the first flush should be would depend on the situation; the type and location of the roof have an effect, but also the length of the antecedent dry period, as during this period contaminants accumulate at the roof. In general the first 0.1 to 3.8 mm (for horizontal roofs covered with gravel) have to be disposed of to reach a good quality [14,17,20,21,45,48,50,55,56].

As literature data from Europe and densely populated areas show large similarities, it can be assumed that the quality of harvested rainwater in the Netherlands would be similar to the qualities described in literature. This means that for drinking water applications, a robust treatment, especially disinfection, would be required.

3.2. Quantity of Rainwater

Data of 25 meteorological weather stations and 325 stations for deposition measurements across the Netherlands, gathered between 2006 and 2016, were studied [27]. During this period the yearly amount of rain increased from 814 mm to 856 mm due to the occurrence of more heavy showers. The amount of water that can be harvested depends on the run-off coefficient: the ratio of rainwater that can be harvested to the total amount of rainwater that falls on a roof. This factor depends on the type and angle of the roof, the dominant wind direction, the intensity of the showers and the amounts of water that are "lost" as a result of evaporation or leakage. The run-off coefficient varies between 0.7 and 0.95, with an average value of about 0.8 [20]. If both the run-off coefficient and a first flush of 2 mm are applied to the deposition data, the percentage of rainwater that can actually be harvested appears to be about 50%, as shown in Table 2. The data in Table 2 refer to the weather stations shown in Figure 1. For these calculations, all showers with rainfall below 2 mm (the first flush), were not taken into account. The total amount of rainfall appeared to be practically the same over the whole country. In some parts of the country, like Nieuw Beerta, however, relatively more small showers occurred, as a result of which a larger part of the total rainfall was discarded. However, in general it was concluded that the differences in type of rainfall over the country were small.

Table 2. Part of rainwater that could effectively be harvested, calculated for weather stations across the Netherlands, as shown in Figure 1.

Location in the Netherlands (City)	% of Rainwater that Could Effectively Be Harvested
Vlissingen	48
De Bilt	51
Maastricht	49
Twente	48
De Kooy	48
Nieuw Beerta	46
Average	48

Figure 1. Weather stations across the Netherlands.

In the Netherlands, the average roof surface is 60 m^2, and the average house is inhabited by 2.2 persons [57,58]. On average, a Dutch person uses 119 L of drinking water per day, or 95.6 m^3 per family per year (see Supplementary Information Table S9) [28]. If the first flush is disregarded, about 41 m^3 may be harvested on 60 m^2 of roof surface, which is far from enough to cover the total water demand of about 95.6 m^3. Besides, part of the summer rainfall occurs in large showers (requiring relatively large collecting tanks), which alternate with periods of up to six weeks of drought. This means that it is important that the capacity of the pond or basin is large enough to collect the large showers, but also that the storage basins for treated water are large enough to bridge periods or drought, depending on the season. Water saving showerheads, taps and toilet cisterns are common in the Netherlands and already contribute to the reduction of water use. Further decrease of the water demand could be achieved by active systems such as vacuum toilets and recirculation showers, but these are still very expensive. To come into the range of a water supply fully based on rainwater harvesting, a further radical reduction of water consumption is necessary. For instance, by applying a waterless toilet, a saving of approximately 29% of the water demand could be achieved, and another 25% reduction of water demand for showering would be necessary to enable self-sufficiency. Both measures would lead to a total water demand of 72 L/p/day, which could be covered by RWH.

From the above it is concluded that rainwater harvesting for an average Dutch dwelling is not providing enough water to realize a self-sufficient drinking water supply system.

3.3. Sustainability and Cost-Effectiveness of Rainwater Harvesting in the Netherlands

The preceding paragraph showed that the amount of rainwater that can be harvested on Dutch roof tops is not sufficient to cover the drinking water demand in general. However, to explore the possibilities in more detail, costs and environmental impact of rainwater harvesting were calculated for two cases:

1. A new city district, being developed in urban Amsterdam, considering all rainwater from paved and built surfaces, in order to also decrease negative effects from heavy showers, like flooding, and overcharge of the sewer system.
2. An individual house in the peri-urban area of Amsterdam, assuming that in this case the roof area would be large enough to cover the drinking water demand of the inhabitants.

For the city district the calculations were based on city government's plans for layout of the area [59]. The surface area of the new district, which is to be located on an artificial island in the IJ lake, is 13,000 m^2, and will comprise 1300 unit (partly single houses, partly apartment buildings). As probably the total roof area would be too small due to the presence of multi-story apartment buildings, it was assumed that rainwater from all built and paved surface areas could be harvested (a best case scenario). According to literature, the quality of this rainwater still should be better than the quality of surface water, which may contain wastewater treatment plant effluent [18]. This effluent in general still contains pharmaceutical residues, microbial contaminations, etc. The total built and paved surface area in this district is expected to be 93,600 m^2. Based on the meteorological data and an average run-off coefficient of 0.8, approximately 685 mm of rainfall could be harvested. Thus, it can be calculated that a maximum of 64,000 m^3 of water may be harvested in this district, collecting all rainfall on paved and built surfaces. This amount would cover about 51% of the drinking water demand of the planned number of inhabitants at the current rate of water use.

Combining rainwater harvesting with the regular central drinking water production and distribution as a backup system would solve this problem. However, in order to be able to deliver sufficient water at any moment (including periods with a shortage of rain and empty rainwater storage tanks or reservoirs), the capacity of the treatment process and network would have to be identical to a regular system. As a result, no savings could be realized on investments for central drinking water treatment, but the water volume produced by the central system on average would be smaller due to the use of rainwater harvesting, resulting in higher costs per m^3 for the regular drinking water. A negative side-effect of this system would be that the residence time of water in the drinking water network would increase because the demand for centrally produced drinking water is low in times that decentralized rainwater harvesting could be used. This may result in a lower water quality [60]. Thus, a combination of a "regular" central network with decentralized rainwater harvesting system results in higher drinking water costs per m^3 and possibly lower qualities of drinking water.

For the second case, the individual house in the rural area, it was assumed the total roof area would be large enough to be able to harvest sufficient rainwater for the residents. Houses in the rural areas in general are larger, and often there are also outbuildings like barns and stables.

For both the city district and the single house, TCO were calculated (see Table S1 in the Supplementary Information). Two treatment processes, based on either reverse osmosis (RO) or advanced oxidation (UV/H$_2$O$_2$) were taken into account. This was done to have a double disinfection step, and to be able to remove any micropollutants that may have been present due to industry, traffic and possibly agricultural emissions. An additional requirement was that the produced drinking water had to be supplied without chlorine disinfection, similar to the current Dutch drinking water supply. Analyses results only become available after 24 h after potential contamination, meaning that, as water is consumed immediately after production, there is a risk that contaminated water would be consumed prior to the detection of the contamination. Therefore, it was decided that a robust treatment system is needed to deal with any pollutants that may occur, especially since water harvested from parking lots, pavements and roads would have to be treated. A detailed description of the total processes is given in Table 1.

For the city district, it was suggested to collect rainwater in an open pond. In a closed basin, water quality would become anaerobic and deteriorate quickly, as a result of which the water would have to be treated shortly after collection, resulting in a large treatment and storage capacity. With an open pond this would be less important, as water could be treated continuously over a longer period. Naturally, the water quality here too would deteriorate as a result of dust and contaminants from the surroundings, but the treatment process could be adjusted to this, as is the case with surface water used as a source of drinking water. In order to be able to harvest the maximum amount of water, and to prevent nuisance from heavy rainfall, the volume of this pond should be 14,000 m^3, or, at a depth of 4 m, it would require an area of about 3500 m^2, equaling about half the area of all sports fields and parks planned in the district.

For the single house, an open pond of 20 m^3 would be required, but it can be doubted whether the presence of such a large pond, which would be nearly empty during most of the year, would be desirable in the vicinity of a house. Therefore, we also calculated a situation in which water is collected in a closed tank. As untreated water cannot be stored for a longer period, and as the tank would have to be emptied within a short time (in order to be able to collect the next rain shower), the treatment capacity of the process would have to be relatively large in this case, although it would only be used occasionally. This would result in relatively high investment costs and operational costs of the system.

In total, six scenarios were studied, as shown in Table 1. In a case in which RO is applied as the main treatment process, the permeate would have to be conditioned in order to meet drinking water standards. As microorganisms may grow on the calcite filter used for conditioning, a second disinfection by UV would be required. For treatment processes based on a UV/H$_2$O$_2$ process, first rapid sand filtration is applied in order to obtain a first disinfection step and to remove particles and NOM (Natural Organic Matter), in order to improve the UV transmittance and turbidity of the water. Filtration over activated carbon is applied to remove the excess of H$_2$O$_2$ and any byproducts that may have been formed during the oxidation process. Conditioning is required to meet the drinking water standards for calcium. In order to remove any microorganisms originating from the carbon or calcite filters, a UV disinfection is applied afterwards.

For all scenarios, TCO and LCA were calculated. The results were compared to costs and environmental impact of centrally produced drinking water, with the use of surface water as a raw water source for the city of Amsterdam. Details of the TCO calculations are shown in Table 3.

Table 3. Details of TCO calculations of all six scenarios.

Scenario	1	2	3	4	5	6
Building costs ($€$)	1.44×10^6	1.43×10^6	6.02×10^4	4.54×10^4	1.83×10^4	1.83×10^4
Investment costs ($€$)	1.98×10^6	1.97×10^6	8.30×10^4	6.27×10^4	2.52×10^4	2.53×10^4
Interest & depreciation ($€/y$)	1.15×10^5	1.15×10^5	4837	3650	1469	1473
Operation & maintenance ($€/y$)	4.95×10^4	4.93×10^4	2076	1567	630	632
Energy ($€/y$)	713	3.96×10^4	3	123	2	123
Chemicals ($€/y$)	13	4355	0	1189	0	33
Membrane replacement ($€/y$)	1200	0	160	0	210	0
Filter ($€/y$)	50	5050	5	507	76	99
Lamps ($€/y$)	32	480	3	0	0	0
TCO ($€/m^3$)	2.71	3.43	85.24	84.76	38.27	38.02

The environmental impact of all scenarios was also calculated both in ecopoints and in CO_2 equivalents. This CO_2 footprint is often used, but it doesn't take into account all effects. As the LCA calculations are based only on consumables, the environmental impact for scenarios 5 and 6 equals the impact for scenarios 3 and 4, as these scenarios only differ in the type of collection tank (either a closed tank or an open pond).

In order to calculate the positive effects of rainwater harvesting (preventing nuisance and damage due to heavy rainfall) the situation in Berlin was taken as a starting point. Here, taxes are levied to compensate for the costs for water treatment and nuisance caused by (heavy) showers when the water cannot be drained because of the presence of hard surfaces. These taxes amount to $€1.84/m^2$ of paved surface area [61]. Therefore, it was assumed that a similar amount of money per m^2 could be saved in the Amsterdam area if rainwater were harvested and used as a drinking water source, instead of being discharged into the sewer. So for the calculation of the total costs, $€1.84/m^2$ was deducted from the production costs. The results are shown in Table 4. In order to be able to guarantee drinking water safety, regular analyses would be required. The yearly costs for water quality monitoring are on the average $€2500$. These costs would have to be made for every production plant, as a result of which the costs are very high for a single house ($€25.93/m^3$), but low for a district with more houses ($€0.04/m^3$). These costs have been included in the production costs in Table 4.

Table 4. TCO and LCA for scenario 1–6.

Scenario	Production Costs ($€/m^3$)	Analyses Costs ($€/m^3$)	Savings ($€/m^3$)	Net Costs ($€/m^3$)	Impact (mPt/m^3)	Impact ($kg\ CO_2/m^3$)
1	2.71	0.04	1.60	1.15	14.7	0.003
2	3.43	0.04	1.60	1.87	11.8	0.004
3	85.24	25.93	4.48	106.69	32.5	0.002
4	84.76	25.93	4.48	106.21	24.1	0.004
5	38.27	25.93	2.69	61.51	32.5	0.002
6	38.02	25.93	2.69	61.26	24.1	0.004
Centrally treated drinking water			0	1.63 [*]	36.4	0.130

[*] Price includes taxes, administration, etc. These costs would have to be added to the net price of scenarios 1–6.

The relatively high costs for small drinking water treatment systems are in accordance with literature findings. Roebuck et al. [62] studied 3840 domestic systems and concluded that harvesting rainwater was significantly less cost-effective than using only centrally produced drinking water. None of the RWH systems were able to demonstrate a return on investment. Although the operation of RWH appeared to be cheaper than drinking water, the periodic recurring costs for maintenance proved to be greater in magnitude than drinking water savings, resulting in a larger total rate. Domènech and Saurí [63] evaluated the use of RWH systems in the metropolitan area of Barcelona, Spain. In their study, they investigated social aspects, drinking water savings and costs of single- and multi-family buildings. For the economic modeling they also used the RainCycle model Roebuck used. For single-family homes, the harvested rainwater was used for toilet flushing, cleaning, filling the swimming pool or washing the car. In multi-family buildings, only garden irrigation was assumed. Again, in this study long payback times were found, up to 60 years with the main cause being high capital costs. Farreny et al. [64] investigated RWH on a larger scale in dense Mediterranean urban neighborhoods. The research compared cost-efficiency at two scales (single building and neighborhood) and implementation (new construction areas and existing area retrofits). However, the case study was limited to the use of rainwater for laundry washing only. The authors concluded that cost-efficiency of RWH strategies may be in doubt as long as local water prices are low. Furthermore, they concluded that RWH systems should be preferably installed at the neighborhood level, because of economy of scale. Installations should be realized in new construction areas to be cost-effective. Morales-Pinzón et al. [65] investigated 87 scenarios in a number of Spanish cities consisting of RWH systems of various sizes, ranging from two single houses to a group of apartment buildings connected to a single RWH system. They concluded that the material type used for a storage tank was not a fundamental financial factor, but planning on a neighborhood scale was. The costs per functional unit ranged from 0.94 to 10.59 €/m^3, with the lowest cost for the category "group of apartment buildings". RWH systems have a better financial fit for large-scale and high-density constructions. According to these authors, the best strategy was implementation at a neighborhood level. An example of such a system can be found in Ringdansen, Norrköping (Sweden) [66]. In these studies, too, it was concluded that the variability of rainfall is an important factor to be considered in detail during design because it has a direct impact on the RWH tank size.

In Table 4 it also can be seen that by only taking into account the CO_2 footprint, the difference between the scenarios is very small. Differences in effects, like those on human health or ecosystems, may not be fully accounted for when looking only at the CO_2 footprint, which makes up only part of the footprint in ecopoints. For example, scenarios 3 and 5 seem to have a low CO_2 footprint, whereas they have the highest footprint in ecopoints. The CO_2 footprint of the centrally produced drinking water is significantly higher, as the water has to be transported over a distance of about 60 km from intake to the treatment plant, and softening is applied. Furthermore, it is likely that the impact of filtration over activated carbon is higher in a large-scale process, as in such a scenario surface water is used as a source, which probably would contain more micropollutants, resulting in a higher reactivation frequency for the activated carbon.

These results are in accordance with literature data, where it was also found that the overall generated impact of water treatment is driven by the consumption of energy. When the impact of the installations is also included, a significantly higher total impact would be the result [33]. Some previous LCA studies at water utility Waternet showed that the most significant impact contributors of the centralized treatment process are the use of conventional energy, coagulation, softening and filtration over granular activated carbon [32,35]. As softening and coagulation don't have to be applied with rainwater, this lowers the environmental impact of the treatment process. Garfí, Cadena, Sanchez-Ramos and Ferrer [31] compared the environmental impacts caused by drinking water consumption in Barcelona, comparing centrally treated tap water from a conventional plant and from a plant based on RO, tap water treated with point-of-use RO, mineral water in plastic bottles and mineral water in glass bottles. The results showed that the centrally treated drinking water caused the smallest impact, the impact of domestic RO being 10–24% higher.

As with other sources of drinking water, rainwater can only be used as a source for drinking water when sufficient purification is applied, especially when water from all paved and built surfaces is used. The microbial safety of the collected rainwater is a point of concern in decentralized treatment systems, and would require robust water treatment and frequent and expensive analyses to guarantee that the purification system was still working properly. The same applies to centralized drinking water treatment processes (e.g., based on surface water), but, as less water would be produced, the costs per m^3 would be higher for decentralized processes. Apart from the high analysis costs, it would be very difficult to guarantee safe drinking water in decentralized systems, as the enforcement of measures that must be taken based on analytical data would be very difficult. Besides, monitoring would only give results after at least 24 h, while problems could occur immediately in decentralized systems. At present, no online monitoring is possible.

Taking into account the cost savings related to RWH as a result of less problems caused by heavy showers, the costs of decentralized drinking water production based on rainwater harvesting would be in the same order of magnitude as costs for centrally produced drinking water. For a single house, costs would be much higher. This is due to relatively high investment and analyses costs. Especially when water is collected in a closed tank, costs become very high due to the additional spare capacity required for enabling a rapid treatment of harvested water, which is necessary for emptying the collection tank for the next rain event.

Although the environmental impact seems to decrease by using rainwater as a source for the production of drinking water instead of surface water, the relative savings are very small. As one Dutch person yearly on average uses 43.4 m^3 of drinking water, a maximum of $43.4 \times 24.6 = 1068$ mPts could be saved, which is about 1‰ of the total environmental impact of this person. Besides, only consumables were taken into account for this investigation, not the impact of the installations and networks required. If these were also included, the impact of decentralized systems would increase compared to the impact of a centralized system, as a large number of small installations requires more material than one large scale installation.

Although from this study it can be concluded that production of drinking water from harvested rainwater in the Netherlands is far more expensive and doesn't really have a positive effect on the environment in comparison with the centralized production and distribution of drinking water, this doesn't mean that rainwater harvesting should not be applied. It is a proven tool for storm water management, and when rainwater has been collected, it might as well be used for certain applications, like industrial applications or maybe household water. However, using it as a source for drinking water is not recommended.

4. Conclusions

Based on literature data it is expected that the quality of harvested rainwater in general would not meet (Dutch) drinking water standards, and thus a robust treatment is required. The quality is strongly affected by the surface used to harvest the rainwater, and the microbiological quality often requires a robust disinfection in order to produce safe drinking water. For decentralized treatment systems, the required analyses result in high costs per m^3, and even then, it would be very difficult to guarantee water safety, as enforcement of required maintenance would be hardly possible on a small scale.

By considering only consumables, using rainwater as a source for decentralized drinking water production results in a slightly smaller ecological footprint compared to the use of surface water in a central system. However, this difference only results in a decrease of about 1‰ in the total environmental impact of a person per year. When the impact of the installations is also included, a significantly higher total impact results. The use of steel has an especially large contribution to the total impact. Besides, for the RO processes, the disposal and treatment of concentrate is another factor that should be considered in costs and environmental impact.

The costs of a decentralized drinking water production are (much) higher than the costs of a centralized water supply system. In order to make rainwater harvesting economically interesting, water should be collected on a neighborhood scale and not per individual building. To create a fully self-sufficient system, water not only from roof tops but also from paved surfaces should be harvested. By doing this, water nuisance during extreme weather events may be reduced, lowering the total societal costs involved (rainwater harvesting and treatment plus costs of storm-water management).

In densely populated areas, like the city of Amsterdam, the amount of rainwater that could be harvested is insufficient to cover the water demand of the inhabitants, even if water-saving measures are being taken. As a result, a centralized drinking water treatment system and network would still be required, in addition to the RWH system.

Supplementary Materials: The following are available online at http://www.mdpi.com/2073-4441/11/3/511/s1: Table S1—TCO calculations for six scenarios; Table S2—comparison of composition of harvested rainwater and Malaysian and Dutch drinking water standards; Table S3—inorganic parameters in directly harvested rainwater; Table S4—heavy metals in directly harvested rainwater; Table S5—physico-chemical parameters in harvested rainwater; Table S6—inorganic parameters in harvested rainwater after first flush; Table S7—heavy metals in harvested rainwater after first flush; Table S8—microbiological parameters in harvested rainwater after first flush; Table S9—use of drinking water in the Netherlands in L/person/day in 2016.

Author Contributions: Conceptualization, R.H.-C., J.H. and J.P.v.d.H.; methodology, R.H.-C., J.H. and C.B.; software, L.d.W. and T.v.d.B.; validation R.v.d.A. and J.P.v.d.H.; investigation, R.H.-C., C.B., L.d.W., T.v.d.B. and J.H.; writing—original draft preparation, R.H.-C. and J.H.; writing—review and editing, J.P.v.d.H.; visualization, J.H.; supervision, J.P.v.d.H. and R.v.d.A.

Funding: This research was funded by a joint research programme (BTO) for the Dutch water companies De Watergroep (BE) and the branch association Vewin.

Acknowledgments: The authors would like to thank Marcel Paalman (KWR) for his contribution to the literature research, and Oasen and Waternet (Dutch drinking water utilities) for their financial contribution and input for this research.

Conflicts of Interest: The authors declare no conflict of interest.

References

1. Van den Hurk, B.; Tank, A.K.; Lenderink, G.; Van Ulden, A.; Van Oldenborgh, G.J.; Katsman, C.; Van de Brink, H.; Keller, F.; Bessembinder, J.; Burger, G.; et al. *KNMI Climate Change Scenarios 2006 for the Netherlands*; Royal Netherlands Meteorological Institute: De Bilt, The Netherlands, 2006.

2. IPCC WG2. AR6 Climate Change 2021: Impacts, Adaptation and Vulnerability. Available online: https://www.ipcc.ch/report/sixth-assessment-report-working-group-ii/ (accessed on 29 December 2018).

3. Rovers, V.; Bosch, P.; Albers, R.E. *Climate Proof Cities*; Final Report; Climate Proof Cities Consortium: Amsterdam, The Netherlands, 2014.

4. Ringelstein, O. Now we can shower with Rain Water. *GWF Wasser-Abwasser* **2015**, *156*, 58–61.

5. Bellu, A.; Sanches Fernandes, L.F.; Cortes, R.M.V.; Pacheco, F.A.L. A framework model for the dimensioning and allocation of a detention basin system: The case of a flood-prone mountainous watershed. *J. Hydrol.* **2016**, *533*, 567–580. [CrossRef]

6. Terêncio, D.P.S.; Sanches Fernandes, L.F.; Cortes, R.M.V.; Moura, J.P.; Pacheco, F.A.L. Rainwater harvesting in catchments for agro-forestry uses: A study focused on the balance between sustainability values and storage capacity. *Sci. Total Environ.* **2018**, *613*, 1079–1092. [CrossRef] [PubMed]

7. Terêncio, D.P.S.; Sanches Fernandes, L.F.; Cortes, R.M.V.; Pacheco, F.A.L. Improved framework model to allocate optimal rainwater harvesting sites in small watersheds for agro-forestry uses. *J. Hydrol.* **2017**, *550*, 318–330. [CrossRef]

8. Wang, H.; Mei, C.; Liu, J.; Shao, W. A new strategy for integrated urban water management in China: Sponge city. *Sci. China Technol. Sci.* **2018**, *61*, 317–329. [CrossRef]

9. Xia, J.; Zhang, Y.; Xiong, L.; He, S.; Wang, L.; Yu, Z. Opportunities and challenges of the Sponge City construction related to urban water issues in China. *Sci. China Earth Sci.* **2017**, *60*, 652–658. [CrossRef]

10. Han, M.Y.; Mun, J.S. Operational data of the Star City rainwater harvesting system and its role as a climate change adaptation and a social influence. *Water Sci. Technol.* **2011**, *63*, 2796–2801. [CrossRef]

11. Sanches Fernandes, L.F.; Terêncio, D.P.S.; Pacheco, F.A.L. Rainwater harvesting systems for low demanding applications. *Sci. Total Environ.* **2015**, *529*, 91–100. [CrossRef]

12. Kim, T.; Lye, D.; Donohue, M.; Mistry, J.H.; Pfaller, S.; Vesper, S.; Kirisits, M.J. Harvested rainwater quality before and after treatment and distribution in residential systems. *J. Am. Water Works Assoc.* **2016**, *108*, E571–E584. [CrossRef]

13. Dallman, S.; Chaudhry, A.M.; Muleta, M.K.; Lee, J. The Value of Rain: Benefit-Cost Analysis of Rainwater Harvesting Systems. *Water Resour. Manag.* **2016**, *30*, 4415–4428. [CrossRef]

14. Mendez, C.B.; Klenzendorf, J.B.; Afshar, B.R.; Simmons, M.T.; Barrett, M.E.; Kinney, K.A.; Kirisits, M.J. The effect of roofing material on the quality of harvested rainwater. *Water Res.* **2011**, *45*, 2049–2059. [CrossRef] [PubMed]

15. Hamilton, K.A.; Ahmed, W.; Toze, S.; Haas, C.N. Human health risks for Legionella and Mycobacterium avium complex (MAC) from potable and non-potable uses of roof-harvested rainwater. *Water Res.* **2017**, *119*, 288–303. [CrossRef] [PubMed]

16. Leong, J.Y.C.; Chong, M.N.; Poh, P.E.; Hermawan, A.; Talei, A. Longitudinal assessment of rainwater quality under tropical climatic conditions in enabling effective rainwater harvesting and reuse schemes. *J. Clean. Prod.* **2017**, *143*, 64–75. [CrossRef]

17. Leong, J.Y.C.; Oh, K.S.; Poh, P.E.; Chong, M.N. Prospects of hybrid rainwater-greywater decentralised system for water recycling and reuse: A review. *J. Clean. Prod.* **2017**, *142*, 3014–3027. [CrossRef]

18. Angrill, S.; Petit-Boix, A.; Morales-Pinzón, T.; Josa, A.; Rieradevall, J.; Gabarrell, X. Urban rainwater runoff quantity and quality—A potential endogenous resource in cities? *J. Environ. Manag.* **2017**, *189*, 14–21. [CrossRef]

19. Angrill, S.; Segura-Castillo, L.; Petit-Boix, A.; Rieradevall, J.; Gabarrell, X.; Josa, A. Environmental performance of rainwater harvesting strategies in Mediterranean buildings. *Int. J. Life Cycle Assess.* **2017**, *22*, 398–409. [CrossRef]

20. Farreny, R.; Morales-Pinzón, T.; Guisasola, A.; Tayà, C.; Rieradevall, J.; Gabarrell, X. Roof selection for rainwater harvesting: Quantity and quality assessments in Spain. *Water Res.* **2011**, *45*, 3245–3254. [CrossRef] [PubMed]

21. Lee, J.Y.; Bak, G.; Han, M. Quality of roof-harvested rainwater—Comparison of different roofing materials. *Environ. Pollut.* **2012**, *162*, 422–429. [CrossRef] [PubMed]

22. García-Montoya, M.; Sengupta, D.; Nápoles-Rivera, F.; Ponce-Ortega, J.M.; El-Halwagi, M.M. Environmental and economic analysis for the optimal reuse of water in a residential complex. *J. Clean. Prod.* **2016**, *130*, 82–91. [CrossRef]

23. Medema, G.J. *Microbiologische Veiligheid van Huishoudwater. voor Toepassing van Toilet, Wassen Kleding en Buitenkraan*; Kiwa Water Research: Nieuwegein, The Netherlands, 1999.

24. Versteegh, J.F.M.; Evers, E.G.; Havelaar, A.H. *Gezondheidsrisico's en Normstelling voor Huishoudwater*; Rijksinstituut voor Volksgezondheid en Milieu: Bilthoven, The Netherlands, 1997.

25. Oesterholt, F. *Beleidsonderbouwende Monitoring Huishoudwater*; Kiwa Water Research: Nieuwegein, The Netherlands, 2003.

26. Oesterholt, F.; Sluijs, A.; Mons, M.N.; Medema, G.J. Evaluatie van praktijkervaringen met huishoudwater. In *H₂O*; KNW, 2003; Volume 16, pp. 22–24. Available online: http://library.wur.nl/WebQuery/hydrotheek/1692729 (accessed on 30 December 2018).

27. KNMI. Overview of Precipitation and Evaporation in The Netherlands. Available online: https://www.knmi.nl/nederland-nu/klimatologie/gegevens/monv (accessed on 8 August 2018).

28. Vewin. Water Supply Statistics 2017. Available online: http://www.vewin.nl/SiteCollectionDocuments/Publicaties/Cijfers/Drinkwaterstatistieken-2017-NL.pdf (accessed on 8 August 2018).

29. DHV Water. Handboek Kosten Kleinschalige Waterbehandeling. 2000. Available online: https://www.kostenstandaard.nl/ (accessed on 30 December 2018).

30. Loubet, P.; Roux, P.; Loiseau, E.; Bellon-Maurel, V. Life cycle assessments of urban water systems: A comparative analysis of selected peer-reviewed literature. *Water Res.* **2014**, *67*, 187–202. [CrossRef] [PubMed]

31. Garfí, M.; Cadena, E.; Sanchez-Ramos, D.; Ferrer, I. Life cycle assessment of drinking water: Comparing conventional water treatment, reverse osmosis and mineral water in glass and plastic bottles. *J. Clean. Prod.* **2016**, *137*, 997–1003. [CrossRef]

32. Mohapatra, P.K.; Siebel, M.A.; Gijzen, H.J.; Van der Hoek, J.P.; Groot, C.A. Iproving eco-efficiency of Amsterdam water supply: A LCA appraoch. *J. Water Supply Res. Technol. Aqua* **2002**, *51*, 217–227. [CrossRef]

33. Igos, E.; Dalle, A.; Tiruta-Barna, L.; Benetto, E.; Baudin, I.; Mery, Y. Life Cycle Assessment of water treatment: What is the contribution of infrastructure and operation at unit process level? *J. Clean. Prod.* **2014**, *65*, 424–431. [CrossRef]

34. Bonton, A.; Bouchard, C.; Barbeau, B.; Jedrzejak, S. Comparative life cycle assessment of water treatment plants. *Desalination* **2012**, *284*, 42–54. [CrossRef]

35. Barrios, R.; Siebel, M.; van der Helm, A.; Bosklopper, K.; Gijzen, H. Environmental and financial life cycle impact assessment of drinking water production at Waternet. *J. Clean. Prod.* **2008**, *16*, 471–476. [CrossRef]

36. Goedkoop, M.J.; Huijbregts, M.A.J.; Heijungs, R.; Stuijs, J.; van Zelm, R. *ReCiPe 2008 a Life Cycle Impact Assessment Method Which Comprises Harmonised Category Indicators at the Midpoint and the Endpoint Level*; Report I: Characterisation; Ministerie van Volkshuisvesting, ruimtelijke ordening en milieubeheer: The Hague, The Netherlands, 2009.

37. Baayen, H. *Eco-Indicator 99; Manual for Designers; A Damage Oriented Method for Life Cycle Impact Assessment*; Ministry of Housing, Spatial Planning and the Environment: The Hague, The Netherlands, 2000.

38. UNFCCC. Adoption of the Paris Agreement. Available online: https://unfccc.int/resource/docs/2015/cop21/eng/l09r01.pdf (accessed on 30 December 2018).

39. Hofman-Caris, C.H.M.; Huiting, H.; Snip, L.; Brand, T.V.D.; Palmen, L. *Ontharding 2.0 bij Waternet*; Productielocatie Leiduin; KWR Watercycle Research Institute: Nieuwegein, The Netherlands, 2016.

40. Grömping, A.H.J.; Ostapczuk, P.; Emons, H. Wet deposition in Germany: Long-term trends and the contribution of heavy metals. *Chemosphere* **1997**, *34*, 2227–2236. [CrossRef]

41. Bhaskar, V.V.; Rao, P.S.P. Annual and decadal variation in chemical composition of rain water at all the ten GAW stations in India. *J. Atmos. Chem.* **2017**, *74*, 23–53. [CrossRef]

42. Rao, P.S.P.; Tiwari, S.; Matwale, J.L.; Pervez, S.; Tunved, P.; Safai, P.D.; Srivastava, A.K.; Bisht, D.S.; Singh, S.; Hopke, P.K. Sources of chemical species in rainwater during monsoon and non-monsoonal periods over two mega cities in India and dominant source region of secondary aerosols. *Atmos. Environ.* **2016**, *146*, 90–99. [CrossRef]

43. Deusdará, K.R.L.; Forti, M.C.; Borma, L.S.; Menezes, R.S.C.; Lima, J.R.S.; Ometto, J.P.H.B. Rainwater chemistry and bulk atmospheric deposition in a tropical semiarid ecosystem: The Brazilian Caatinga. *J. Atmos. Chem.* **2017**, *74*, 71–85. [CrossRef]

44. Beysens, D.; Mongruel, A.; Acker, K. Urban dew and rain in Paris, France: Occurrence and physico-chemical characteristics. *Atmos. Res.* **2017**, *189*, 152–161. [CrossRef]

45. Zobrist, J.; Müller, S.R.; Ammann, A.; Bucheli, T.D.; Mottier, V.; Ochs, M.; Schoenenberger, R.; Eugster, J.; Boller, M. Quality of roof runoff for groundwater infiltration. *Water Res.* **2000**, *34*, 1455–1462. [CrossRef]

46. Vet, R.; Artz, R.S.; Carou, S. A global assessment of precipitation chemistry and deposition of sulfur, nitrogen, sea salt, base cations, organic acids, acidity and pH, and phosphorus. *Atmos. Environ.* **2014**, *93*, 3–100. [CrossRef]

47. Cukurluoglu, S. Sources of trace elements in wet deposition in Pamukkale, Denizli, western Turkey. *Environ. Forensics* **2017**, *18*, 83–99. [CrossRef]

48. Campisano, A.; Butler, D.; Ward, S.; Burns, M.J.; Friedler, E.; DeBusk, K.; Fisher-Jeffes, L.N.; Ghisi, E.; Rahman, A.; Furumai, H.; et al. Urban rainwater harvesting systems: Research, implementation and future perspectives. *Water Res.* **2017**, *115*, 195–209. [CrossRef]

49. Dutch Government. Drinkwaterbesluit. Available online: https://wetten.overheid.nl/BWBR0030111/2018-07-01#BijlageA (accessed on 30 December 2018).

50. Förster, J. Variability of roof runoff quality. In *Proceedings of the 1998 International Congress on Options for Closed Water Systems—Sustainable Water Management*; Elsevier Science Ltd.: Amsterdam, The Netherlands, 1999; Volume 39, pp. 137–144.

51. Cindoruk, S.S.; Ozturk, E. Atmospheric deposition of organochlorine pesticides by precipitation in a coastal area. *Environ. Sci. Pollut. Res.* **2016**, *23*, 24504–24513. [CrossRef]

52. Göbel, P.; Dierkes, C.; Coldewey, W.G. Storm water runoff concentration matrix for urban areas. *J. Contam. Hydrol.* **2007**, *91*, 26–42. [CrossRef] [PubMed]

53. Lamprea, K.; Ruban, V. Micro pollutants in atmospheric deposition, roof runoff and storm water runoff of a suburban catchment in Nantes, France. In Proceedings of the 11th international conference on urban drainage, Edingburgh, UK, 31 August–5 September 2008; pp. 1–8.

54. Boogaard, F.C.; Lemmen, G.B. *Achtergrondrapport Database Regenwater*; 2007-WO09; Stowa: Zwijndrecht, The Netherlands, 2007; ISBN 978.90.5773.378.9.

55. Yaziz, M.I.; Gunting, H.; Sapari, N.; Ghazali, A.W. Variations in rainwater quality from roof catchments. *Water Res.* **1989**, *23*, 761–765. [CrossRef]

56. Gikas, G.D.; Tsihrintzis, V.A. Assessment of water quality of first-flush roof runoff and harvested rainwater. *J. Hydrol.* **2012**, *466*, 115–126. [CrossRef]

57. Rijksoverheid. Compendium voor de Leefomgeving. Available online: http://www.clo.nl/indicatoren/nl2114-huishoudens (accessed on 30 December 2018).

58. Weber, D. Zonnepanelen-Weetjes. Available online: https://www.zonnepanelen-weetjes.nl/blog/afmetingen-van-zonnepanelen/ (accessed on 2 January 2019).

59. Amsterdam, M.O. Centrumeiland: Zelfbouw en Duurzaamheid. Available online: https://www.amsterdam.nl/projecten/ijburg/centrumeiland (accessed on 8 August 2018).

60. Agudelo-Vera, C.; Blokker, M.; Vreeburg, J.; Vogelaar, H.; Hillegers, S.; Van der Hoek, J.P. Testing the robustness of two water distribution system layouts under changing drinking water demand. *J. Water Resour. Plan. Manag.* **2016**, *142*, 05016003. [CrossRef]

61. Welt. Regensteuer. Available online: https://www.welt.de/print/die_welt/hamburg/article10416092/Regensteuer-tritt-2012-in-Kraft.html (accessed on 8 August 2018).

62. Roebuck, R.M.; Oltean-Dumbrava, C.; Tait, S. Whole life cost performance of domestic rainwater harvesting systems in the United Kingdom. *Water Environ. J.* **2011**, *25*, 355–365. [CrossRef]

63. Domènech, L.; Saurí, D. A comparative appraisal of the use of rainwater harvesting in single and multi-family buildings of the Metropolitan Area of Barcelona (Spain): Social experience, drinking water savings and economic costs. *J. Clean. Prod.* **2011**, *19*, 598–608. [CrossRef]

64. Farreny, R.; Gabarrell, X.; Rieradevall, J. Cost-efficiency of rainwater harvesting strategies in dense Mediterranean neighbourhoods. *Resour. Conserv. Recycl.* **2011**, *55*, 686–694. [CrossRef]

65. Morales-Pinzón, T.; Lurueña, R.; Rieradevall, J.; Gasol, C.M.; Gabarrell, X. Financial feasibility and environmental analysis of potential rainwater harvesting systems: A case study in Spain. *Resour. Conserv. Recycl.* **2012**, *69*, 130–140. [CrossRef]

66. Villarreal, E.L.; Dixon, A. Analysis of a rainwater collection system for domestic water supply in Ringdansen, Norrköping, Sweden. *Build. Environ.* **2005**, *40*, 1174–1184. [CrossRef]

water

MDPI

Article

Understanding the Costs of Inaction–An Assessment of Pluvial Flood Damages in Two European Cities

Harry Nicklin [1,*], Anne Margot Leicher [2], Carel Dieperink [1] and Kees Van Leeuwen [1,3]

[1] Copernicus Institute for Sustainable Development and Innovation, Utrecht University, Heidelberglaan 2, 3584 CS Utrecht, The Netherlands; c.dieperink@uu.nl (C.D.); kees.van.leeuwen@kwrwater.nl (K.V.L.)

[2] Nelen & Schuurmans, Zakkendragershof 34-44, 3511 AE Utrecht, The Netherlands; margot.leicher@nelen-schuurmans.nl

[3] KWR Watercycle Research Institute, Groningenhaven 7, 3433 PE Nieuwegein, The Netherlands

* Correspondence: harrynicklin@gmail.com; Tel.: +31(6)1133-9871

Received: 2 April 2019; Accepted: 15 April 2019; Published: 17 April 2019

Abstract: Today, over 50% of the global population lives near water. Due to population growth, ongoing economic development, and extreme weather events, urban areas are growing more susceptible to flood risks, and the costs of inaction of failing to manage flood risks are high. Research into the benefits of pluvial flood-risk management is needed to spread awareness and motivate investments in pluvial flood-risk reduction. So far, such research is lacking. This research therefore assesses pluvial flood damage from a single 60 mm/1-h rainfall event in the cities of Rotterdam and Leicester using 3Di flood modelling and the flood damage estimation tool (waterschadeschatter; WSS). The results demonstrate that potential pluvial flood damages exceed €10 million in each city. From this research, inhabitants and authorities of Leicester and Rotterdam can learn that preparing for upcoming pluvial floods can save millions of euros resulting from future damages. The application of these tools also makes clear that data availability is a highly relevant bottleneck to the pluvial flood damage assessment process. By addressing data shortages, flood damage estimates can be strengthened, which improves decision support and enhances the chance actions are taken in reducing pluvial flood risks.

Keywords: cost of inaction; urban pluvial flooding; flood damage assessment; flood risk

1. Introduction

Throughout history, cities have sprouted in proximity to freshwater sources, as water is vital for drinking, agriculture, transportation and domestic use [1]. However, due to urban expansion and climate change, cities are growing increasingly prone to floods with serious socio-economic and environmental consequences. Flooding is the number one most frequently occurring natural disaster, causing over $20 billion in economic damage and claiming over 3300 lives worldwide in 2017 [2]. In Europe, annual flood losses are expected to increase five-fold by 2050 and as much as seventeen-fold by 2080, highlighting the need for cities to build flood resilience [3,4].

There are various types of floods, for example, river, coastal, groundwater, and pluvial floods, each requiring different techniques to prepare for. Coastal and river floods receive the most attention as they are generally the largest and longest-lasting flood types, while pluvial floods-caused by heavy rainfall that urban drainage systems are unable to cope with-are relatively underrepresented in research [5]. Recent research has suggested that due to the frequent nature of pluvial floods, cumulative direct damage to property from pluvial floods equals or may even exceed damage from river and coastal floods [6]. Continued urbanization accompanied by the intensification of rainfall patterns due to climate change will likely exacerbate pluvial flood risks, so research is needed into how to manage pluvial floods to minimize damage to our economy, environment, and society [6,7].

Financial resources for managing pluvial flood risks are limited, so decision-makers need convincing that investments in flood-risk reduction are worthwhile. The occurrence of a flood disaster is often used as motivation, for example, the city of Copenhagen initiated the Cloudburst Management Plan in response to a July 2011 pluvial flood disaster that cost upwards of €800 million [8]. Such a reactive mindset does little to reduce the cost of the initial flood disaster. The concept of the cost of inaction (COI), defined as the total cost due to climate change in the absence of adaptation and mitigation measures, can be used to present the consequences of disasters that have not yet occurred [9]. By estimating and considering the COI, decision-makers may see that it is costlier to wait than act now to reduce flood risks.

To understand the COI, it is necessary to assess the amount of flood damage that occurs in the absence of any further investments in flood-risk management. Since flooding can cause a wide array of economic, environmental, and societal impacts, a distinction is often made between tangible/intangible and direct/indirect flood damage. Direct damage occurs in the flooded area due to immediate physical contact with floodwater, while indirect damages arise with a time lag or outside the flooded area [9]. For example, if a flooded business halts production, the physical damage to the building and contents within is direct damage, while induced losses to supply and demand suffered later in time outside of the flooded area are indirect. Tangible flood damage is damage to assets that can be easily monetized with a market price, whereas non-market priced damage (e.g., health loss, environmental damage) that cannot be immediately given a monetary value is intangible [10,11]. Some examples are shown below in Figure 1 [10–13].

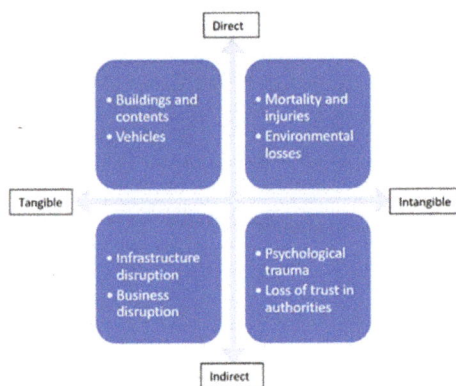

Figure 1. Distinction between tangible, intangible, direct, and indirect flood damages.

1.1. Direct Damage Assessments

Direct damage to property is considered the dominant type of pluvial flood damage and is the subject of most pluvial flood damage research [7]. This research also focuses on direct pluvial flood damage as such physical damage to structures is the most relevant damage type for densely built urban areas, and there are established assessment methods to draw upon. Potential indirect and intangible pluvial flood impacts are touched upon later in the discussion section.

The common framework for assessing direct flood damage consists of three steps: simulation of the flood hazard (hazard analysis), identification of the types of assets exposed to flooding (exposure analysis), and the translation into monetary flood damage based on the characteristics of the exposed objects (vulnerability analysis). This framework (illustrated below in Figure 2) has mainly been applied to assessments of river and coastal flood damages [14–16], but recent studies have used the same methods to assess pluvial flood damage.

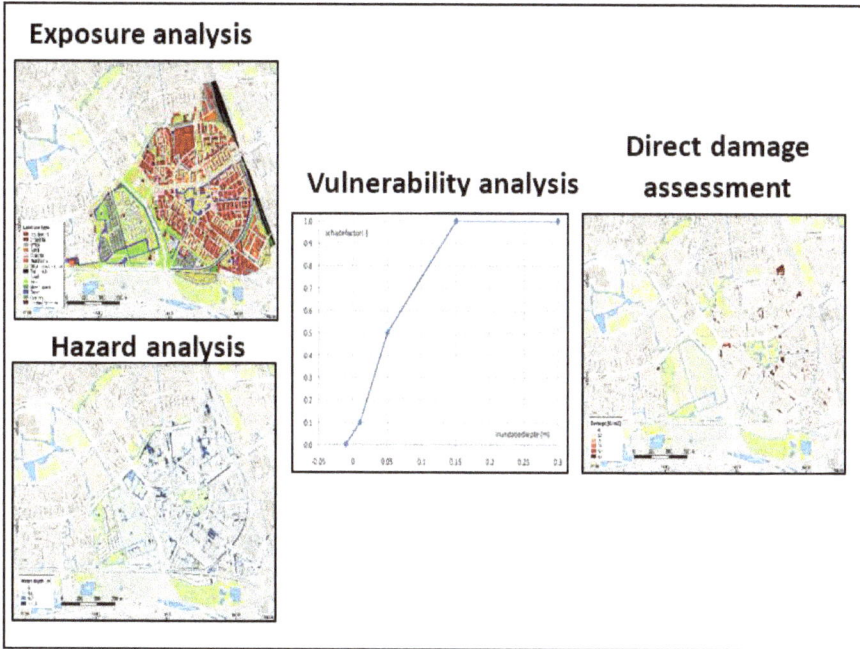

Figure 2. Standard direct flood damage assessment framework.

1.1.1. Hazard Analysis

The magnitude of direct flood damage is partly determined by the characteristics of the flood, known as impact parameters. Impact parameters such as flood depth, flood duration, flow velocity, contamination, and rise rate may determine damage to building structure and contents, but such detailed information is often not available about the flood hazard. In practice, the flood depth is the dominant and often only parameter used to represent the flood hazard. This is especially the case for pluvial floods, which usually do not occur over long durations or at high enough flow velocities to incorporate these parameters in the hazard analysis. To simulate pluvial floods, coupled 1D/2D hydrological models, which integrate both 1D sewer flow and 2D surface water flow to solve the shallow water equations, are used [17–19]. Despite representing the best option for accounting for realistic drainage processes, these models suffer from data limitations as they require hourly precipitation data and localized information on flow paths, as well as impediments to flow and interactions with drainage components [17,20]. Uncertainties in this process include extreme value statistics used, stationary and homogeneity of data series, consideration of physical properties (e.g., dikes and drainage systems) of a location, and calibration and validation of model output etc. [20,21]. As complete data are not always available, GIS-based digital elevation models (DEM) can be used to approximate pluvial flood flow [17].

1.1.2. Exposure Analysis

Flood exposure could be gauged at the object-level, but there are so many types of assets and buildings that can be exposed to flooding that assets are usually aggregated into groups based on land use classification. For city-wide flood damage assessments, land use maps displaying residential, commercial/industrial, environmental, infrastructural, and other land use/building types are used to group exposed assets. Average asset and content values for each land use type can be surmised using aggregated national or regional statistics, real estate data, or expert consultation to analyze the total value of objects exposed to flooding. However, uncertainty regarding asset types and approximate values can be critical in many cases, especially when models developed for a specific location at a specific time are not validated and transferred across spatial or temporal boundaries [14]. The exposure analysis for pluvial floods can be carried out in the same way as has been established for river and coastal floods, as described in Merz et al. [14].

1.1.3. Vulnerability Analysis

A loss model is a central element of flood damage estimation and the most common way of estimating direct damage amount is the use of depth–damage functions, often called susceptibility or vulnerability functions [14,15]. Some loss models are multi-parameter models based on several impact parameters and resistance parameters including building function, type, age, size, presence of mitigation measures, profiles of inhabitants, etc. Prominent examples of multi-parameter flood damage assessment techniques within Europe include the UK multi-colored manual, and German FLEMOps and FLEMOcs models [16,22–25]. These models were developed specifically for fluvial (river) flood damage assessments, concentrating mainly on damage to residential and commercial buildings. The same damage functions relating fluvial flood depth to monetary damage should not be used for pluvial floods, as flooding from rainfall is dictated by different flow properties and the magnitude of flood damage to building structure and contents is likely different [26]. However, the central idea of a loss model is also applicable to pluvial flood damage assessments, for example, Zhou et al. [7] described a framework for economic pluvial flood risk assessment considering future climate change which quantifies flood risk in monetary terms as expected annual damage in different return periods of rainfall. Susnik et al. [27] used a threshold method in which pluvial flood depths above 30cm are attributed to a fixed damage amount to assess pluvial flood damages from a heavy rainfall event in Eindhoven. Efforts have also been made to relate pluvial flood damage to various parameters based on insurance data or surveys distributed to pluvial flood victims [28–33]. Such research is important for distinguishing how parameters other than flood depth, such as preparation, prior flood experience, and presence of mitigation measures, can determine pluvial flood losses. Outcomes of these studies illustrate how non-structural measures, for example, increasing the flood warning times or spreading awareness on how to adequately respond to a flood event, can significantly reduce pluvial flood risks [30–33]. However, the flood damage assessment models to date contain a number of uncertainties in both the hazard and damage models. The largest sources of uncertainties in damage modelling are associated with prescribed depth–damage functions [11,13,14]. A reason for uncertainty in many loss models is the crude assumption of the relationship between damage and flood depth only. Optimally, other parameters that impact flood damage like building age, presence of basements, and preparedness to respond to flooding would be included in the damage assessment, yet data on these factors are often incomplete [28,34]. There is still a need to develop a stronger understanding of different parameters impacting pluvial flood damage to develop stronger and more rigorous flood damage assessment methods [9,10,30]. Equally important is for decision-makers to recognize the simplifications and uncertainties present in flood damage assessment models, so the results of imperfect damage assessments do not misguide pivotal policy choices in flood-risk management.

1.2. Objective of This Research

In this paper, the process of estimating the COI of urban pluvial flooding is illustrated through pluvial flood damage assessments in two selected European cities. Using a combination of a state-of-the-art flood simulation model and a flood damage estimation tool developed by a consortium of Dutch water companies, flood damages are estimated for a single rainfall event in the cities of Rotterdam (NL) and Leicester (UK). The purpose of this research is two-fold: to illustrate potential flood costs that could arise if these cities fail to prepare for a pluvial flood event, and to shine a light on the key limitations of the flood damage assessment process so decision makers are better prepared to translate results into tangible policy action towards reducing pluvial flood risks. If estimates of the

COI are transparent and trustworthy enough, they can be used to convince urban flood-risk managers to reduce flood risks proactively rather than responsively, thus saving the cost of the initial flood disaster [13].

In the next section, the flood damage assessments carried out for neighborhoods in the cities of Leicester (United Kingdom) and Rotterdam (the Netherlands) are described. In Section 3, the results of the flood damage assessments for both study areas are presented, followed by a discussion of the values and limitations of the research in Section 4. This paper finishes with a short summary of the research and concluding remarks in Section 5.

2. Materials and Methods

A flood damage assessment is conducted for a 60 mm/h rainfall event (constant rainfall) in Belgrave (Figure 3), a part of Leicester (UK) and Lombardijen (Figure 4), a part of Rotterdam (NL) using 3Di flood modelling software and the Dutch 'Waterschadeschatter' (WSS) flood damage assessment tool.

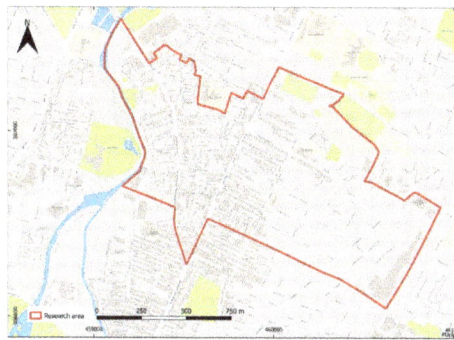

Figure 3. Belgrave location.

Figure 4. Lombardijen location.

2.1. Study Areas

These areas were both selected for analysis, firstly because they are both low-lying areas that are identified as flood prone [35–38]. Through consultation with members of the Leicester City Council, Belgrave was identified as a neighborhood especially at risk to pluvial flooding. Belgrave is also of a comparable size, population and climate to Lombardijen, so it is included as a point of comparison for Lombardijen. The Netherlands contains many low-lying, flood-prone areas, but Lombardijen was chosen in particular because of the availability of a pre-calibrated and validated high-quality 3Di model for the area. The use of a high-quality 3Di increases the trust in the results of the 3Di modelling conducted in this research.

2.2. Flood Modelling with 3Di

The 3Di flood modelling software was developed by a combination of Stelling Hydraulics, Deltares, TU Delft, and Nelen & Schuurmans. It is a physically-based model designed to simulate the passage of water through urban areas during flood events. According to Van Dijk [17], the sewer and surface water systems should be coupled in dual drainage models for realistic flood simulations. The governing equations of 3Di can be accessed on the 3Di Water Management website (https: //3diwatermanagement.com/3di-start). 3Di flood modeling is used because it represents state-of-the art hydrodynamic modelling. It uses a sub-grid method for 2D surface water flow, 3Di provides fast and accurate results [39]. This is one of the best currently available ways to link 2D surface flow to 1D drainage flow to simulate the process of flooding in urban areas. In the UK, 3Di software is also regarded as one of the best currently available methods for dual drainage modeling based on a benchmark developed last year.

For the Lombardijen study area, data accessed were from AHN2 digital elevation map (DEM) (2008, 25 m^2), BAG (building register), TOP10 (topography), OSM (open streetmap), and CBS land use datasets. A map of the urban drainage network was provided by the municipality of Rotterdam. For Belgrave, a soil type map was accessed on the Cranfield Soil and Agriculture Institute website [40], a DEM (2015, 25 m^2) was obtained from the UK governmental environmental data online portal [41], and a land use map was created in a GIS environment based on CDRC Open Map Survey data (https://data.cdrc.ac.uk/dataset/cdrc-2015-os-geodata-pack-leicester-e06000016). The parking lots were manually added based on Open Street Map data. In the figures below, the water depth maps developed for Belgrave (Figure 5) and Lombardijen (Figure 6) are displayed. It is important to note that a digital map of the urban drainage system was unavailable for Belgrave because of the privatization of the water supply industry in the UK and concerns regarding confidentiality. Therefore, the interactions between surface water flow and the sewer systems are left out of the Belgrave flood model.

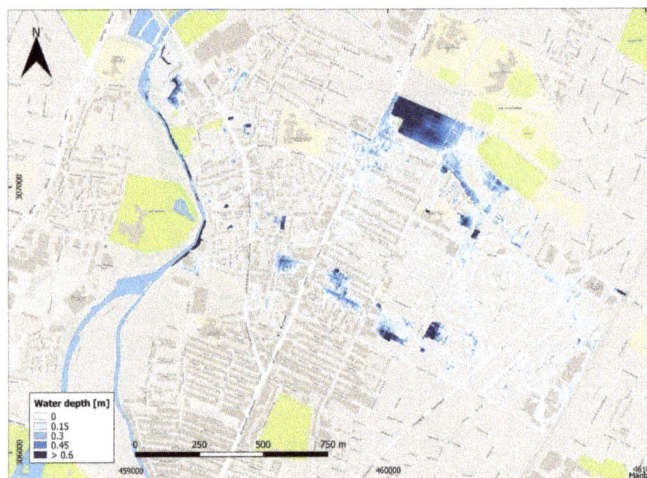

Figure 5. Belgrave water depth map.

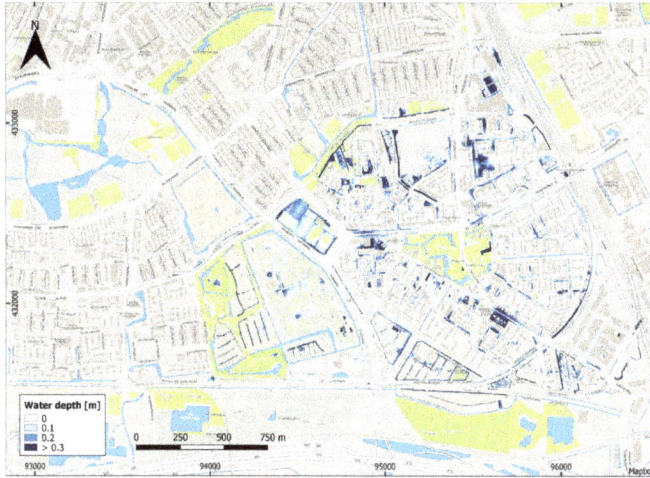

Figure 6. Lombardijen water depth map.

From the figures above, it is visible that the rainfall event causes pooling of water in the fields to the North of Belgrave and in residential areas in the Center/South (Figure 5). In Lombardijen (Figure 6) the water pooling appears more spread out, but there are still significant flood depths over 0.5 m in the Northwest, and just Southeast of the center. To assess flood damage, the water level map (maps above show water depth, water level = water depth + elevation) for each study area is combined with land use information and damage functions in the online WSS damage estimation tool. Below, the land use maps developed for Belgrave (Figure 7) and Lombardijen (Figure 8) are displayed. It should be mentioned that due to difficulties accessing land use and building register information in the UK, the land use classification is less detailed (7 land use classes) in Belgrave than Lombardijen (14 classes).

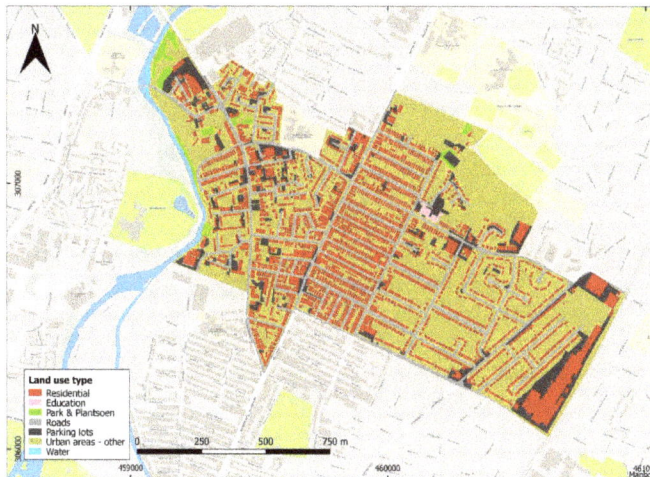

Figure 7. Belgrave land use map.

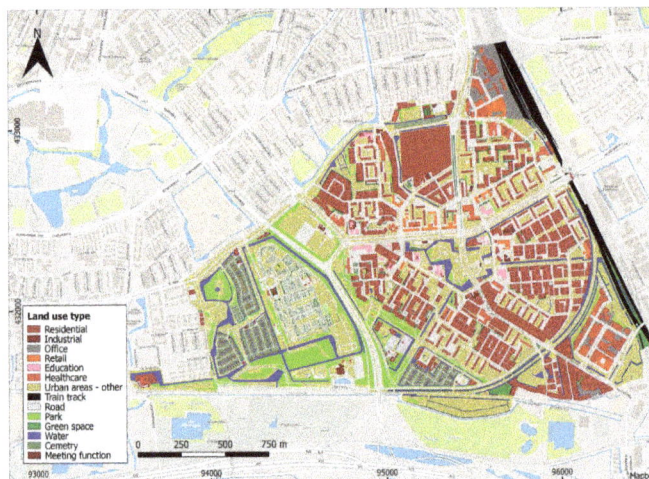

Figure 8. Lombardijen land use map.

2.3. Damage Assessment with WSS

The WSS is a freely accessible cloud-based flood damage estimation tool. It is owned by the STOWA consortium of Dutch water companies and was designed to estimate damages from flood depths up to 2.5 m. The flood damage estimation tool can be accessed at the website: https://www.waterschadeschatter.nl/damage/. WSS software was used for this damage estimation because it is a free, entirely web-based tool that does not require heavy computing power which has been developed to estimate pluvial flood damage in the Netherlands. It is simple enough that the functions used to estimate damage can be understood and used to calculate flood damage for areas outside the Netherlands.

Within the Netherlands, the only required input for the WSS is a water level map because land use and DEM data are already included in the WSS tool. With the water level map, the WSS subtracts the DEM to calculate water depth in each grid cell. Land use information is then used to identify the flooded land use classes, and damage functions are applied to relate the flood depth to monetary direct flood damage values for each land use class. Since the WSS is a tool developed for Dutch flood damage assessments, the water level and land use raster with a UK projection cannot be processed by the WSS cloud. Therefore, the damage estimation for Belgrave was conducted using a raster calculator in Python with the default damage functions used in the WSS tool and Lombardijen case study.

After uploading a water level map to the WSS website, the user is asked to select the flood duration (h), recovery time for buildings and roads (h), month of flood event, and whether to use minimum, average, or maximum damage values. In this research of direct damage from a one- hour rainfall event of 60 millimeters, the flood duration was set to one hour, recovery times for buildings and roads were set to zero (direct damage only), the month was set to September, and average damage values were selected. Below, Tables 1 and 2 display the average damage values for all land use classes considered in the Belgrave and Lombardijen case studies. Figure 9a,b shows the damage functions that were applied for the different land use classes.

Table 1. Average damage values for each land use class, Belgrave.

Land Use Type	Average Damage (€/m²)
Residential	271
Educational	271
Parks & greens	0.1086
Roads	0.076
Parking lots	0.076
Urban areas-other	0
Water	0

Table 2. Average damage values for each land use class, Lombardijen.

Land Use Type	Average Damage (€/m²)
Residential	271
Industrial	271
Office	271
Retail	271
Educational	271
Healthcare	271
Meeting	271
Sport	54
Parks & greens	0.1086
Train track	0.076
Roads	0.076
Urban areas–other	0
Cemetery	0
Water	0

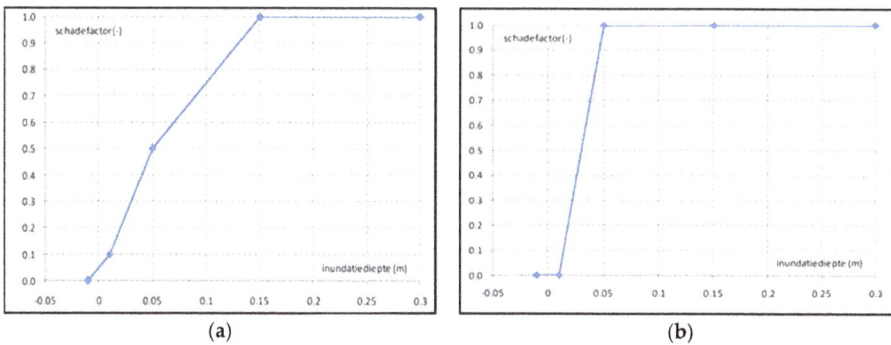

Figure 9. Damage function for residential, educational, industrial, retail, office, healthcare, meeting, urban area – other, and sport land use classes (**a**); Damage function for train tracks, roads, parks and greens, parking lots, water, and cemetery land use classes (**b**). On the *X*-axis, the flood (inundation) depth is displayed (up to 30 cm) and on the *Y*-axis, the damage factor, total share of the asset value that is damaged, (0–1) is shown.

Tables 1 and 2 display the average maximum value that is susceptible to flood damage per m² of each land use type. Figure 9a,b shows the relative damage curves used to translate water depth to a share of the average maximum value that can be damaged in each land use class. This results in a monetary estimate of direct flood damage for each land use type, as presented in the next section.

3. Results

Below, the results of the flood damage assessments for a one-hour 60 mm rainfall event in Belgrave and Lombardijen are shown.

3.1. Direct Flood Damage in Belgrave

Figure 10 shows that flood damage in Belgrave is concentrated in a couple of hotspots in the North and South. Comparing this with the water depth map of Belgrave (Figure 5), it is noticeable that flood damage is not always highest in areas with the greatest flood depths and is highly dependent on land use. For example, the field to the North of Belgrave that experiences heavy pooling in Figure 5 does not show flood damage in Figure 10. On the other hand, the less-flooded residential area surrounding the field in Figure 5 does show heavy flood damage. Direct flood damage can only be as high as the value of the asset that is damaged, so it is important that areas identified as flood-prone on flood hazard maps do not become overdeveloped. As more areas in the Netherlands and UK are expected to become flood-prone in the future, it is crucial that anticipated future rainfall patterns be incorporated in present-day decisions regarding development and urban spatial planning. Below, Table 3 breaks down the flood damage in Belgrave per land use type.

Figure 10. Belgrave flood damage map (dark red signifies high flood damage).

Table 3. Belgrave damage estimate per land use type.

Damage Type	Damage (€)
Residential	10,895,788
Educational	87,912
Roads	2833
Parking lot	2717
Parks & greens	473
Water	0
Urban areas-other	0
Total	10,989,723

As shown in Table 3, the total direct damage from the one-hour pluvial flood event is nearly €11 million. Over 98% of total damage comes from the residential sector, which is likely because all buildings were considered residential except for one school. Other studies show building damage of 60–95% of total direct damage [42], whereas in this research the total share of building damage is over 99%. This could be because other damage types (e.g., infrastructure damage) were neglected and the average damage values for water and urban areas–other were set to zero. Thus, the >99% of total damage associated with property damage yielded in this study is likely an overestimation. However, the total damage estimate of €10.99 million could be an underestimation as flooding of urban areas–other and water land use classes would realistically cause some monetary damage.

3.2. Direct Flood Damage in Lombardijen

Figure 11 displays the distribution of flood damage throughout the Lombardijen neighborhood. There are several areas of concentrated high damage to the east and southeast of Spinozapark (note the dark red blotches). Comparing this with the water depth map (Figure 6), it is clear that water pooling does not always correlate with high flood damage. More important is the land use class of the area that is flooded. In the case of Lombardijen, the damage hotspots exist in the densely packed residential/industrial areas where there are more buildings exposed to flooding.

As shown in Table 4, residential damage makes up most of the pluvial flood damage, followed by damage to industrial and meeting function buildings. The total share of building damage (99%) is on the high end of what is reported in literature [12,42,43]. This is likely because damage to water utilities, power stations, and other infrastructural components was not part of this assessment, resulting in an overstatement of the total share of building damage. It is important to also mention that although direct damages to roads and train tracks were minimal in this simulation, traffic delays and diversions would be expected to cause high indirect and intangible damages like inconvenience and lost time. Although the urban drainage system was accounted for in the Lombardijen damage estimation, the results still show greater flood damages in Lombardijen than Belgrave. This could be due to the inclusion of a more detailed land use map for Lombardijen (Figure 8). Much of the Belgrave area was classified as urban area – other, which had a maximum damage value of zero, whereas the more detailed land use classification for Lombardijen considered comparatively much less area as urban area - other. Damage estimates are sensitive to the level of detail in the land use classification, so it is difficult to compare the damage estimates between the two areas due to the difference in the land use maps used.

Table 4. Lombardijen damage estimate per land use type.

Damage Type	Damage (€)
Residential	7,558,742
Industrial	2,283,720
Meeting	1,224,713
Educational	672,239
Office	487,369
Sport	137,291
Retail	13,025
Roads	645
Train track	220
Parks & greens	135
Healthcare	0
Urban areas-other	0
Total	12,378,099

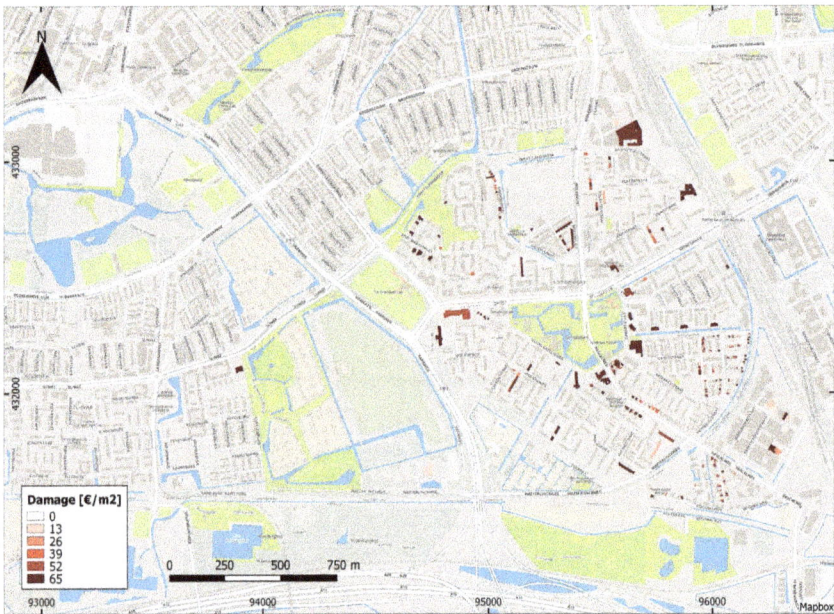

Figure 11. Lombardijen flood damage map (dark red signifies high flood damage).

4. Discussion

4.1. Data Limitations

The two case studies are contrasting since all Lombardijen information was readily accessible, while the Belgrave damage estimation was hampered by data availability issues. Consequently, critical elements for the flood damage assessment like a sewer network map displaying locations of drains, sewers, pipes, and manholes, and average asset values specific for buildings in Belgrave were not available. Because of this, the damage assessment for Belgrave was conducted with the best available data to the knowledge of the authors and the damage estimate of €10.9 million should only be considered a ballpark estimate. It has been noted that even with state-of-the-art 1D/2D hydrodynamic flood models, there is still a poor understanding of the mechanisms of urban flooding and high uncertainties in flood depth simulations [13,17–20]. Data on real flood events are needed, not just

228

to validate the results of flood level maps, but also to better our understanding of the relationship between flood depth and flood damage among different land use classes and to develop damage functions tailored to each class.

Due to input data shortages for model validation and damage function development, the damage estimates presented here have not been tested, thus caution should be taken if the outcomes are used to inform decisions in urban flood-risk management. It is also important to note that this damage estimate focuses only on direct damages to physical property, neglecting indirect and intangible damage to health and the environment that is also likely to occur. By omitting these damage types, the full spectrum of flood damage is not included in the damage assessment, resulting in an estimate of direct damage only. It is pivotal to include all damage types in flood damage assessments to build a complete understanding of flood risks and ensure the benefits of flood-risk reduction are not underestimated [44]. In fact, some studies comparing the benefits of coastal flood-risk reduction measures to implementation costs have shown that investments in flood-risk reduction are only net-beneficial if intangible damages like loss of life are included in the damage assessment [45,46]. Without the inclusion of the diverse types of potential flood damage, the estimated benefits of flood-risk reduction are understated, thus the case for implementation is weakened. Even though pluvial floods are likely to be smaller and cause less significant intangible and indirect damage than coastal floods, all damage types should still be considered to ensure nothing is left out and the assessment of flood risk is complete. It could also be unethical to base decisions in flood risk reduction only on direct damage estimates as it may lead to the prioritization of reducing flood risks in areas with the most economically valuable building assets, in other words, the richest areas. Although the WSS attributes constant average damage values to all buildings regardless of building quality, the use of these damage estimates could result in the prioritization of flood damage mitigation for the largest buildings, not necessarily the most vulnerable. Recent research has highlighted that the less fortunate and socially vulnerable encounter greater flood damage due to information and resource shortages, yet this is ignored by damage assessments based only on asset damage [47,48]. Although there may be less chance of heavy intangible or indirect damage compared to river or coastal floods, pluvial floods can result in casualties, injuries, long-term trauma, and significant environmental damage [9,49]. As climate change is expected to bring about more frequent and intense pluvial flood hazards, it is prudent to continue to dedicate research to the assessment of not only direct, but all types of pluvial flood risks.

4.2. Looking Towards the Future

Despite the simplicity and imperfections of the damage assessments conducted here, the results are still valuable for raising awareness of the absolute minimum potential flood damage that communities can expect to face, as well as pinpointing where future research is urgently required. This can be helpful for inspiring people to start a (very necessary) conversation about what flood risks we are willing to face, and how to be prepared to reduce flood risks in the future. Many cities are either sitting atop outdated drainage systems that do not have the capacity for future rainfall patterns or are developing rapidly without a centralized drainage system in place [50,51]. In any case, decisions need to be made on how to decrease pluvial flood risks to avoid future disasters. Options for reducing pluvial flood risks exist, from conventional 'grey' options like increasing the capacity of existing sewer systems to 'green' and 'blue' innovations such as green roofs, sustainable urban drainage systems, and designated floodwater reservoirs in cities, which serve to delay or divert rainfall from over-stressed sewers [8,50,51]. Since each approach comes at a cost, it is necessary to estimate pluvial flood damage in cities to determine whether investments in flood risk reduction are economically justifiable. Flood damage estimates also provide valuable information for insurance firms, local businesses, spatial planning authorities, emergency planners, and households.

Data collection and accessibility need to be improved for water depth and damage model validation, so results of flood damage assessments can be trusted to better represent reality. Flood depth and flood damage data are collected for a variety of purposes, but methods and standards for data collection are rarely aligned. This is problematic since consistent and accurate data are needed to validate the outcomes of both water level and flood damage estimates. Without a consistent and reliable source of flood data, it becomes more difficult to reinforce and improve flood damage estimates. It is understandable that drainage network and infrastructure maps are difficult to attain due to safety, security, and strategic concerns, but there are tangible benefits to increasing accessibility of this information for dedicated research purposes. Not only would this allow for more accurate simulation of the flood routing and urban drainage process, but maps of other infrastructural components like utility stations and power networks can identify further components at risk to flood exposure. Severe pluvial floods can incapacitate infrastructure networks, and the interdependencies and knock-on effects should be further investigated [52–54]. Without drainage and infrastructural network maps in the hands of researchers, it is difficult to trust that the simulated flood depths or estimated value of assets exposed to flooding match with reality.

The assumption of this research that all building types have the same damage characteristics and average values can be called into question, for example, industrial buildings often include heavy machinery and equipment that is more valuable, thus susceptible to greater flood damage than residential buildings. By associating the same damage functions and average asset values to all buildings in Lombardijen and Belgrave, the damage estimates of €10.9–12.2 million to buildings in the two cities merely reflect the fact that they are comparably sized with similar building densities. It is imperative to dedicate continued research to addressing data deficiencies so the role of flood damage assessments as decision-support mechanisms is strengthened. Flood damage assessments are also key for raising awareness of flood risks and closing the gap between perceived and actual risks [55]. Flood damage assessment therefore plays the important role of increasing awareness of flood risks prior to disasters so societies can be better prepared to reduce future flood risks. Flood defense via structural or technical means is regarded as the cornerstone of European flood risk management, but risks can also be addressed with other paradigms in flood risk management such as flood prevention via spatial planning, risk mitigation, preparation, and recovery [55,56]. Research has shown that non-structural means like installation of early warning systems or adaptation of water sensitive cities master plans can repay setup costs within a matter of years, especially when cleverly combined with structural options [46,57]. Key to raising awareness and hastening action flood risk reduction is for people to understand, as well as believe, the results of flood damage assessments [44,58]. Thus, it is important to continue to research and dissect existing flood damage assessment methods to increase our understanding of the methods used as well as the limitations. By studying urban pluvial flooding, the risks appear more real and awareness can be raised on potential flood impacts. Raising awareness is a crucial pre-cursor for inspiring action to reduce flood risks and cope with the challenges of climate change in cities.

5. Conclusions

The significance of expanding knowledge about pluvial flood risks and the costs of inaction cannot be understated. Climate change is steering society into unchartered territory, and urban conglomerations only serve to aggregate flood risks. People are becoming increasingly aware of flood risks after catastrophic events that have plagued society in recent decades. This backward-looking attitude is not well-suited for a future where we face uncertain conditions yet near-certain intensifications of flood risk. Instead, it is urgent to act now to reduce risks before disasters occur. Awareness precludes action, and to raise awareness, flood risks need to be assessed.

This research used a combination of 3Di flood modelling and the WSS flood damage estimation tool to assess direct flood damage from a 60 mm/1-h pluvial flood event in two urban areas: Belgrave (Leicester, United Kingdom) and Lombardijen (Rotterdam, the Netherlands). For Belgrave, direct damage was estimated at roughly €11 million, while for Lombardijen direct damage was €12.4 million. Due to a lack of pluvial flood damage data, identical average asset values and damage functions were applied for both neighborhoods. Thus, the comparable damage estimates yielded in this research

Water **2019**, *11*, 801

merely reflect the fact that the areas are of similar size and building densities. This research could be improved by using locally-tailored asset values, land use maps, and damage functions to account for differences between building types as well as study areas. Furthermore, the Belgrave damage assessment was hampered by the failure to include a map of the urban drainage infrastructure, resulting in an overly-simplified portrayal of the flood propagation. Urgent research is needed to address data bottlenecks, especially for validating flood depth simulations and developing damage values and functions for the wide variety of land use classes that exist in modern cities [59]. With more complete data of past flood events, work can be put into improving pluvial flood damage assessments that can play a stronger role supporting decisions in flood-risk reduction. This can be vital for reducing future flood risks, as most studies demonstrate investments in flood-risk reduction at least break even with implementation costs [46,60].

Pluvial flood damage assessments are essential for supporting arguments for building flood resilience, raising awareness of flood risks, and determining how flood-risk management could best be implemented. Ideally, such damage assessments are based on sound data. Decisions should not be based on damage assessments alone without recognizing input data limitations, which result in some significant uncertainties underpinning the damage assessment process. A way forward is to keep studying flood damages and developing solid databases to validate and improve models for assessing potential future flood damages. Stronger flood damage assessments will be essential for building the case that the costs of inaction are too great to ignore flood risks any longer. To borrow a quote from John F. Kennedy, "There are risks and costs to action. But they are far less than the long-range risks of comfortable inaction".

Author Contributions: Conceptualisation: H.N., K.V.L. and A.M.L.; Methodology: H.N., A.M.L., C.D. and K.V.L.; Validation: H.N., A.M.L., C.D. and K.V.L.; Formal analysis: H.N., A.M.L., C.D. and K.V.L.; Investigation: H.N., A.M.L., C.D. and K.V.L.; Writing—original draft preparation: H.N. and K.V.L.; Writing—review and editing: H.N., A.M.L., C.D. and K.V.L.; Visualization: M.L. and H.N. Supervision: C.D. and K.V.L. Funding acquisition: K.V.L. and C.D.

Funding: Part of this research took place in the context of Watershare®, a worldwide network of water research organisations and utilities dedicated to applying global expertise to master local water challenges, funded by KWR Watercycle Research Institute. Part of this research was funded by POWER (Political and social awareness on water environmental challenges). The European Commission is acknowledged for funding POWER in H2020-Water under Grant Agreement No. 687809.

Acknowledgments: The authors would like to acknowledge the researchers that were involved in this research project for their contributions to the mapping and digitalization efforts, as well as for the in-depth discussions that have helped to interpret the empirical material and translate these damage estimates. The support of the city of Leicester is greatly acknowledged, and especially Janet Riley, for her practical support as well as the comments on the manuscript. Last but not least, for allowing Lombardijen as a case study, the support of the city of Rotterdam is gratefully acknowledged.

Conflicts of Interest: The authors declare no conflict of interest.

References

1. Kummu, M.; De Moel, H.; Ward, P.J.; Varis, O. How close do we live to water? A global analysis of population distance to freshwater bodies. *PLoS ONE* **2011**, *6*, e20578. [CrossRef] [PubMed]

2. CRED. Cred Crunch 50: Natural Disasters in 2017—Lower mortality, Higher Cost. Available online: http://www.emdat.be/publications (accessed on 20 October 2018).

3. European Environment Agency. *Flood Risks and Environmental Vulnerability: Exploring the Synergies between Floodplain Restoration, Water Policies and Thematic Policies*; European Environment Agency, Publications Office of the European Union: Luxembourg, Luxembourg, 2016; ISBN 978-92-9213-716-8.

4. World Economic Forum. *The Global Risks Report 2018*, 13th ed.; World Economic Forum: Geneva, Switzerland, 2018; ISBN 978-1-944835-15-6.

5. Olesen, L.; Lowe, R.; Arnbjerg-Nielsen, K. *Flood Damage Assessment: Literature Review and Recommended Procedure*; Cooperative Research Centre for Water Sensitive Cities: Melbourne, Australia, 2017; ISBN 978-1-921912-39-9.

6. Szewranski, S.; Chruscinski, J.; Kazak, J.; Swiader, M.; Tokarczyk-Dorociak, K.; Zmuda, R. Pluvial Flood Risk Assessment Tool (PFRA) for Rainwater Management and Adaptation to Climate Change in Newly Urbanised Areas. *Water* **2018**, *10*, 386. [CrossRef]

7. Zhou, Q.; Mikkelsen, P.S.; Halsnæs, K.; Arnbjerg-Nielsen, K. Framework for economic pluvial flood risk assessment considering climate change effects and adaptation benefits. *J. Hydrol.* **2012**, *414*, 539–549. [CrossRef]

8. City of Copenhagen. *Cloudburst Management Plan 2012*; The City of Copenhagen Technical and Environmental Administration: Copenhagen, Denmark, 2012.

9. European Environment Agency. *Climate Change: The Cost of Inaction and the Cost of Adaptation*; European Environment Agency, Publications Office of the European Union: Luxembourg, Luxembourg, 2007; ISBN 978-92-9167-974-4.

10. Hammond, M.J.; Chen, A.S.; Djordjević, S.; Butler, D.; Mark, O. Urban flood impact assessment: A state-of-the-art review. *Urban Water J.* **2015**, *12*, 14–29. [CrossRef]

11. De Moel, H.; Jongman, B.; Kreibich, H.; Merz, B.; Penning-Rowsell, E.; Ward, P.J. Flood risk assessments at different spatial scales. *Mitig. Adapt. Strat. Glob. Chang.* **2015**, *20*, 865–890. [CrossRef] [PubMed]

12. Green, C.; Viavattene, C.; Thompson, P. *Guidance for Assessing Flood Losses*; CONHAZ Report WP06, D6.1; Middlesex University: London, UK, 2011.

13. Apel, H.; Thieken, A.H.; Merz, B.; Blöschl, G. Flood risk assessment and associated uncertainty. *Nat. Hazards Earth Syst. Sci.* **2004**, *4*, 295–308. [CrossRef]

14. Merz, B.; Kreibich, H.; Schwarze, R.; Thieken, A. Review article "Assessment of economic flood damage". *Nat. Hazards Earth Syst. Sci.* **2010**, *10*, 1697–1724. [CrossRef]

15. Smith, D.I. Flood damage estimation—A review of urban stage-damage curves and loss functions. *Water* **1994**, *20*, 231–238.

16. Penning-Rowsell, E.C.; Priest, S.; Parker, D.; Morris, J.; Tunstall, S.; Viavattene, C.; Chatterton, J.; Owen, D. *Flood and Coastal Erosion Risk Management: A Manual for Economic Appraisal*; Routledge: London, UK, 2013; ISBN 978-0415815154.

17. Van Dijk, E.; van der Meulen, J.; Kluck, J.; Straatman, J.H.M. Comparing modelling techniques for analysing urban pluvial flooding. *Water Sci. Technol.* **2014**, *69*, 305. [CrossRef] [PubMed]

18. Maksimović, Č.; Prodanović, D.; Boonya-Aroonnet, S.; Leitão, J.P.; Djordjević, S.; Allitt, R. Overland flow and pathway analysis for modelling of urban pluvial flooding. *J. Hydraul. Res.* **2009**, *47*, 512–523. [CrossRef]

19. Leandro, J.; Chen, A.S.; Djordjević, S.; Savić, D.A. Comparison of 1D/1D and 1D/2D coupled (sewer/surface) hydraulic models for urban flood simulation. *J. Hydraul. Eng.* **2009**, *135*, 495–504. [CrossRef]

20. Freni, G.; La Loggia, G.; Notaro, V. Uncertainty in urban flood damage assessment due to urban drainage modelling and depth-damage curve estimation. *Water Sci. Technol.* **2010**, *61*, 2979–2993. [CrossRef]

21. Apel, H.; Aronica, G.T.; Kreibich, H.; Thieken, A.H. Flood risk analyses—How detailed do we need to be? *Nat. Hazards* **2009**, *49*, 79–98. [CrossRef]

22. Thieken, A.H.; Müller, M.; Kreibich, H.; Merz, B. Flood damage and influencing factors: New insights from the August 2002 flood in Germany. *Water Resour. Res.* **2005**, *41*. [CrossRef]

23. Thieken, A.H.; Olschewski, A.; Kreibich, H.; Kobsch, S.; Merz, B. Development and evaluation of FLEMOps—A new Flood Loss Estimation Model for the private sector. *Wit Trans. Ecol. Environ.* **2008**, *118*, 315–324.

24. Elmer, F.; Thieken, A.H.; Pech, I.; Kreibich, H. Influence of flood frequency on residential building losses. *Nat. Hazards Earth Syst. Sci.* **2010**, *10*, 2145–2159. [CrossRef]

25. Kreibich, H.; Seifert, I.; Merz, B.; Thieken, A.H. Development of FLEMOcs—A new model for the estimation of flood losses in the commercial sector. *Hydrol. Sci. J.* **2010**, *55*, 1302–1314. [CrossRef]

26. Kellens, W.; Vanneuville, W.; Verfaillie, E.; Meire, E.; Deckers, P.; De Maeyer, P. Flood risk management in Flanders: Past developments and future challenges. *Water Resour. Manag.* **2013**, *27*, 3585–3606. [CrossRef]

27. Sušnik, J.; Strehl, C.; Postmes, L.A.; Vamvakeridou-Lyroudia, L.S.; Savić, D.A.; Kapelan, Z.; Mälzer, H.J. Assessment of the effectiveness of a risk-reduction measure on pluvial flooding and economic loss in Eindhoven, The Netherlands. *Procedia Eng.* **2014**, *70*, 1619–1628. [CrossRef]

28. Spekkers, M.H.; Rozer, V.; Thieken, A.; ten Veldhuis, M.C.; Kreibich, H. A comparative survey of the impacts of extreme rainfall on two international case studies. *Nat. Hazards Earth Syst. Sci.* **2017**, *17*, 1337–1355. [CrossRef]

29. Spekkers, M.H.; Kok, M.; Clemens, F.H.L.R.; ten Veldhuis, J.A.E. Decision-tree analysis of factors influencing rainfall-related building structure and content damage. *Nat. Hazards Earth Syst. Sci.* **2014**, *14*, 2531–2547. [CrossRef]

30. Van Ootegem, L.; Verhofstadt, E.; Van Herck, K.; Creten, T. Multivariate pluvial flood damage models. *Environ. Impact Assess. Rev.* **2015**, *54*, 91–100. [CrossRef]

31. Rözer, V.; Müller, M.; Bubeck, P.; Kienzler, S.; Thieken, A.; Pech, I.; Schröter, K.; Buchholz, O.; Kreibich, H. Coping with pluvial floods by private households. *Water* **2016**, *8*, 304. [CrossRef]

32. Grahn, T.; Nyberg, L. Assessment of pluvial flood exposure and vulnerability of residential areas. *Int. J. Disaster Risk Reduct.* **2017**, *21*, 367–375. [CrossRef]

33. Spekkers, M.H.; Ten Veldhuis, J.A.E.; Clemens, F.H.L.R. Collecting data for quantitative research on pluvial flooding. In Proceedings of the 12th International Conference on Urban Drainage, Porto Alegre, Brazil, 11–15 September 2011; IWA—International Water Association: London, UK, 2011.

34. Merz, B.; Kreibich, H.; Lall, U. Multi-variate flood damage assessment: A tree-based data-mining approach. *Nat. Hazards Earth Syst. Sci.* **2013**, *13*, 53–64. [CrossRef]

35. Leicester City Council. *Surface Water Management Plan—Part 1 Report*; Leicester City Council: Leicester, UK, 2012.

36. Dai, L.; Wörner, R.; van Rijswick, H.F. Rainproof cities in the Netherlands: Approaches in Dutch water governance to climate-adaptive urban planning. *Int. J. Water Resour. Dev.* **2018**, *34*, 652–674. [CrossRef]

37. Hooimeijer, F.; van der Toorn Vrijthoff, W. *More Urban Water: Design and Management of Dutch Water Cities*; CRC Press: London, UK, 2014; ISBN 9781482288551.

38. Knowledge for Climate. *Exploring Opportunities for Green Adaptation in Rotterdam*; Knowledge for Climate: Utrecht, The Netherlands, 2012; ISBN 978-94-90070-62-5.

39. Nelen & Schuurmans. 3Di Docs: Introduction. Available online: https://docs.3di.lizard.net/en/stable/a_introduction.html (accessed on 1 February 2018).

40. Cranfield Soil and Agriculture Institute. Soilscapes. Available online: http://www.landis.org.uk/soilscapes/ (accessed on 8 May 2018).

41. UK Government Online Environmental Data Portal. Available online: https://data.gov.uk/ (accessed on 8 May 2018).

42. Schröter, K.; Kreibich, H.; Vogel, K.; Riggelsen, C.; Scherbaum, F.; Merz, B. How useful are complex flood damage models? *Water Resour. Res.* **2014**, *50*, 3378–3395. [CrossRef]

43. Fenn, T.; Clarke, C.; Burgess-Gamble, L.; Harding, E.; Ogunyoye, F.; Hick, E.; Dawks, S.; Morris, J.; Chatterton, J. The costs and impacts of the winter 2013/14 floods in England and Wales. In *E3S Web of Conferences*; EDP Sciences: Les Ulis, France, 2016; Volume 7, p. 05004.

44. Bradford, R.A.; O'Sullivan, J.J.; Van der Craats, I.M.; Krywkow, J.; Rotko, P.; Aaltonen, J.; Bonaiuto, M.; De Dominicis, S.; Waylen, K.; Schelfaut, K. Risk perception–issues for flood management in Europe. *Nat. Hazards Earth Syst. Sci.* **2012**, *12*, 2299–2309. [CrossRef]

45. Brouwer, R.; Van Ek, R. Integrated ecological, economic and social impact assessment of alternative flood control policies in the Netherlands. *Ecol. Econ.* **2004**, *50*, 1–21. [CrossRef]

46. Shreve, C.M.; Kelman, I. Does mitigation save? *Reviewing cost-benefit analyses of disaster risk reduction. Int. J. Disaster Risk Reduct.* **2014**, *10*, 213–235. [CrossRef]

47. Koks, E.E.; Bočkarjova, M.; de Moel, H.; Aerts, J.C. Integrated direct and indirect flood risk modeling: Development and sensitivity analysis. *Risk Anal.* **2015**, *35*, 882–900. [CrossRef]

48. Rufat, S.; Tate, E.; Burton, C.G.; Maroof, A.S. Social vulnerability to floods: Review of case studies and implications for measurement. *Int. J. Disaster Risk Reduct.* **2015**, *14*, 470–486. [CrossRef]

49. Van Ootegem, L.; Verhofstadt, E. Well-being, life satisfaction, and capabilities of flood disaster victims. *Environ. Impact Assess. Rev.* **2016**, *54*, 134–138. [CrossRef]

50. OECD. *Water and Cities: Ensuring Sustainable Futures*; OECD Studies on Water; OECD Publishing: Paris, France, 2015.

51. OECD. *Implementing the OECD Principles on Water Governance: Indicator Framework and Evolving Practices*; OECD Studies on Water; OECD Publishing: Paris, France, 2018.

52. Emanuelsson, M.A.E.; Mcintyre, N.; Hunt, C.F.; Mawle, R.; Kitson, J.; Voulvoulis, N. Flood risk assessment for infrastructure networks. *J. Flood Risk Manag.* **2014**, *7*, 31–41. [CrossRef]

53. Koks, E.E.; Thissen, M. The economic-wide consequences of natural hazards: An application of a European interregional inputoutput model. In Proceedings of the 22nd Input Output Conference, Lisboa, Portugal, 14–18 July 2014.

54. Koks, E.E.; Thissen, M. A multiregional impact assessment model for disaster analysis. *Econ. Syst. Res.* **2016**, *28*, 429–449. [CrossRef]

55. Driessen, P.; Hegger, D.; Kundzewicz, Z.; van Rijswick, H.; Crabbé, A.; Larrue, C.; Matczak, P.; Pettersson, M.; Priest, S.; Suykens, C.; Raadgever, G.; Wiering, M. Governance strategies for improving flood resilience in the face of climate change. *Water* **2018**, *10*, 1595. [CrossRef]

56. Driessen, P.P.; Hegger, D.L.; Bakker, M.H.; van Rijswick, H.F.; Kundzewicz, Z.W. Toward more resilient flood risk governance. *Ecol. Soc.* **2016**, *21*. [CrossRef]

57. Löwe, R.; Urich, C.; Domingo, N.S.; Mark, O.; Deletic, A.; Arnbjerg-Nielsen, K. Assessment of urban pluvial flood risk and efficiency of adaptation options through simulations–A new generation of urban planning tools. *J. Hydrol.* **2017**, *550*, 355–367. [CrossRef]

58. Ten Veldhuis, J.A.E. How the choice of flood damage metrics influences urban flood risk assessment. *J. Flood Risk Manag.* **2011**, *4*, 281–287. [CrossRef]

59. Gerl, T.; Kreibich, H.; Franco, G.; Marechal, D.; Schröter, K. A review of flood loss models as basis for harmonization and benchmarking. *PLoS ONE* **2016**, *11*, e0159791. [CrossRef]

60. Poussin, J.K.; Botzen, W.W.; Aerts, J.C. Effectiveness of flood damage mitigation measures: Empirical evidence from French flood disasters. *Glob. Environ. Chang.* **2015**, *31*, 74–84. [CrossRef]

water **MDPI**

Article

Governance Strategies for Improving Flood Resilience in the Face of Climate Change

Peter P. J. Driessen [1], Dries L. T. Hegger [1,*], Zbigniew W. Kundzewicz [2,3],
Helena F. M. W. van Rijswick [4], Ann Crabbé [5], Corinne Larrue [6], Piotr Matczak [7],
Maria Pettersson [8], Sally Priest [9], Cathy Suykens [4,10], Gerrit Thomas Raadgever [11] and
Mark Wiering [12]

[1] Copernicus Institute of Sustainable Development, Utrecht University, Princetonlaan 8a, 3584 CB Utrecht,
     The Netherlands; p.driessen@uu.nl
[2] Institute for Agricultural and Forest Environment, Polish Academy of Sciences, Bukowska 19,
     60-809 Poznań, Poland; kundzewicz@yahoo.com
[3] Potsdam Institute for Climate Impact Research, Telegrafenberg A 31, 14473 Potsdam, Germany
[4] Utrecht Centre for Water, Oceans and Sustainability Law, Utrecht University School of Law, Newtonlaan 201,
     3584 BH Utrecht, The Netherlands; h.vanrijswick@uu.nl (H.F.M.W.v.R.); c.b.r.suykens@uu.nl or
     cathy.suykens@kuleuven.be (C.S.)
[5] Research Group Environment & Society, Sociology Department, Faculty of Social Sciences, University of
     Antwerp, Sint-Jacobsstraat 2, S.M.380, 2000 Antwerp, Belgium; ann.crabbe@uantwerpen.be
[6] Paris School of Planning, Lab'Urba, Paris Est University, 14-20 Bd Newton, 77454 Marne La Vallée CEDEX 2,
     France; corinne.larrue@u-pec.fr
[7] Institute of Sociology, Adam Mickiewicz University, ul. Szamarzewskiego 89c, 60-568 Poznań, Poland;
     matczak@amu.edu.pl
[8] Law Unit, Luleå University of Technology, SE-971 87 Luleå, Sweden; maria.pettersson@ltu.se
[9] Flood Hazard Research Centre, Middlesex University, The Burroughs, Hendon, London NW4 4BT, UK;
     s.priest@mdx.ac.uk
[10] Institute for Environmental and Energy Law, KU Leuven, Oude Markt 13-bus 5500, 3000 Leuven, Belgium
[11] Sweco Netherlands B.V., De Holle Bilt 22, 3743 HM De Bilt, The Netherlands; tom.raadgever@sweco.nl
[12] Institute for Management Research, Radboud University Nijmegen, Heyendaalseweg 141,
     6525 AJ Nijmegen, The Netherlands; m.wiering@fm.ru.nl
*   Correspondence: d.l.t.hegger@uu.nl; Tel.: +31-(0)30-253-7829

Received: 8 October 2018; Accepted: 4 November 2018; Published: 7 November 2018

Abstract: Flooding is the most common of all natural disasters and accounts for large numbers of casualties and a high amount of economic damage worldwide. To be 'flood resilient', countries should have sufficient capacity to resist, the capacity to absorb and recover, and the capacity to transform and adapt. Based on international comparative research, we conclude that six key governance strategies will enhance 'flood resilience' and will secure the necessary capacities. These strategies pertain to: (i) the diversification of flood risk management approaches; (ii) the alignment of flood risk management approaches to overcome fragmentation; (iii) the involvement, cooperation, and alignment of both public and private actors in flood risk management; (iv) the presence of adequate formal rules that balance legal certainty and flexibility; (v) the assurance of sufficient financial and other types of resources; (vi) the adoption of normative principles that adequately deal with distributional effects. These governance strategies appear to be relevant across different physical and institutional contexts. The findings may also hold valuable lessons for the governance of climate adaptation more generally.

Keywords: flood risk management; flood resilience; governance strategies; climate change

1. Introduction

Flood risks are increasing in coastal areas and river basins due to worsening hazards related to urbanization and the effects of climate change [1–6]. Metropolitan regions such as Bangkok, Mumbai, Jakarta, Shanghai, Dhaka, London, Hamburg, and Rotterdam are extremely vulnerable to flooding. Extreme events in the Caribbean, the United States, and Bangladesh in 2017 and Japan, the US, and the Philippines in 2018 show how devastating the effects of floods can be. In response to these risks, an abundance of natural science and technical knowledge has been developed. In scientific literature, the concept of 'resilience' has been introduced as the ultimate and desirable outcome of flood risk management. A well-known definition of resilience is that of Walker et al. (p. 6): "the capacity of a system to absorb disturbances and reorganize while undergoing change, so as to still retain essentially the same function, structure, identity and feedback" [7]. A flood resilient system has in place the capacity to resist floods (e.g., by flood defenses), the capacity to absorb and recover from floods (e.g., by spatial planning, disaster management, insurance), and the capacity to adapt and transform (in order to moderate potential damages, to take advantage of opportunities, and to cope with the consequences of floods and respond in a flexible way) [8,9].

Flood risk management literature is well-grounded in risk assessments and future scenarios and is based on profound insights into the effectiveness of technical measures [10,11]. In contrast, the question of which governance actions should be taken to improve flood resilience has received lesser attention, despite the strong barriers that social and institutional factors often present to successful flood risk management [11–13].

To address this knowledge gap, a comparative analysis and evaluation of flood risk governance was conducted. This took place under the auspices of the STAR-FLOOD project (2012–2016) which was funded through the Seventh Framework program of the European Commission. This was the first large integrated European research project in the floods domain that focused almost entirely on governance and legal issues. The studied countries, namely Belgium, England, France, The Netherlands, Poland, and Sweden, are all European Union (EU) countries in which governments take a significant level of responsibility. This is partly due to the fact that there is EU legislation in place and is also partly for historical and political reasons. The flood risk management practices in these countries were studied, as detailed in the materials and methods section.

The current paper presents the six main governance strategies that the authors identified based on the research that was carried out. After providing the materials and methods in Section 2, Section 3 will subsequently present each of the strategies. We hold that these strategies are relevant across different contexts and have a wider meaning internationally.

2. Materials and Methods

In the framework of the STAR-FLOOD project, systematic and detailed analyses and evaluations of flood risk governance and associated legal frameworks in Belgium, England, France, The Netherlands, Poland, and Sweden were conducted between October 2013 and September 2015, which have been laid down, amongst others, in six country reports [14–19]. All partners conducted empirical analyses and evaluations of flood risk governance in their country, both at the national level and at the level of three specific urban areas that were used to illustrate and further explore developments at the national level. The research was based on a common analytical framework. We applied four different research methods in all countries: desk research (analysis of policy documents, legal texts, case law, literature); semi-structured interviews (70 in Belgium, 61 in England, 64 in France, 45 in The Netherlands, 54 in Poland, and 19 in Sweden), legal comparison, and at least one workshop with stakeholders in each country ([20,21], p. 14). Similarities and differences between the countries as well as lessons to be derived from them were discussed in plenary meetings and meetings in sub-groups in which the full consortium (>30 policy analysts and legal scholars) participated. These discussions have led to the first syntheses of the findings [20,21]. The author team of the current paper consists of at least one representative from each consortium partner, representing each country and the policy and legal disciplines. This team engaged in intensive discussions aimed at further condensing the findings and drawing out the key lessons. Each of the key lessons that are presented in the next sections are backed-up by solid empirical evidence [14–19]. We recognize that the paper's focus on the European context brings limitations in terms of potential generalization of findings to geographical regions other than Europe. Important issues that make these regions different from Europe are the general degree of risk awareness, linked to the population's alphabetization rate, policymakers' skills, and the rate of democracy. These and other issues are in many cases far removed from the European and north American situation. Moreover, a lack of financial resources and the poorness of infrastructure are important complicating factors to establishing efficient flood risk prevention or to build flood defense. Nevertheless, as we will show in Section 3 and conclude upon in Section 4, we believe that our paper's six governance strategies can serve as an important source of inspiration for improving flood risk governance more internationally, even though some of the lessons might need some degree of translation to other geographical contexts. To validate the lessons, the author team reassessed the empirical data and, in some cases, consulted researchers who were not directly involved in writing this paper. The steps above, in our view, ensure a certain robustness, however they also ensure transparency of the findings.

3. Results: Key Governance Strategies for Enhancing Flood Resilience

This section subsequently presents each of the six governance strategies that were identified based on the country comparison. The lessons pertain to diversification of flood risk management approaches (3.1), alignment of these approaches (3.2), involvement of private actors in flood risk governance (3.3), the role of formal rules (3.4), the assurance of sufficient resources (3.5), and the presence of adequate normative principles to deal with distributional effects (3.6).

3.1. Diversification of Flood Risk Management Approaches, Taking into Account the Physical and Institutional Context

It is increasingly argued that optimizing the capacity to resist floods, absorb and recover from floods, and to adapt and transform requires a diversified portfolio of flood risk management approaches [9,22,23]. We identify five modes of flood risk management: flood prevention; flood defense; flood mitigation; flood preparation and flood recovery (see Table 1).

Table 1. Five modes of flood risk management, adapted from [21,23]. Note that the main distinguishing feature in this categorization is the focus of each mode on reducing the probability of floods, reducing the consequences of floods, or promoting recovery. This is an entry point that differs to the classical one by Thampapillai and Musgrave who use the term flood damage mitigation to refer to all strategies and further distinguish between structural and non-structural measures. In our categorization, structural measures mainly relate to the flood defense strategy [24].

Mode	Characteristics
1. Flood risk prevention	
	Prevention measures aim to **decrease the exposure of people/property** etc. via methods that prohibit or discourage development in areas that are at risk of flooding (e.g., spatial planning, re-allotment policy, expropriation policy etc.). The main focus of this approach is on "keeping people away from water" by only building outside flood-prone areas. This is a pro-active strategy that focuses both on probability reduction (if people do not live in flood-prone areas, then it is unlikely that they will get flooded) and the consequences of flooding (an area may still be flooded, however in the absence of large-scale urban developments, the consequences are relatively small).

Table 1. *Cont.*

Mode	Characteristics
2. Flood defense	
	Flood defense measures aim to **decrease the probability of flooding** areas through infrastructural works, such as dikes, dams, embankments and weirs, mostly referred to as "flood defense" or "structural measures" through options that increase the capacity of **existing** channels for water conveyance (natural or man-made) to increase the space for water (laterally or vertically) or the creation of **new** spaces for water retention **outside of the area to be defended**. In brief, the focus is on "keeping water away from people".
3. Flood risk mitigation	
	Flood risk mitigation focuses on **decreasing the consequences of floods through measures inside the vulnerable area**. Consequences can be moderated by a smart design of the flood-prone area. Measures include smart spatial ordering inside urban areas, water retention within the protected area, or (regulations for) flood-proof building. Flood risk mitigation thus includes all measures to flood-proof the built environment as well as measures to retain or store water.
4. Flood preparation	
	Consequences of floods can also be alleviated by **preparing for a flood event**. Measures include developing flood forecasting and warning systems, as well as preparing disaster management and evacuation plans.
5. Flood recovery	
	This approach **facilitates an effective recovery after a flood event**. Measures include reconstruction or rebuilding plans (that may also provide a window of opportunity for making the new buildings more flood-proof or relocate these to safer areas) as well as compensation or insurance systems.

Each of these modes differs in its focus on the reduction of flood probability, flood exposure, and flood consequences, however it is assumed that together these modes will effectively decrease the flood risk [9,23]. A diversified portfolio of effectively functioning flood risk management approaches is needed to be able to address the three resilience capacities. The flood defense approach contributes to the capacity to resist, while the other four strategies contribute to the capacity to absorb and recover and the capacity to adapt and transform [9,23].

Countries have their own strengths and weaknesses in terms of the implemented portfolio of flood risk management approaches. This has repercussions for the way in which diversification should be pursued. For instance, Poland and especially The Netherlands show a dominant focus on flood defense. These countries are now facing the challenge of broadening their portfolio to better address the potential consequences of flooding. Further development of the complementary strategies to increase flood resilience has proven to be challenging. The Netherlands have gained several centuries of experience fighting against the water. The country's track record regarding flood defense is excellent. Moreover, the flood-defense approach is highly institutionalized [16]. However, over the years, new ideas have started to have more of an impact. From the 1990s onwards, the dominant 'keeping the water out' paradigm has been complemented with Room for the River approaches that can be seen as a broadening *within* the flood defense strategy. As a next step, in more recent years, the further development of flood mitigation and recovery—and to a lesser extent flood risk prevention—have gained more prominence in policy discourses, while experiments with the new approaches have started [16,25]. In the Polish situation, the dominant tendency was to sort out flood defense first, despite the fact that Polish water management agencies suffer from a severe lack of resources [18]. Amongst other events, the floods of 1997 and 2010 gave an impetus to the development of improved flood forecasting and warning systems. However, the further development of the other strategies is proceeding very gradually. Both in The Netherlands and Poland, there is a relative dominance of flood defense which contributes to path dependency. To some extent, the flood defense strategy is seen as 'inescapable' because developments in flood prone areas—in The Netherlands, a low-lying country, more than half of its surface—have already taken place and are unlikely to be reversed in the short to medium term.

At the other extreme, flood risk management in England has been diversified for several decades and England's portfolio of management approaches has been diagnosed as balanced, as in: showing a high degree of diversity, although the effectiveness of specific approaches was found to need improvement [9,14]. This relatively positive evaluation of England's portfolio might come across as counter-intuitive, given that England has experienced several floods in recent years, most notably in 2000, 2007, and 2014. However, a strength of the English flood risk governance arrangement is that it has a range of approaches at its disposal, each of them tailored to specific circumstances in terms of the physical characteristics of an area and the potential for damage and economic loss [14].

Belgium, France, and Sweden are positioned on a spectrum between these two extremes. Both in Belgium and France, there used to be a focus on flood defense, however to a lesser extent than in The Netherlands and Poland. Belgium's prevention and mitigation strategies overtook those of The Netherlands when it started to implement designated policies that aimed to put restrictions on urban development in flood prone areas (including signal areas and the water test) [19]. At the same time, in France, a discursive shift can be observed towards an increased emphasis on flood prevention [17]. Swedish flood risk governance is characterized by a very localized approach with a dominant role for the municipalities as well as the civil contingency agencies. While there are important local variations, there is a strong reliance on flood preparation [15].

Overall, the findings show that a diversification of modes of flood risk management contributes to flood resilience in that it helps to address a broader range of resilience capacities: that to resist, to absorb and recover, and to adapt and transform [9]. However, the extent to which approaches can be diversified depends on physical and institutional features, as well as general levels of development, flood risk management legacies, culture, and politics.

3.2. Alignment of Flood Risk Management Approaches to Overcome Fragmentation

To diversify flood risk management approaches, different sectors need to be aligned. Besides water management, spatial planning and emergency management, amongst other sectors, are also necessary to cover the range of approaches from prevention to recovery by providing adequate policy instruments. Hence, enhancing connectivity between policy sectors and administrative levels is needed to overcome the fragmentation that a diversification of flood risk management approaches may create.

Such connectivity requires the employment of bridging mechanisms or processes that broadly entail information exchange, coordination of policies, and cooperation mechanisms [26]. These mechanisms can be anchored through laws and policy instruments and can be promoted through finances and by putting knowledge infrastructures in place. Bridging mechanisms may help to establish links between different flood risk management approaches, thus ensuring that these approaches reinforce each other and, hence, help to strengthen all resilience capacities. Two examples include:

- Recovery schemes to compensate victims of a flood event, e.g., through private insurance schemes (possibly in cooperation with the nation state) or the activation of a public fund, can enhance the success of the risk prevention and mitigation strategies and vice versa [26]. Compensation and insurance schemes may incentivize or dis-incentivize citizens and public authorities to undertake flood risk reduction and promote the implementation of property-level measures through risk differentiation, structured premium increases or premium benefits, or resilient reinstatement. For instance, the French system is lauded for its comprehensive recovery scheme through the CAT-NAT ('Natural Disaster' (CATastrophe NATurelle in French) fund, however it was found that this scheme forms a dis-incentive for taking preventative measures [17,27], hampering ongoing efforts to strengthen flood prevention approaches by limiting urban development in flood prone areas. On the other hand, in England, experience has been gained with risk-based insurance premiums, which in some cases have been complemented with premium reductions in case property owners take mitigating measures. In this situation, recovery schemes serve as an incentive for strengthening flood risk prevention and mitigation [14]. Hence, policy instruments that are used in one strategy in a specific country cannot be applied in another jurisdiction without looking at the overall picture of the several strategies and the policy instruments that are related to them [28].

- By analogy, effective flood emergency management requires tailor-made spatial planning policies and probabilistic decision-making processes to improve the effectiveness of evacuation [26]. For instance, the effective evacuation of citizens out of flood-prone areas puts requirements on spatial preconditions such as the presence of elevated evacuation routes and the realization of shelters. In some cases, existing spatial configurations make horizontal evacuation impossible and necessitate the realization of plans for vertical evaluation. This is the case for several low-lying areas in The Netherlands [29].

Besides establishing links between different sectors, bridging mechanisms can also help to link different governance levels. Top-down and bottom-up policy processes should be combined and coordinated. At a strategic level, acceptable levels of risk and the division of responsibilities in dealing with these risks should be discussed [20]. Countries show a different track record in this respect. In England, we see that the Environment Agency is playing a key role in facilitating discussions at the strategic level [14]. The Netherlands have established a national 'Delta program' which stimulates the exchange of ideas and experiences as well as strategic planning with the inclusion of governmental actors at the national level (Rijkswaterstaat, the Dutch Office for Public Works), the Regional Water Authorities, municipalities, and knowledge institutes, among other actors [16]. Such discussions are also held in Belgium, although coordination takes place mostly at the level of the three regions, being the Flanders, Walloon, and Brussels Capital regions, while inter-regional exchange is still in an embryonic stage [19]. In Poland, the new administration body: the National Water Management Holding Polish Waters, was established in 2018 with the intention to consolidate water management, including flood risk management. In Sweden, a lack of strategic discussions at the national level was explicitly acknowledged as an important weakness of the national flood risk governance arrangement [15].

On the other hand, more room should be created for bottom-up activities: local and regional stakeholders (preferably at the river basin level) draft flood risk reduction plans together that are based on their objectives and are hereby supported with funding and expertise from the higher governments. Moreover, regions and municipalities should coordinate upstream and downstream measures. Multi-level governance is necessary because diversification of flood risk management approaches makes it very unlikely that a single governmental actor can oversee and implement complete portfolios of approaches. Multi-level governance is also necessary because often experiences at the local level, derived through pilots and experiments as in The Netherlands [16], hold valuable lessons that deserve upscaling. This requires some degree of coordination of experiments as well as translation and dissemination of their findings.

A key example of a bridging mechanism is the Flemish and Dutch water assessment, which intends to bridge water management and spatial planning [16,19]. Through the assessment, the impact of spatial planning actions on the water system, be it permits or plans and programs, are systematically scrutinized. Hence, flood risk managers can ensure that the strategies of flood risk prevention and flood mitigation are routinely considered in spatial planning processes, bringing together actors at the local level. Another prominent example is the designation of properties as at risk properties. For instance, England knows the duty of property owners to inform prospective buyers of flood risks. Belgium and France have procedures to designate flood zones, e.g., signal areas in Belgium and areas falling under the auspices of the PPRI (the national Plan de Prevention de Risk d'Inondation) in France. The designation of high-risk zones leads to additional requirements and restrictions in the Belgium case, while it might lead to the prohibition of urban development or at least severe restrictions to it in the case of France [17,19].

3.3. Increased Involvement of Private Actors, Including Businesses, NGOs, and Citizens in Addition to Strong Public Involvement

In the European Union context, governmental actors are playing a strong role in establishing and maintaining vital infrastructure. These actors alone, however, cannot implement a comprehensive portfolio of flood risk management approaches. Hence, the involvement of private parties in flood risk governance, including businesses, Non Governmental Organisations (NGOs), and citizens, is necessary to enhance flood resilience. First, from a substantive viewpoint, the input of a diverse set of resources and capacities that are embodied in the expertise and knowledge of stakeholders is needed to implement a diverse set of flood risk management approaches as governmental institutions do not have all the necessary resources to cope with complex flood risks [9]. Second, from a normative perspective, participation in decision-making is considered important because participation enhances the legitimacy of measures that are taken to decrease flood risks and contributes to representative deliberation, procedural and distributive justice, and socio-political acceptability [30].

Private sector involvement is needed to strengthen all three capacities. Measures that enhance the capacity to absorb and recover cannot be taken by governmental actors alone, while enhancing the capacity to adapt and transform inherently necessitates the involvement of a broad set of societal actors. Think of house-owners who need to make their houses flood-proof and insurance companies that need to activate flood prevention more.

Companies can contribute to flood risk governance through public-private cooperation. A good practice in this respect is the English Partnership Funding Scheme which obliges local authorities, the private sector, or civil society to match national government funding. Thus, actors with a financial stake in flood risk management that have not been previously actively involved can enter into the governance strategies at the project level [14]. Countries which have private flood insurance mechanisms require cooperation between public and private actors, since public actors are often the reinsurers of the last resort.

Citizens in their capacity of residents can take actions in and around their homes. First, new houses should be built to be flood-proof. The implementation of property level measures such as the (dry or wet) flood-proofing of houses or the creation of capacity for rainwater retention (through green roofs and other infiltration capacities) can complement actions that are also taken by governmental actors. Citizen involvement in these cases may be necessary for substantive reasons (e.g., because measures are taken on private properties), however also for normative reasons (e.g., because citizens should have a say in issues in which they have a stake or out of the 'beneficiary pays' principle, i.e., the belief that those benefitting should implement measures [31]). That is why co-production between citizens and governmental actors is increasingly called for.

In a European context, however, we see that public authorities have traditionally taken the lead in flood risk management and are struggling with how to best engage citizens. There is a tendency for citizens to adopt a paternalistic "the-government-should-take-care-of-me" attitude, while communicating flood risks to citizens and companies is made difficult by the highly technical language of flood managers, which is poorly understood by the public or is poorly communicated [32]. A prime example is the involvement of citizens in fluvial flood risks in The Netherlands. While flood consequences in The Netherlands are relatively large, there is a serious lack of flood awareness and corresponding action being possible or taken by residents [16,33] that is only changing gradually. However, the lack of action being taken can to some extent be explained out of a lack of action perspectives, e.g., in areas in which possible inundation depths are high and flooding occurs fast. In their efforts to co-produce with citizens, governmental actors are focusing mainly on pluvial flooding [31]. Engagement of residents in flood risk governance seems to be highest in England [14] and Poland where Voluntary Fire Brigades have an important role in flood preparation [18]. In England, this seems to be directly related to the fact that the country has already diversified its portfolio of flood risk management strategies, combined with dedicated measures that lead to flood consciousness such as awareness campaigns by the Environment Agency (Bristol, UK). The frequent floods of the past years have undoubtedly contributed to this level of flood awareness. In Poland, the relatively high level of flood awareness seems to be more directly attributable to the occurrence of devastating floods, most notably that of 1997 [18]. In terms of involving residents in co-producing flood risk management, Belgium, France, and Sweden seem to rank between these extremes.

3.4. Adequate Formal Rules that Balance Legal Certainty and Flexibility

Flood risk governance requires a delicate balancing act between legal certainty—that is the assurance that one's rights are clear and will be respected—and flexibility which might be necessary for adaptive governance [34,35]. To establish the different resilience capacities, both are needed to some extent.

Legal certainty means that decisions are based in law and are possible to foresee. There are differences in the degree to which individuals in the different countries are sure of their rights and to what extent actions/decisions in respect of these rights can be predicted. Legal certainty is required to achieve clarity about responsibilities amongst public authorities and communities, for instance about what level of flood protection can be ensured or aimed for and who is responsible for this. It furthermore provides clarity on the scope of formal powers based on the several relevant policy fields, the legal instruments that are available, and the legal basis for taxes, insurance schemes, and compensation to be used for flood risk management. Hence, legal certainty provides much needed structure and guidance, which is a necessary precondition for achieving sufficient capacity to resist as well as to absorb and recover. Land-use planning can, for instance, be used to keep people away from the water, i.e., steer development away from flood prone zones. However, the legal certainty provided by land-use plans differs between the various countries. In Sweden, only detailed plans (zoning) are legally binding for individuals and there are no mechanisms to ensure their implementation, while in The Netherlands, land-use or spatial plans are legally binding and provide certainty regarding the possibilities to use land and property for diverse functions. It is, for example, not possible to change the function of agricultural land towards a flood retention area without a formal change of the designated function that the land use may have in a spatial plan and often also involves related compensation for the land owner. The same goes for Room for the River projects and dike reinforcement projects, as rivers and dikes are being designated as such in a spatial plan. This protects citizens in that they know when and where they can expect changes in the possibilities to use their land. Regulating restrictions in land use—especially of private property—need explicit changes in land use plans (by means of a formal change of the plan itself or a formal permission to deviate from the plan) and related regulations in order to provide private actors legal certainty and to enable them to take investment decisions based on adequate information of their rights and responsibilities.

On the other hand, to achieve a high capacity to adapt and transform, formal rules should also provide enough flexibility to enable tailored local approaches and adaptive pathways whereby a change in course is possible if dynamics require it, e.g., due to urbanization or climate change patterns. At the same time, we face the challenge of not sacrificing legal certainty, including the possibility to hold public authorities accountable [34]. Flexibility and the possibility to adapt because of changing circumstances is generally provided by multiple year planning cycles, for example, the six year planning cycle of water plans in The Netherlands. A new planning period should take new circumstances into account and should mention the proposed measures for the coming years, thus providing insight into new policies and possible restrictions for citizens. As the preparation of new plans takes place with participation of the public, more adaptive and still legitimate policies can be introduced. Long-term strategic planning combined with short-term operational planning provides citizens with clarity on what can be expected in the near future. The issue of legal certainty versus flexibility is ubiquitous and is typically "solved" on a case-to-case basis. Some laws and legal rules are said to be "better" than others in this respect, for example, performance based rules, since these imply both requirements and flexibility in implementation. The possibility to appeal and revoke (plan) decisions is, however, an important feature of a flexible system that still adheres to the principle of (procedural) legal certainty. In Sweden, this is legally possible, however it is unusual due to the municipal planning monopoly.

In Europe, the establishment of formal rules is taking place both at the European Union (EU) and at the national level. At the former level, in 2007, the Directive 2007/60/EC on the assessment and management of flood risks (Floods Directive or FD) was introduced. This is widely seen as an important step towards establishing a minimum common flood management framework. The FD obliges EU Member States to establish flood risk assessments; draw flood hazard and flood risk maps and make flood risk management plans that should be revised every six years. The main consideration that drove the establishment of a European approach was to ensure international solidarity amongst other rules by mandating that measures do not significantly increase risk to societies elsewhere [36]. The FD has been effective in establishing greater discussion and consideration of flood issues, particularly in those countries where flood management was underdeveloped. There are, furthermore, instances whereby the FD has stimulated improvements in regional, national, and transboundary flood risk management and cooperation [36–38].

The FD takes a procedural approach and leaves the Member States a large amount of policy discretion about how they want to design flood risk management. It takes flexibility and subsidiarity as the main points of departure. However, both the nature and implementation of the FD lead to concerns about its ability to achieve its aims. An important lesson is that its lack of substantive requirements and strong reliance on procedural obligations might undermine its ability to lead to increased flood resilience [34–36]. The FD may serve as a good practice to develop a comprehensive overview of flood risks, which enables decision-makers and the broad public to be aware of the risk in their living environment. The cyclic planning and programming approach with plans and programs that must be revised every six years makes it possible to adapt and learn. The FD may help to decrease external effects, despite the differing political, legal, and policy contexts that are present. It is a policy choice whether the legal framework for flood risk governance should embody only a procedural or also a substantive approach. The advantage of a purely procedural approach is that it respects the different geographical conditions and leaves policy discretion to the individual states in terms of how much and in what ways they want to improve flood risk governance. However, clear definitions are necessary to ensure the coherent implementation of flood risk governance frameworks, especially in the case of transboundary river basins [39,40]. The advantage of a combined procedural and substantive approach is that there is also more clarity about the level of protection that will be offered and about who is responsible for the design, development, and implementation, adding to legal certainty. As of now, this is not the path taken in the approach that is laid down in the Floods Directive, which through its procedural approach, can be seen as non-committal. The lack of substantive requirements implies that individuals cannot rely on the Directive directly, however they are referred to the discretion of the national legislation and the way that the national courts deal with this [36]. The differences between the several Member States that are part of the research are significant. The Netherlands has legally binding safety standards that provide clarity on the scope of the governmental responsibilities for flood protection. The other countries do not have these standards and citizens should determine the level of flood protection and the scope of their private responsibilities in a more indirect way by interpreting policy documents.

3.5. The Assurance of Sufficient Resources

Diversification of flood risk management strategies requires resources, including finances, knowledge, and capacities, that are possibly formalized in legislation, for example, when flood protection is paid from general or designated taxes [4–44]. Flood resilience will be stimulated when sufficient and satisfactory resources are mobilized and dedicated to multiple flood risk management approaches. Resources channeled into one approach should not lead to underinvestment in other approaches. This will to some extent entail a reallocation of resources [20]. In this regard, we see two important developments that, from a resilience perspective, provide opportunities and risks: private investments and decentralization. These may help to broaden the resource base of flood risk management.

In Europe, typically the nation state and general tax income play the principal role in flood risk management. The nation state is also responsible for the education of experts and providing legal frameworks for management. In many countries, financial resources for flood risk management have been under pressure, in particular after the 2008 crisis, due to central budget constraints. Involvement of local governments, businesses (e.g., insurance), communities, and household owners is, however, increasing. At the same time, an increasing appeal is being made to local governments as well as private actors to take up more responsibility in flood risk management [9]. From a resilience perspective, this has, in principle, the advantage that the resource base is broadened and that locally tailored approaches are enabled, strengthening the capacity to adapt and transform and the capacity to absorb and recover. However, decentralization as well as reliance on private investments is not without risks. A lack of funds and expertise at the level of local communities can limit the development of approaches [42]. It may also have distributional effects and justice implications in that it is often economically marginalized groups and territories that suffer from floods. There may also be a financial incentive for local governments to permit spatial developments in at-risk areas [9,42].

In The Netherlands, for example, protection against floods is strongly based on the solidarity principle, as every person in a certain area has the same level of protection. Resources are provided by taxes from the regional water authorities combined with funding by general taxes. Investments that have to be made for dike reinforcements or flood retention areas include the costs to compensate those who suffer more than others because the measures will make use of their private property [28].

In terms of resources, in The Netherlands, the division of resources mirrors the relative importance that is attached to flood defense, with many financial resources going to large dike strengthening projects for which a special fund has been designated (https://www.infomil.nl/onderwerpen/lucht-water/handboek-water/wetgeving/waterwet/financiele/deltafonds/), while maintenance is paid from tax revenues of regional water authorities. This notwithstanding, as part of the Delta program, pilots at the local level are also taking place, leading to the additional development of knowledge and expertise and the exchange of experiences. England has had a long-standing approach to funding flood risk management from central government funds through a dedicated budget. This commitment has been a stabilizing factor and consistent investment has led to the incremental improvement of flood risk management (FRM) which was further strengthened in 2015 through the adoption of a 6-year investment program [45]. Funds are allocated to areas using economic prioritization and it recognizes the management principle that resources are limited and not all risks can be avoided. From 2012, a new Partnership Funding approach was adopted [46] which aims to raise a proportion of the funds locally, thereby placing a higher burden of those benefitting from FRM (and questions remain about fairness and the success of delivery) [14], however increasing the overall investment for FRM and permitting more management measures to be funded.

3.6. Appropriate Normative Principles for Dealing with Distributional Effects

The governance of risk bestows benefits and imposes burdens in several ways [42] as virtually all measures and policies have distributional effects [28,47–49]. The development towards a diversification of flood risk management approaches will probably be no exception. A consequence might for instance be an increased financial burden for at-risk households [48,49] or for agricultural businesses that are confronted with restrictions in crop use because of Room for the River projects based on new restrictions in land-use plans or the designation and use of flood retention areas. Another consequence might be a transfer of safety/money between upstream and downstream communities and 'privatization' of responsibilities to non-stake actors and stakeholders [50,51]. While in principle inequity in the distribution of the benefits and burdens can be dealt with through targeted actions [28], in reality it might prove difficult to identify vulnerable groups and to design instruments with the desired effect. Because of this, resilient flood risk governance requires mechanisms to ensure social equity and to address 'unfair' distributional effects. The Netherlands has a strong history in dealing with distributional effects, while, for example, France and Belgium show different approaches depending on whether measures are based on spatial planning or water management [49]. This might lead to a choice for a flood risk management measure that increases the unequal distribution of burdens or, on the contrary, a choice that best suits fairness in flood risk management. Examples of different mechanisms with different underlying normative principles are: (i) the partnership funding in England which encourages local actors and stakeholders to engage in FRM and facilitates flood defense schemes; (ii) the constitutional right to flood protection in The Netherlands which is enacted through legal safety standards based on risk type [30]; (iii) compensation regimes/schemes which increase equity, and mechanisms for compensating losses. These are present in most of the researched countries [52].

To put such mechanisms in place, inclusive societal debates are needed that lead to the establishment of adequate normative principles that are seen as fair and legitimate [9]. Depending on the (national) context, different principles may be preferred. Whereas from a solidarity perspective it would be considered fair that people in low risk areas also contribute to flood protection measures, if fairness is interpreted as 'beneficiary pays', the situation will be perceived as fair if contributions are based on risks or on benefits [53]. Another principle that could be applied is to give priority to measures that protect the most vulnerable people [54] in those countries that do not take solidarity and protection for all as the basis for their flood risk policies. It seems that approaches presuppose different normative systems. Flood defenses which have a public good character seem to be based on the solidarity principle. At the same time, a market-based system can still be based on the solidarity principle [55], such as in France. To increase the acceptability of flood risk management measures, debates on what is seen as desirable and how and to what extent different strategies may help to realize these normative ideals should be held in the open rather than be restricted to a selective number of public officials and experts [9,53]. Diversification of flood risk management strategies might serve as a mechanism to open up such debates, since it implies that flood managers have to make a shift from highly technical jargon towards more lay language.

4. Discussion and Conclusions

The current paper has addressed an important knowledge gap in that governance and legal knowledge on flood risk management was relatively limited and fragmented until recently. While the number of studies in this field has increased over the years, the step towards the identification of overall governance strategies that can be deemed applicable in multiple contexts had not yet been taken. The six governance strategies laid down in this paper have addressed this gap. Besides that, we further elaborated a visualized categorization of five modes of flood risk management (flood risk prevention, flood defence, flood mitigation, flood preparation, and flood recovery). The distinction between these five modes might form a precious source for teaching and for the improvement of the awareness of policy makers and possibly also citizens in order to enhance their involvement in flood risk governance.

In a fully flood-resilient society, the three resilience capacities (capacity to resist, to absorb and recover, and to adapt and transform) are sufficiently present and well-aligned to increase effectiveness. In practice, however, countries make trade-offs and focus flood risk governance efforts on specific capacities, leading to differences in countries' strengths and weaknesses regarding flood resilience. For instance, England has all three capacities developed to a large extent, whereas The Netherlands is focusing much on the capacity to resist. Trade-offs are made based on normative starting points, physical circumstances (e.g., rainfall patterns, altitude and gradient of physical terrains, the degree of complexity of river systems), institutional and social contexts, and available resources [9]. However, well-elaborated choices are essential, taking diversification and alignment of flood risk management approaches as an important starting point.

Flood resilience also means the efficient use of resources (economic, human, technological), maximizing desired outputs and minimizing required inputs. (Societal) Cost Benefit Analysis [56] is considered to be a helpful tool to improve resource efficiency by selecting cost-efficient measures. Recent research also points to the usability of other financial tools that use Life Cycle Cost and Return on Investment as important decision variables [57].

Flood resilience can only be achieved when the legitimacy of governance strategies is not questioned: the input, process, and output should be legally legitimate and societally acceptable. Access to information and transparency, procedural justice, and accountability is well organized in the analyzed European countries, however there is room for improvement on the criteria of social equity, public participation, and acceptability by all actors that are involved [21]. These issues appear to be important in all countries that face huge flood risks.

The lessons derived from the STAR-FLOOD project have clearly shown that there is no concrete 'one size fits all' solution for achieving flood resilience. In that sense, there clearly is a limit to decentralization policies. Governance strategies should be specified and tailored to physical circumstances as well as institutional and social contexts. Nevertheless, the six governance strategies proposed in the current paper are a fruitful starting point for achieving such context-specific flood resilience. We maintain that these strategies have a wider meaning internationally. For instance, Tullos' [58] findings on the weaknesses of flood risk management in the USA strikingly apply to the European Union (EU), despite all the political, economic, and social differences between the two entities. Her call for reducing reliance on the structural measures resonates with the main finding of the STARFLOOD project that flood risk management strategies have to be diversified. The strategies proposed in this paper could also be relevant beyond Europe [59], for example, in several metropoles such as Bangkok, Mumbai, Jakarta, and Mexico City. These metropoles are in the front line of flood risk due to their overwhelming rate of urbanization and their lack of resources as compared to European metropolises.

In terms of the next steps, the six governance strategies point to important research themes that deserve to be considered further. Additional detailed case studies as well as studies in geographical regions other than Europe are necessary. These may help to further assess the relationship between governance and legal approaches and mechanisms and their outcomes in terms of flood risk reduction (the 'what works?' question). Issues to be taken up in this strand of research include the social and institutional mechanisms that facilitate or hamper the diversification of flood risk management strategies; the precise functioning and effectiveness of certain bridging mechanisms in different contexts; the mechanisms that may facilitate citizen co-production; and the types of resources that are present in different circumstances. At the same time, normative evaluations of flood risk governance are in order (the 'is it fair?' question), including the question of whether public debate on these questions is taking place. Assessments of these debates (e.g., through stakeholder analyses or argumentative discourse analyses) need to be related to institutionalization, e.g., in legal frameworks. We invite other scholars to join us in this exciting endeavor.

Author Contributions: The authors made the following contributions to the paper. Conceptualisation: P.P.J.D. and D.L.T.H. Methodology: all. Validation: all. Formal analysis: all. Investigation: all. Writing—original draft preparation: P.P.J.D. and D.L.T.H. Writing—review and editing: all. Visualization: G.T.R. Supervision: P.P.J.D. Funding acquisition: P.P.J.D., D.L.T.H., Z.W.K., H.F.M.W.v.R., A.C., C.L., M.P., G.T.R. and M.W.

Funding: This research was funded by the EUROPEAN COMMISSION within its Seventh Framework Programme through the grant to the budget of the Integrated Project STAR-FLOOD, Contract 308364. The APC was covered by the institution of the lead author.

Acknowledgments: The authors would like to acknowledge the researchers that were involved in the STAR-FLOOD project for the in-depth discussions that have helped to interpret the empirical material and translate these into six key governance strategies. Earlier drafts of this paper have been presented at the FLOODRISK 2016 conference in Lyon, the European Climate Change Adaptation (ECCA) conference in Glasgow (2017), and the Lund Conference on Earth System Governance (2017). The audiences of these conferences are acknowledged for their feedback. We finally thank Tina Newstead for her language corrections in earlier drafts of this paper.

Conflicts of Interest: The authors declare no conflict of interest. The funders had no role in the design of the study; in the collection, analyses, or interpretation of the data; in the writing of the manuscript, or in the decision to publish the results.

References

1. Kundzewicz, Z.W.; Krysanova, V.; Dankers, R.; Hirabayashi, Y.; Kanae, S.; Hattermann, F.F.; Huang, S.; Milly, P.C.D.; Stoffel, M.; Driessen, P.P.J.; et al. Differences in flood hazard projections in Europe–their causes and consequences for decision making. *Hydrol. Sci. J.* **2017**, *62*, 1–14. [CrossRef]

2. Kundzewicz, Z.W.; Kanae, S.; Seneviratne, S.I.; Handmer, J.; Nicholls, N.; Peduzzi, P.; Mechler, R.; Bouwer, L.M.; Arnell, N.; Mach, K.; et al. Flood risk and climate change: Global and regional perspectives [Le risque d'inondation et les perspectives de changement climatique mondial et régional]. *Hydrol. Sci. J.* **2014**, *59*, 1–28. [CrossRef]

3. Winsemius, H.C.; Aerts, J.C.J.H.; Van Beek, L.P.H.; Bierkens, M.F.P.; Bouwman, A.; Jongman, B.; Kwadijk, J.C.J.; Ligtvoet, W.; Lucas, P.L.; Van Vuuren, D.P.; et al. Global drivers of future river flood risk. *Nat. Clim. Chang.* **2016**, *6*, 381–385. [CrossRef]

4. Hansen, J.; Sato, M.; Ruedy, R. Perception of climate change. *Proc. Natl. Acad. Sci. USA* **2012**, *109*, E2415–E2423. [CrossRef] [PubMed]

5. Arthur, R.; Boulton, C.A.; Shotton, H.; Williams, H.T.P. Social sensing of floods in the UK. *PLoS ONE* **2018**, *13*, e0189327. [CrossRef] [PubMed]

6. Messeri, A.; Morabito, M.; Messeri, G.; Brandani, G.; Petralli, M.; Natali, F.; Grifoni, D.; Crisci, A.; Gensini, G.; Orlandini, S. Weather-Related Flood and Landslide Damage: A Risk Index for Italian Regions. *PLoS ONE* **2015**, *10*, e0144468. [CrossRef] [PubMed]

7. Walker, B.; Holling, C.S.; Carpenter, S.R.; Kinzig, A. Resilience, adaptability and transformability in social-ecological systems. *Ecol. Soc.* **2004**, *9*, 5. [CrossRef]

8. Folke, C.; Carpenter, S.R.; Walker, B.; Scheffer, M.; Chapin, T.; Rockström, J. Resilience thinking: Integrating resilience, adaptability and transformability. *Ecol. Soc.* **2010**, *15*, 4. [CrossRef]

9. Hegger, D.L.T.; Driessen, P.P.J.; Wiering, M.; Van Rijswick, H.F.M.W.; Kundzewicz, Z.W.; Matczak, P.; Crabbé, A.; Raadgever, G.T.; Bakker, M.H.N.; Priest, S.J.; et al. Toward more flood resilience: Is a diversification of flood risk management strategies the way forward? *Ecol. Soc.* **2016**, *21*, 4. [CrossRef]

10. Hirabayashi, Y.; Mahendran, R.; Koirala, S.; Konoshima, L.; Yamazaki, D.; Watanabe, S.; Kim, H.; Kanae, S. Global flood risk under climate change. *Nat. Clim. Chang.* **2013**, *3*, 816–821. [CrossRef]

11. Dieperink, C.; Hegger, D.L.T.; Bakker, M.H.N.; Kundzewicz, Z.W.; Green, C.; Driessen, P.P.J. Recurrent Governance Challenges in the Implementation and Alignment of Flood Risk Management Strategies: A Review. *Water Resour. Manag.* **2016**, *30*, 4467–4481. [CrossRef]

12. Van den Brink, M.; Termeer, C.; Meijerink, S. Are Dutch water safety institutions prepared for climate change? *J. Water Clim. Chang.* **2011**, *2*, 272–287. [CrossRef]

13. Jongman, B.; Winsemius, H.C.; Aerts, J.C.J.H.; Coughlan De Perez, E.; Van Aalst, M.K.; Kron, W.; Ward, P. Declining vulnerability to river floods and the global benefits of adaptation. *Proc. Natl. Acad. Sci. USA* **2015**, *112*, E2271–E2280. [CrossRef] [PubMed]

14. Alexander, M.; Priest, S.; Micou, A.P.; Tapsell, S.; Green, C.; Parker, D.; Homewood, S. *Analysing and Evaluating Flood Risk Governance in England—Enhancing Societal Resilience through Comprehensive and Aligned Flood Risk Governance*; STAR-FLOOD Consortium: Utrecht, The Netherlands, 2016; ISBN 978-94-91933-07-3.

15. Ek, K.; Goytia, S.; Pettersson, M.; Spegel, E. *Analysing and Evaluating Flood Risk Governance in Sweden—Adaptation to Climate Change?* STAR-FLOOD Consortium: Utrecht, The Netherlands, 2016; ISBN 978-94-91933-10-3.

16. Kaufmann, M.; Van Doorn-Hoekveld, W.J.; Gilissen, H.K.; Van Rijswick, H.F.M.W. *Drowning in Safety. Analysing and Evaluating Flood Risk Governance in The Netherlands*; STAR-FLOOD Consortium: Utrecht, The Netherlands, 2016; ISBN 978-94-91933-11-0.

17. Larrue, C.; Bruzzone, S.; Lévy, L.; Gralepois, M.; Schellenberger, T.; Trémorin, J.B.; Fournier, M.; Manson, C.; Thuillier, T. *Analysing and Evaluating Flood Risk Governance in France: From State Policy to Local Strategies*; STAR-FLOOD Consortium: Utrecht, The Netherlands, 2016; ISBN 978-94-91933-08-0.

18. Matczak, P.; Lewandowski, J.; Choryński, A.; Szwed, M.; Kundzewicz, Z.W. *Flood Risk Governance in Poland: Looking for Strategic Planning in a Country in Transition*; STAR-FLOOD Consortium: Utrecht, The Netherlands, 2016; ISBN 978-94-91933-09-7.

19. Mees, H.; Suykens, C.; Beyers, J.C.; Crabbé, A.; Delvaux, B.; Deketelaere, K. *Analysing and Evaluating Flood Risk Governance in Belgium. Dealing with Flood Risks in an Urbanised and Institutionally Complex Country*; STAR-FLOOD Consortium: Utrecht, The Netherlands, 2016; ISBN 978-94-91933-06-6.

20. Hegger, D.L.T.; Driessen, P.P.J.; Bakker, M.H.N. *A View on More Resilient Flood Risk Governance: Key Conclusions of the STAR-FLOOD Project*; STAR-FLOOD Consortium: Utrecht, The Netherlands, 2016; ISBN 978-94-91933-13-4.

21. Raadgever, G.T.; Hegger, D.L.T. *Flood Risk Management Strategies and Governance*; Springer: Berlin, Germany, 2018; ISBN 978-331967699-9.

22. Aerts, J.C.J.H.; Botzen, W.; Van der Ven, A.; Krywkow, J.; Werners, S. Dealing with uncertainty in flood management through diversification. *Ecol. Soc.* **2008**, *13*, 41. [CrossRef]

23. Hegger, D.L.T.; Driessen, P.P.J.; Dieperink, C.; Wiering, M.; Raadgever, G.T.; Van Rijswick, H.F.M.W. Assessing stability and dynamics in flood risk governance: An empirically illustrated research approach. *Water Resour. Manag.* **2014**, *28*, 4127–4142. [CrossRef]

24. Thampapillai, D.J.; Musgrave, W.F. Flood damage mitigation: A review of structural and nonstructural measures and alternative decision frameworks. *Water Resour. Res.* **1985**, *21*, 411–424. [CrossRef]

25. Van Buuren, A.; Ellen, G.J.; Warner, J.F. Path-dependency and policy learning in the Dutch delta: Toward more resilient flood risk management in The Netherlands? *Ecol. Soc.* **2016**, *21*, 43. [CrossRef]

26. Gilissen, H.K.; Alexander, M.; Beyers, J.-C.; Chmielewski, P.; Matczak, P.; Schellenberger, T.; Suykens, C.B.R. Bridges over troubled waters: An interdisciplinary framework for evaluating the interconnectedness within fragmented domestic flood risk management systems. *J. Water Law* **2015**, *25*, 12–26.

27. Suykens, C.; Priest, S.J.; Van Doorn-Hoekveld, W.J.; Thuillier, T.; Van Rijswick, M. Dealing with flood damages: Will prevention, mitigation, and ex post compensation provide for a resilient triangle? *Ecol. Soc.* **2016**, *21*, 1. [CrossRef]

28. Van Doorn-Hoekveld, W.J. Distributional Effects of EU Flood Risk Management and the Law. Ph.D. Thesis, Utrecht University, Utrecht, The Netherlands, 2018.

29. Kolen, B.; Helsloot, I. Decision-making and evacuation planning for flood risk management in The Netherlands. *Disa* **2014**, *38*, 610–635. [CrossRef] [PubMed]

30. Alexander, M.; Doorn, N.; Priest, S. Bridging the legitimacy gap—Translating theory into practical signposts for legitimate flood risk governance. *Reg. Environ. Chang.* **2018**, *18*, 397–408. [CrossRef]

31. Mees, H.; Crabbé, A.; Alexander, M.; Kaufmann, M.; Bruzzone, S.; Lévy, L.; Lewandowski, J. Coproducing flood risk management through citizen involvement: Insights from cross-country comparison in Europe. *Ecol. Soc.* **2016**, *21*, 7. [CrossRef]

32. Klijn, F.; Samuels, P.; Van Os, A. Towards flood risk management in the EU: State of affairs with examples from various European countries. *Int. J. River Basin Manag.* **2008**, *6*, 307–321. [CrossRef]

33. OECD. *Water Governance in The Netherlands: Fit for the Future?* OECD Studies on Water; OECD Publishing: Paris, France, 2014.

34. Goytia, S.; Pettersson, M.; Schellenberger, T.; Van Doorn-Hoekveld, W.J.; Priest, S. Dealing with change and uncertainty within the regulatory frameworks for flood defense infrastructure in selected European countries. *Ecol. Soc.* **2016**, *21*, 23. [CrossRef]

35. Green, O.O.; Garmestani, A.S.; Van Rijswick, H.F.M.W.; Keessen, A.M. EU water governance: Striking the right balance between regulatory flexibility and enforcement? *Ecol. Soc.* **2013**, *18*, 10. [CrossRef]

36. Priest, S.J.; Suykens, C.; Van Rijswick, H.F.M.W.; Schellenberger, T.; Goytia, S.; Kundzewicz, Z.W.; Van Doorn-Hoekveld, W.J.; Beyers, J.-C.; Homewood, S. The European union approach to flood risk management and improving societal resilience: Lessons from the implementation of the Floods Directive in six European countries. *Ecol. Soc.* **2016**, *21*, 50. [CrossRef]

37. Suykens, C.B.R. The Law of the River. The Institutional Challenge for Transboundary River Basin Management and Multi-Level Approaches to Water Quantity Management. Ph.D. Thesis, Utrecht University, Utrecht, The Netherlands, 2017.

38. Suykens, C. EU water quantity management in international river basin districts: Crystal clear? *Eur. Energy Environ. Law Rev.* **2015**, *24*, 134–143.

39. Keessen, A.M.; Van Rijswick, H.F.M.W. Adaptation to climate change in European water law and policy. *Utrecht Law Rev.* **2012**, *8*, 38–50. [CrossRef]

40. Keessen, A.M.; Van Kempen, J.J.H.; Van Rijswick, M.; Robbe, J.; Backes, C.W. European river basin districts: Are they swimming in the same implementation pool? *J. Environ. Law* **2010**, *22*, 197–221. [CrossRef]

41. Gralepois, M.; Larrue, C.; Wiering, M.; Crabbé, A.; Tapsell, S.; Mees, H.; Ek, K.; Szwed, M. Is flood defense changing in nature? Shifts in the flood defense strategy in six European countries. *Ecol. Soc.* **2016**, *21*, 37. [CrossRef]

42. Fournier, M.; Larrue, C.; Alexander, M.; Hegger, D.; Bakker, M.; Pettersson, M.; Crabbé, A.; Mees, H.; Chorynski, A. Flood risk mitigation in Europe: How far away are we from the aspired forms of adaptive governance? *Ecol. Soc.* **2016**, *21*, 49. [CrossRef]

43. Gilissen, H.K.; Alexander, M.; Matczak, P.; Pettersson, M.; Bruzzone, S. A framework for evaluating the effectiveness of flood emergency management systems in Europe. *Ecol. Soc.* **2016**, *21*, 27. [CrossRef]

44. Thieken, A.H.; Kienzler, S.; Kreibich, H.; Kuhlicke, C.; Kunz, M.; Mühr, B.; Müller, M.; Otto, A.; Petrow, T.; Pisi, S.; et al. Review of the flood risk management system in Germany after the major flood in 2013. *Ecol. Soc.* **2016**, *21*, 51. [CrossRef]

45. Environment Agency. *Flood and Coastal Erosion Risk Management in England: Investment Programme 2015 to 2021*; EA: Bristol, UK, 2018.

46. Department for Environment, Food and Rural Affairs (Defra). *Flood and Coastal Resilience Partnership Funding. Defra Policy Statement on an Outcome-Focused, Partnership Approach to Funding Flood and Coastal Erosion Risk Management*; Department for Environment, Food and Rural Affairs (Defra): London, UK, 2011.

47. Lazarus, R. Pursuing Environment Justice: The Distributional Effects of Environmental Protection. *Nw. UL Rev.* **1992**, *87*, 787.

48. Van Doorn-Hoekveld, W. Transboundary flood risk management: Compatibilities of the legal systems of flood risk management in The Netherlands, Flanders and France—A comparison. *Eur. Energy Environ. Law Rev.* **2017**, *26*, 81–96.

49. Van Doorn-Hoekveld, W. Equal distribution of burdens in flood risk management—The application of the 'égalité principle' in the compensation regimes of The Netherlands, Flanders and France. *Rev. Eur. Adm. Law* **2017**, *10*, 81–110. [CrossRef]

50. Thaler, T.; Hartmann, T. Justice and flood risk management: Reflecting on different approaches to distribute and allocate flood risk management in Europe. *Nat. Haz.* **2016**, *83*, 129–147. [CrossRef]

51. Van Rijswick, H.F.M.W.; Havekes, H.J.M. *European and Dutch Water Law*; Europa Law Publishing: Groningen, The Netherlands, 2012; ISBN 9789089521071.

52. Van Doorn-Hoekveld, W.J.; Goytia, S.B.; Suykens, C.; Homewood, S.; Thuillier, T.; Manson, C.; Chmielewski, P.; Matczak, P.; Van Rijswick, H.F.M.W. Distributional effects of flood risk management—A cross-country comparison of preflood compensation. *Ecol. Soc.* **2016**, *21*, 26. [CrossRef]

53. Penning-Rowsell, E.; Pardoe, J. The distributional consequences of future flood risk management in England and Wales. *Environ. Plan. C* **2015**, *33*, 1301–1321. [CrossRef]

54. Doorn, N. Resilience indicators: Opportunities for including distributive justice concerns in disaster management. *J. Risk Res.* **2017**, *20*, 711–731. [CrossRef]

55. Pettersson, M.; Van Rijswick, M.; Suykens, C.; Alexander, M.; Ek, K.; Priest, S. Assessing the legitimacy of flood risk governance arrangements in Europe: Insights from intra-country evaluations. *Water Int.* **2017**, *42*, 929–944. [CrossRef]

56. Rojas, R.; Feyen, L.; Watkiss, P. Climate change and river floods in the European Union: Socio-economic consequences and the costs and benefits of adaptation. *Glob. Environ. Chang.* **2013**, *23*, 1737–1751. [CrossRef]

57. De Risi, R.; De Paola, F.; Turpie, J.; Kroeger, T. Life Cycle Cost and Return on Investment as complementary decision variables for urban flood risk management in developing countries. *Int. J. Disaster Risk Reduct.* **2018**, *28*, 88–106. [CrossRef]

58. Tullos, D. How to achieve better flood-risk governance in the United States. *Proc. Natl. Acad. Sci. USA* **2018**, *115*, 3731–3734. [CrossRef] [PubMed]

59. Dai, L.; Van Doorn-Hoekveld, W.J.; Yu Wang, R.; Van Rijswick, H.F.M.W. Dealing with distributional effects of flood risk management in China: Compensation mechanisms in flood retention areas. under review.

water

MDPI

Article

Interdisciplinary Collaboration on Green Infrastructure for Urban Watershed Management: An Ohio Case Study

Shawn Dayson Shifflett [1], Tammy Newcomer-Johnson [2], Tanner Yess [3] and Scott Jacobs [4,*]

[1] Oak Ridge Institute for Science and Education (ORISE) Research Participation Program, U.S. Environmental Protection Agency, 26 W. Martin Luther King Drive, Cincinnati, OH 45268, USA; shifflett.shawn@epa.gov

[2] Office of Research and Development, National Exposure Research Laboratory, U.S. Environmental Protection Agency, 26 W. Martin Luther King Drive, Cincinnati, OH 45268, USA; newcomer-johnson.tammy@epa.gov

[3] Mill Creek Alliance, 1662 Blue Rock St. Cincinnati, OH 45223, USA; tyess@themillcreekalliance.org

[4] Office of Research and Development, National Risk Management Research Laboratory, U.S. Environmental Protection Agency, 26 W. Martin Luther King Drive, Cincinnati, OH 45268, USA

* Correspondence: jacobs.scott@epa.gov; Tel.: +1-513-569-7223

Received: 12 March 2019; Accepted: 3 April 2019; Published: 9 April 2019

Abstract: Many older Midwestern cities of the United States are challenged by costly aging water infrastructure while working to revitalize urban areas. These cities developed much of their water infrastructure before the Clean Water Act became law and have struggled to mitigate contaminant loading to surface waters. An increasingly common approach to resolving these challenges is the integration of green infrastructure with gray infrastructure improvements to manage point and non-point source pollution. Stakeholder engagement and collaboration during green infrastructure planning can help address impairments and promote community involvement through the revitalization process. Mill Creek watershed in Cincinnati, OH, USA has seen improvement in watershed integrity indicators after being impaired for many decades by flashy hydrology, combined sewer overflows, and water quality degradation. A workshop was conducted to examine how integrated green and gray infrastructure has contributed to improvements in Mill Creek over the past several decades. This effort sought to examine internal and external factors that influence a multi-stakeholder watershed approach to planning, implementing, and evaluating green infrastructure techniques. Community investment and physical infrastructure, access to datasets, and skills and knowledge exchange were essential in improving use attainment in the Mill Creek. Strategic placement of green infrastructure has the potential to maximize water quality benefits and ecosystem services. However, green infrastructure deployment has been more opportunistic due to the diversity of stakeholder and decision maker interests. Future work should consider collaborative approaches to address scaling challenges and workforce development to maximize green infrastructure benefits.

Keywords: use-attainment; social network analysis; urban planning; governance; social infrastructure

1. Introduction

Watershed management in aging urban cities of the Midwestern United States is an expensive but critical need [1,2]. These cities have unique challenges including flashy hydrology, channelized and buried streams, and combined sewer overflows (CSOs) from merged stormwater and wastewater infrastructure [3–5]. In addition, these watersheds have impervious surface areas ranging from >10% to >26% [3,6]), inter-watershed transfers to meet water resource demands [7], and changing precipitation patterns that challenge resource managers to meet use attainment goals identified through the Clean Water Act.

Achieving use attainment may be addressed by introducing green infrastructure (GI) in combination with traditional engineering and stormwater management practices [2,8]. The term "green infrastructure" was defined by amendment to section 502 of the Federal Water Pollution Control Act as "the range of measures that use plant or soil systems, permeable pavement or other permeable surfaces or substrates, stormwater harvest and reuse, or landscaping to store, infiltrate, or evapotranspire stormwater and reduce flows to sewer systems or to surface waters [9]." GI can help divert pollutants, mitigate flooding, improve groundwater infiltration, and provide some cost savings to private and public expenditures [8]. Strategically placing GI in urban watersheds may extend the benefits of these installations, but current practices have been more opportunistic due to the complex sociopolitical mosaic in which these strategies are deployed.

Siting GI in degraded urban waterways requires careful design, planning, and negotiation by watershed management groups and communities. In the Mill Creek Watershed of Cincinnati, Ohio, USA, GI strategies have incorporated rain gardens, permeable pavement, bioswales, green roofs, land conservation, cisterns, groundwater infiltration installations, stream restoration, and stream daylighting. Daylighting, also referred to as de-culverting, is a radical type of GI that involves the transformation of a buried or piped stream back to an open channel [10–12]. For some communities, these GI implementations have been intended to improve the aesthetics of waterways, while watershed management organizations seek to improve water quality and stormwater management. Two leading organizations involved in planning and implementing GI in Cincinnati are the Municipal Sewer District of Greater Cincinnati (the Sewer District) and the Mill Creek Alliance (MCA). The Sewer District has sought to mitigate CSOs by combining infrastructure upgrades with stream restoration. MCA, a Groundwork USA Trust and advocate for equity in the watershed, has sought to improve ecosystem resiliency and watershed integrity by introducing rain gardens, bioswales, and investing in stream restoration. Rain barrels have also been implemented as part of a pilot effort to research GI incentives [13]. Community engagement has been a high priority for both groups for successful planning and implementation of GI, but more work is needed to understand how these strategies and techniques feed into a watershed-wide approach for use attainment. Furthermore, it is unclear how monitoring and evaluation is utilized once projects have been implemented to better inform future planning and implementiation of GI.

This study is based on a workshop with stakeholders from the MCA, the Sewer District, University of Cincinnati, United States Geological Survey (USGS), United States Environmental Protection Agency (USEPA) Region 5, and EPA Office of Research and Development to discuss designing natural GI through interdisciplinary collaboration. The purpose of the workshop was to examine green infrastructure in the Mill Creek watershed and glean lessons learned that could be shared with other similar communities to move toward strategic placement of GI. A stakeholder network analysis was used to identify stakeholders and how their roles affect the establishment of GI in the Mill Creek Watershed. Additionally, a strength-weaknesses-opportunities-threats (SWOT) analysis was conducted to determine internal versus external and positive versus negative factors affecting GI, such as daylighting Congress Run, a buried tributary within the Mill Creek Watershed (See Supplementary Material Figure S1). Our objectives were to define and to document lessons learned from the workshop regarding internal strengths and weaknesses as well as external opportunities and threats influencing (1) GI planning, (2) implementation, and (3) evaluation. This article broadly synthesizes findings from our approach and identifies research gaps regarding future strategic placement of GI.

2. Materials and Methods

2.1. Study Area Description

The Mill Creek watershed is a 437 km^2 HUC 10 (0509020301) located in southwestern Ohio, USA (−84°29′24″, 39°15′0″; see Figure 1; [14]). The Great Miami River basin bounds the watershed in the northwest, the Little Miami River basin to the east, and the Ohio River and direct tributary watersheds to the south and west. The total elevation difference from Mill Creek's headwaters in Butler County to the barrier dam in Hamilton County is 107 m. Upper Mill Creek has a wide valley bottom with an approximate width of 2.4 km. Lower Mill Creek narrows considerably and has an extensively modified concrete trapezoidal channel that flows through a highly urbanized, industrial landscape. The four major tributaries are underlain by thinly inter-bedded layers of shale and limestone bedrock, except in the lower confluences of the metropolitan area, where the creek has been armored. Thirty-year average annual precipitation is 1072 mm, but inter-annual precipitation differences have deviated greatly from the average with annual precipitation ranging from 931 mm to 1861 mm in the last 18 years [15]. The watershed is highly urbanized with 21.0% (91.8 km^2) impervious surface and approximately 78.9% (345 km^2) developed land as of 2011 [16]. Soils are distributed heterogeneously throughout the watershed but are dominated by well drained to excessively drained urban-Urdothents complexes [17].

The Mill Creek watershed has a complex mosaic of geographical, political, and jurisdictional boundaries. The watershed is divided amongst Hamilton, Butler, and Warren counties. Thirty-six different municipalities are fully or partially within the Mill Creek watershed (see Supplementary Material Figure S2). Within the largest municipality of Cincinnati, Hamilton county recognizes 47 different jurisdictions represented by 52 different community councils (See Supplementary Material Figure S3; [18]). The watershed is also divided into 11 facility planning areas (FPAs) managed by Ohio-Kentucky-Indiana Regional Council of Governments, a Metropolitan Planning Organization that maintains water quality management plan as required by section 208 of the Clean Water Act. The majority (71%) of watershed falls within the boundary of the Sewer District, which serves a population of >850,000 (Figure 1). The Sewer District's boundary crosses into eight different HUC 10 watersheds. Like many midwestern cities, the Sewer District is under consent decree (Civil Action No. C-1-02-107) with the U.S. EPA, the U.S. Department of Justice, and Hamilton County. This legal agreement requires a CSO discharge reduction of 7.6 billion liters per year (or 66%) by 2019 [19,20]. Before the consent decree, Mill Creek watershed contained over 200 CSOs and sanitary sewer overflows (SSOs), a challenged municipal wastewater treatment plant, numerous industrial point discharges, and deteriorating sewage collection systems [19]. Through the consent decree, the Sewer District developed Project Groundwork, a multi-billion dollar and multi-year initiative to implement hundreds of sewer improvements and stormwater control projects, including implementing green infrastructure [21]. Over the past several decades, these projects and others have helped improve watershed integrity [19,22].

Even with this progress, GI planning and implementation has been challenging due to various social, political, and technical challenges. One example identified was the daylighting of a sub-watershed. Congress Run is a 77.6 km^2 HUC 12 (040509020301) with 79.4% developed land and 35.6% impervious cover [22]. Congress Run has been listed as a priority project in multiple watershed plans [22,23]. A stretch of the stream was buried in a 183-m corrugated metal culvert under a golf course owned by Duke Energy. Multi-year plans were made to daylight the stream by removing the culvert and studying the effects of restoration, but various barriers delayed implementation and complicated evaluation. These barriers included divergent landowner interest, discovery of underground obstacles, and financial cost. Furthermore, effective and consistent monitoring of these sites in the urban environment presents its own challenges; these include the loss of monitoring equipment due to extreme flows, vandalization of equipment, hazardous stream banks, and lack of ideal spots for flow gages. These collective challenges are significant to the strategic placement and evaluation of GI that influence overall watershed integrity. Thus, Congress Run was presented as a case study to highlight challenges in the planning, implementation, and evaluation of GI in Mill Creek Watershed.

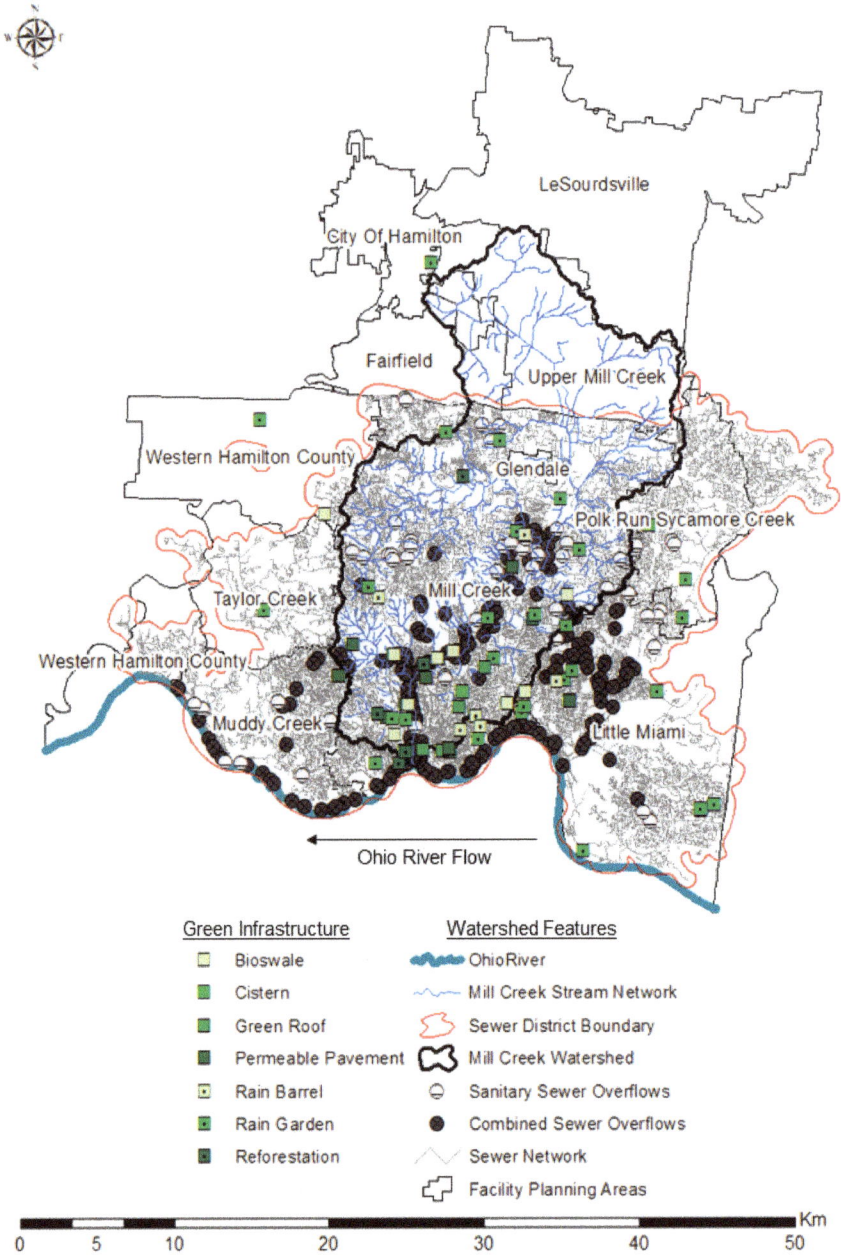

Figure 1. Map layout for Mill Creek Watershed and contributing facility planning areas (FPAs). Locations of combined sewer overflows (CSOs), sanitary sewer overflows (SSOs), and green infrastructure (GI) installations are juxtaposed to demonstrate the complex mosaic of the system's hydrology and its stressors.

2.2. Workshop and Follow-Up Meetings

In September 2017, a meeting was convened in Cincinnati, OH to discuss past GI efforts and identify strengths and weaknesses in the planning, implementation, and evaluation process. The topics covered were both broad and specific. Broadly, attendees discussed some of the regulatory frameworks, public engagement opportunities, and needs for expanding GI. Specifically, Mill Creek and the Congress Run case study combined sewer improvements to meet use-attainment goals with stream restoration and daylighting a failing culvert. Invited participants represented various organizations that were involved in planning, implementation, and evaluation of green infrastructure in the Mill Creek watershed. The participant organizations included EPA Office of Research and Development, EPA Region 5, United States Geological Survey (USGS), the Sewer District, Mill Creek Watershed Council of Communities and Groundwork Cincinnati–Mill Creek (now joined together as the MCA), and the University of Cincinnati.

A stakeholder network analysis was conducted to understand how various stakeholders were affiliated with one another. Researchers and managers are often interested in understanding how social relationships that compose complex governance structures like that of Mill Creek hinder or promote good water resource management practices. Stakeholder network analysis is increasingly being used as a tool in natural resource management to identify critical stakeholders within a network [24–26]. There are many social network analysis techniques [25,27–29], and our effort focused on network affiliation for understanding how different organizations coalesce and diverge from watershed-based goals. Network participants were identified through the workshop and follow-up meetings by creating an inventory of identified partners and collaborators of watershed stakeholders.

A SWOT analysis was conducted to organize relevant concerns raised during the workshop [30]. SWOT is a strategic planning tool for complex situations that aids decision making by condensing information into four categories [30]. In SWOT, internal strengths and weaknesses can include financial resources, efficiency and capacity, structure, image, and others [30]. External opportunities and threats can include regulatory, political, economic, and environmental issues as well as new technologies and social changes, among others [30]. When considering whether a factor was internal or external, participants were asked to share their organization's perspectives. As a result, this SWOT analysis combined the perspectives of multiple individuals from different organizations. Within a year of the workshop, follow-up discussions were held with workshop participants to evaluate the SWOT analysis results and further refine conclusions.

3. Results

3.1. Stakeholder Analysis

Planning, implementing, and evaluating GI projects has required communities to see value in GI projects. Each municipality has unique goals and priorities based on its community makeup and desire to achieve improved water management outcomes. Within the Mill Creek Watershed, the Sewer District and MCA play a leadership role in planning, implementing, and monitoring green infrastructure. Though the motivation for these two organizations is different, both have been successful in adding GI. Figure 2 demonstrates their network affiliation with local municipalities, government agencies, other non-profits working in the watershed, and additional ad hoc partners that may participate in GI efforts. A clear result of this analysis is that the network is somewhat partitioned between government agencies that interact with the Sewer District and non-profit groups that interact with MCA. These two organizations are connected through their relationship with communities in the Mill Creek Watershed. Though these relationships do not inherently lead to strategic placement of GI for improved watershed integrity, GI implementations were more likely when communities had a relationship with both organizations. Thirteen of the 28 (39%) communities served by both MCA and the Sewer District had GI installations, whereas only one out of the eight (13%) communities served strictly by MCA had GI installations (Figure 2).

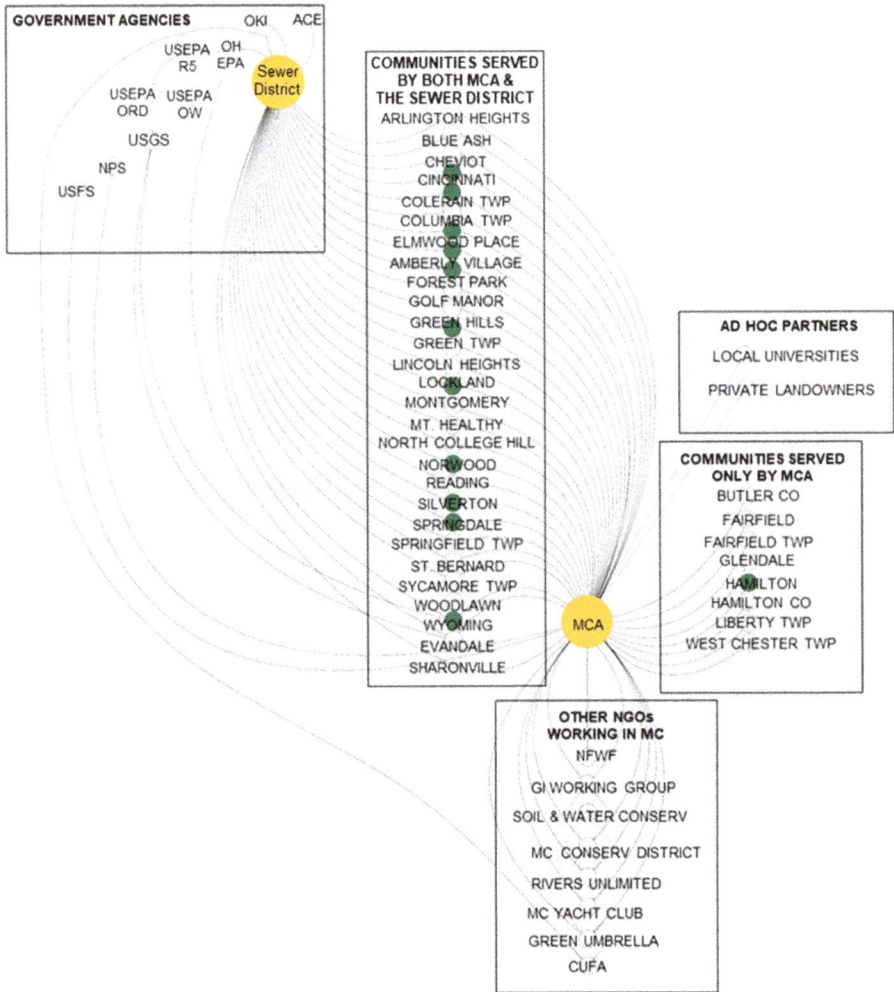

Figure 2. Stakeholder network map for GI planning, implementation, and evaluation in the Mill Creek Watershed. Green nodes are communities with GI installations. Yellow nodes are organizations responsible for planning and implementing GI. Nodes are clustered by organization type (i.e., government agencies versus communities). ACE: Army Corps of Engineers; OKI: Ohio-Kentucky-Indiana Regional Council of Governments; NPS: National Park Service; FWS: US Fish and Wildlife Service; MCA: Mill Creek Alliance; USGS: United States Geological Survey; EPA: United States Environmental Protection Agency; CUFA: Communities United for Action.

For MCA, working in the Mill Creek watershed means partnering with 36 political jurisdictions, interest groups, and community councils, many of whom have different values (Figure 2). Of these 36, only 12 have GI installations, and the largest municipality (Cincinnati) has the most (Figure 2). MCA also partners with other non-profit groups that are not inherently responsible for planning, implementation, or evaluation of GI. These groups tend to have special interest in broadly improving watershed integrity and engaging in river clean-up activities, citizen science water quality monitoring, and building social infrastructure for environmental recreation. To achieve its mission of community empowerment, MCA has several funding sources, including National Park Service (NPS), National Fish and Wildlife Service (FWS), Ohio Environmental Protection Agency, and the Sewer District (Figure 2). For the Sewer District, working in the Mill Creek watershed includes collaborating with a variety of stakeholders, contractors, and subcontractors to meet the requirements of the consent decree as well as community interests. Though the Sewer District partners with 43 municipalities, only 28 are within the Mill Creek Watershed (see Figures 1 and 2). The Sewer District is subject to federal and state oversight via the Ohio-Kentucky-Indiana Council of Governments (OKI), the Army Corps of Engineers (ACE), and Ohio EPA. Ohio EPA also receives support for GI from the US EPA through its Regional Green Infrastructure Coordinator, the Office of Water, and the Office of Research and Development.

In the case of daylighting Congress Run, the stakeholder network has been dynamic. Before the assessment began, stakeholders included Mill Creek Watershed Council of Communities, Groundwork Cincinnati, the municipality of Springfield Township, the private landowner (Duke), the land manager (Duke's Hartwell Golf Course), adjacent landowners (Byer Steel), and researchers from University of Cincinnati and US EPA Office of Research and Development. During this process, Mill Creek Watershed Council of Communities and Groundwork Cincinnati combined to form the Mill Creek Alliance (MCA). Simultaneously, there was concern that this stream daylighting and restoration project would negatively impact the experience of golfers due to the wide corridor needed for stream restoration. During the planning efforts, the golf course closed, and the impact to golfers ceased to be a consideration to the restoration project planning and implementation. Changes observed during the study period captured the fluid nature of a stakeholder network seeking to plan and implement GI within a group demonstrating diverse values.

Identifying the boundary of the stakeholder network affiliation came with some difficulty, as has been documented in several stakeholder network assessments [24,29,31,32]. Through the workshop and subsequent follow-up meetings, GI and watershed stakeholders identified their network and explained how they interact within their network. This approach might have meant that some unidentified but meaningful stakeholders were not considered for this stakeholder network analysis. Thus, this analysis does not preclude that other stakeholders may be playing a role in GI planning, implementation, and evaluation for the Mill Creek Watershed. Similarly, this analysis is limited in capturing and predicting the dynamics of the stakeholder network because social networks are constantly changing as connections between stakeholders sever or form [24,33–35]. To this point, the role of individuals who champion watershed integrity and advocate for GI throughout the watershed is worth noting. Several individuals were identified as key players in establishing relationships among communities and organizations. One outstanding question in this stakeholder network analysis is how the networks will change as these key individuals are no longer able to act as champions in the watershed.

3.2. Green Infrastructure Planning

3.2.1. Inventory of GI Plans

Stakeholders identified comprehensive plans proposing green infrastructure to improve system integrity; however, additional research was needed to identify many of these planning documents (see Table 1). The objectives of these plans included, but were not limited to, expanding stream restoration, improving canopy cover up to 40%, and encouraging the installation of green stormwater management systems to capture, slow, and infiltrate stormwater during moderate and large rain events [22]. Stakeholders broadly identified many strengths, weaknesses, opportunities, and threats to these plans. An internal strength of incorporating GI into formal watershed and other location-based plans is that it can provide a roadmap for strategic placement of GI. An additional strength of formal plans is that they can improve continuity of vision across changes in political administrations as well as local organizational turnover. An external opportunity from these plans is they can enable and strengthen eligibility for funding GI implementation and evaluation (e.g., the Nine-element Nonpoint Source Implementation Strategic Plan for Congress Run–Mill Creek HUC-12 enables eligibility for Clean Water Act (CWA) section 319 grant funding [22,36]). An internal weakness was the lack of formalized accountability in GI plans (e.g., the Green Cincinnati Plan lacks a structure for accountability [37]). Collectively, this SWOT analysis indicated that taking a careful inventory of existing plans at the watershed scale may be critical to achieving improved use-attainment in waterways, but these plans need to be comprehensively reviewed by decision makers for long-term success. Likewise, creating these inventories can better inform where gaps exists in the planning, implementation, and evaluation of GI at the watershed scale.

Table 1. National, regional, and local strategic plans affecting GI in the Congress Run-Mill Creek Watershed.

	Scale	Strategic Plan Title	Lead Organization	Purpose
National	Ohio River Watershed (528,360 km^2)	A Framework for Ecosystem Restoration of the Ohio River and its Watershed [38]	Ohio River Foundation	Restore the Ohio River and its watershed.
Regional	Butler, Clermont, Hamilton, and Warren Counties in Ohio (4532 km^2)	Water Quality Management Plan for Butler, Clermont, Hamilton, and Warren Counties in Ohio [39]	OKI	Manage water quality in the greater Cincinnati region.
Local	Cincinnati Sewershed (751 km^2)	Wet Weather Improvement Plan [40]	Metropolitan Sewer District of Greater Cincinnati	Fulfill consent decree requirements pertaining to submission of the CSO Long Term Control Plan Update and the Capacity Assurance Program Plan.
	Mill Creek Watershed (440 km^2)	Lower Mill Creek Watershed Action Plan [23]	MCA	Improve "water quality and ecological health in the Lower Mill Creek Watershed that will, in turn, create more livable neighborhoods and provide public health, environmental, social, and economic benefits for many years to come."
	Mill Creek Watershed (440 km^2)	Mill Creek Watershed Greenway Trail Master Plan [41]	MCA	Improve water quality and floodplain management and contribute to economic well-being.
	City of Cincinnati (206 km^2)	Plan Cincinnati; A Comprehensive Plan for the Future [42]	Cincinnati City Council	"Redefines our city and what it means to be a thriving urban city."
	City of Cincinnati (206 km^2)	2018 Green Cincinnati Plan [37]	Mayor's Steering Committee	Provides a roadmap guide for transitioning the area into a more sustainable, equitable, and resilient city.
	Congress Run Watershed (77.6 km^2)	Nonpoint Source Implementation Strategic Plan; Congress Run–Mill Creek HUC-12 [22]	MCA	Prioritize nonpoint source pollution reduction strategies and projects in the Congress Run–Mill Creek HUC-12. This plan enables eligibility for Clean Water Act section 319 grant funding.

3.2.2. Social Infrastructure for GI Planning

There was consensus among the workshop participants that an internal strength of GI planning is that it provides communities an opportunity to participate in the planning process. Community-based planning has provided the opportunity for communities to provide input, but it has been met with mixed success [33–46]. The Sewer District has been mandated to conduct regular townhalls to incorporate community values into their GI planning process. However, some local organizations have demanded greater transparency from the Sewer District (e.g., Communities United for Action), and communities regularly express frustration about rising sewer costs. In contrast, MCA is not providing sewer services and is not federally mandated to plan or implement green infrastructure, thus they have more flexibility to invest in their social infrastructure (community relationships, education programs, and green workforce development). MCA has sought to implement GI when communities or individual landowners have expressed interest in protecting and enhancing the value of the Mill Creek and its tributaries. Though the Sewer District and MCA are structurally different, they both depend on social infrastructure investment when seeking to implement GI at the watershed scale. The challenge of effectively building social infrastructure has been identified by other research efforts in the Midwest [47–50]. Poor acceptance of using GI strategies has occurred most frequently in communities with lower socioeconomic status [48,50]. These communities may not have valued GI strategies because they cannot support increased housing costs and property values that come with these implementations [51], or because other environmental justice issues take precedence. Thus, planning GI efforts may need to be reconciled with other community needs.

3.2.3. Funding and Governance in GI Planning

Future funding for GI planning was identified as an external threat in the workshop and follow up discussions. This threat was tied to the challenge in quantifying the benefits for GI. Funding for planning may depend on the structure and efficacy of local governments, sewer districts, metropolitan planning organizations, and non-profits. Funding for planning GI installations in Mill Creek has historically come from multiple sources. The Sewer District has derived some of its funds for GI planning through grants and low-interest construction loans but mainly through customer billing. However, Cincinnati has one of the highest quarterly billing rates of all major cities in the state of Ohio [52]. Furthermore, the Sewer District, like most municipal sewer districts in the Midwest, contracts out design and planning of GI due to a limited in-house workforce capacity (Figure 2). MCA, a non-profit organization, does not receive dedicated funding for GI planning but does receive funding for GI implementations through Ohio EPA, Ohio Department of Natural Resources, National Fish & Wildlife Foundation, National Park Service, US Forest Service, and through the Sewer District. Funding mechanisms were commonly identified as a challenge in GI planning, as many Midwestern cities implement GI under conditions of competing needs. As a result, few funds have been dedicated to helping cities develop good GI plans that lead to meaningful GI installations [53]. Though this phase of the process might be viewed as sweat equity for these organizations, improving funding for the planning process has the potential to improve strategic implementation.

3.2.4. Tools and Models for GI Planning

There are multiple models and datasets available to aid in GI planning, though it is unclear how often these are used in urban Midwestern watersheds. The EPA has developed tools such as the Stormwater Calculator (https://www.epa.gov/water-research/national-stormwater-calculator), which estimates the annual amount of rainwater and frequency of runoff from a specific site to help planners and property owners improve stormwater management. The Stormwater Calculator uses data from national databases that provide soil, topography, rainfall, and evaporation information for a chosen site. The user supplies information about the site's land cover and selects low impact development (LID) controls they would like to use. The LID controls include seven green infrastructure practices. Another EPA tool is the Green Infrastructure Wizard (GIWiz; https://www.epa.gov/sustainability/giwiz), which is a digital repository of GI-related resources and tools. None of the workshop participants were actively using these EPA tools in their day to day work. The Sewer District contracted with the Midwest Biodiversity Institute to develop an Integrated Prioritization System to prioritize their response to the CSO Consent Decree and for Capital Improvement Planning [54]. The Integrated Prioritization System was able to identify sites and stream reaches where use attainment was substantially good and bad, thus allowing the Sewer District to develop plans for sites with high degradation. This approach may be equally as meaningful for other urban Midwestern watersheds seeking to integrate GI to improve use attainment. However, there is a need to improve GI planning tools and models for public use and decision making.

Another challenge to the use of tools and resources is opportunistic pressure. Some of these tools may require substantial technical expertise and time investment, and this can be a limiting factor for organizations seeking to plan strategically. Stakeholders with MCA have used the EJSCREEN Tool (https://www.epa.gov/ejscreen) to identify areas with low-income or minority populations to support educational programs and grant writing. Older cities in the Midwestern United States have unique challenges in improving environmental conditions that may be addressed in part by resilient natural GI in combination with traditional engineering and stormwater management practices. Many community plans will commit to GI efforts for aesthetic improvement, but awareness of how it impacts use attainment remains unclear and under prioritized.

3.2.5. Education and GI Planning

Stakeholders identified the lack of educational programs and workforce development programs in GI planning as a threat and a weakness. University of Cincinnati's School of Planning does offer a Green Roofs Certificate program, which is one of the first in North America (https://daap.uc.edu/academics/sop/overview/green-roofs-cert). There is also the option to become certified as an Accredited Green Roof Professional (GRP) with the organization Green Roofs for Healthy Cities (https://greenroofs.org/grp-faq/). GI planning requires a diverse knowledge base including fundamentals of urban planning, ecology, hydrology, and chemistry. Furthermore, planners may need knowledge of landscape management as well as a capacity to develop social infrastructure with communities. More research is needed to understand how emerging areas of water management and GI can be translated to institutionalized and accredited programs for improved workforce development. Lastly, it is unclear if organizations are positioned to hire and expand their workforce to incorporate trained individuals. This issue is not new but remains a challenge for the planning process.

3.3. Green Infrastructure Implementation

3.3.1. Social Infrastructure for GI Implementation

Stakeholders identified GI implementation as an area for general improvement for Midwestern urban watersheds. Implementation includes three phases: the engineering design, construction, and maintenance of GI projects. Some of the long-term strategic plans were written for ideal circumstances (Table 1), which rarely reflect the sociopolitical and environmental conditions in urban midwestern watersheds. Support for GI has existed at many of the various levels of governance (i.e., municipalities, jurisdictions, counties, state, and federal), but implementations have tended to be opportunistic based on funding and land availability. Land dedicated to GI must either be purchased, donated, occur on public property, or require some agreement between landowners and organizations seeking to install GI. Similarly, many of these sites are highly degraded from historical land use practices and require rehabilitation before GI can be installed. Thus, stakeholders pointed out that threats and weaknesses to GI implementation have been multifold. First, there may be limited funding for GI project completion, which may not include unexpected obstacles such as remediation of buried hazardous materials. Second, implementation of GI is typically achieved by a menagerie of contractors such as engineering firms for design, construction companies for building, and park department employees or other organizations for maintenance. If these diverse teams are not well-coordinated, there is a risk of miscomunication and inefficient GI management, leading to apprehension and a lack of clarity regarding governance. GI maintenance and upkeep may be neglected without a clear plan, funding source, and responsible party. Third, stakeholders noted that GI implementation has required a unique combination of skills, and as a result, creating job descriptions and bids for GI implementation has been challenging. Consequently, this aspect of GI may be an area for growth if watershed scale approaches are desired.

3.3.2. Funding and Governance for GI Implementation

Stakeholders expressed that funding was both a weakness and a threat to GI implementation. These challenges were partly attibuted to difficulties in predicting costs, benefits, and tradeoffs of GI in Midwestern cities. GI tends to be implemented under strict fiscal austerity [53]. As a result, GI implementations that could be effective at the watershed scale may not be installed due to associated costs. In their place have been smaller piecemealed GI efforts that may have impacts with a more limited footprint. For example, during the study period, the Lick Run Project in the lower Mill Creek Watershed was under construction to daylight a one-mile (1.6 km) reach and eliminate 1.5 billion liters of CSO releases annually. The Lick Run stream restoration has been estimated to cost $193 million USD for design and construction [21]. However, this project has not been without its complications. The project included funding for land acquisition, but that funding was insufficient to acquire the number of properties needed to meet the original plan for this project. Similarly, strategic planning may require GI installations ubiquitously throughout the watershed, but not all communities may have the economic capacity to meet the costs of GI. In Mill Creek, community groups must be able to contribute 25% of the total project cost to qualify for certain grants. This cost may be too substantial for communities where the benefits of GI installations and importance of environmental integrity are not a high priority. Therefore, successful implementation is contingent upon key decision makers, landowners, and social networks that bring these individuals together around community goals that can include GI elements. As such, GI has been implemented where spatially possible and where landowners and decision-makers were willing to support environmental resilience and improved water quality downstream.

Stakeholders identified the lack of role clarity in governance of GI implementations as a weakness. Efforts exist to support structured decision making for communities, but the impetus for implementation may need to originate from within local communities to successfully engage in a process that supports GI that is coordinated at a wider scale [55–57]. Watershed-wide implementation coordination exists primarily through regional organizations such as OKI [22,49]. However, implementation on private land has depended upon support from landowners who must weigh the advantages and disadvantages of changes made to the landscape. Strategic planning at a regional level supports a watershed approach but may lack authority and responsibility that rests with ownership and control of property. Furthermore, stakeholders expressed that implementation of GI is frequently site specific due to local soil conditions and existing infrastructure and can be difficult to translate to other sites. A need for streamlining GI implementations could improve this barrier, but other research efforts have noted that this may create divides between GI policy and practice [56,58]. As a result, it is unclear how governance hierarchies would be able to provide guidance on sustainable management [59]; however, the workshop participants advised that a clear plan and common understanding of who is responsible for design, construction, and maintenance is critical.

3.3.3. Education and GI Implementation

GI implementation was identified as an opportunity to create market growth, jobs for workers with a variety of educational backgrounds, and strengthened community engagement. There is a need to improve education for professionals engaged in GI implementation and for communities to understand the function and maintenance of their GI assets. Recent reports have demonstrated looming shortages in GI and water skilled workers capable of engineering design, construction, and maintenance [60]. This gap represents an opportunity for academic institutions and workforce development programs to engage students in multidisciplinary learning programs. Similar efforts have been documented in the assessment of renewable energy [61]. Certification programs for professionals do exist (e.g.,

National Green Infrastructure Certification Program for construction and maintenance [62]), but they are relatively new, and their impact is unclear. The workshop participants also shared the need to better define what skills are needed for the three phases of GI implementation as well as the need to develop language for job descriptions. Much like the planning phase, installing GI implementations can require a diverse skillset including knowledge on hydrology, ecology, engineering, construction, and landscape management. Similarly, investment in community education and outreach regarding the potential benefits of GI is needed to inform citizens who, in turn, have influence over planning and investments in restoration or protection of source and recreational waters. Many workshop participants noted that they themselves had not been well versed in GI terminology (e.g., rain garden, pervious pavement, etc.) until they became engaged through their occupations. Thus, developing language specific to GI implementation has been limited. This may be an artifact of institutional knowledge delay. Stakeholder institutions that are less fluid in their adaptation to modern water management strategies may have to play catch up with GI applications as a strategy for improved watershed integrity.

3.4. Green Infrastructure Evaluation

3.4.1. Inventory of GI Evaluations

Through the workshop and follow-up conversations, GI evaluation was identified as an area where more work is needed, which matches larger multi-decade trends calling for better evaluation of the mult-billion ecological restoration industry [12,63,64]. However, workshop participants considered evaluating GI effectiveness to be a lower priority than implementing additional GI and sewer repairs. GI evaluation is composed of monitoring the site for metrics relevant to project goals, analyzing monitoring results, and potential adaptive management if needed. Though these metrics can be quantified through a variety of experimental designs, workshop participants advised that the evaluation of use attainment in waterways was a high priority [19,22]. This prioritization may be largely attributed to the governance structures surrounding GI. Because regulations specific to GI have only recently been implemented (e.g., H.R. 7279 [9]), surface water regulations through the CWA have been the focus of evaluating GI impacts. In Mill Creek, multiple water quality management practices (CSO reduction and GI addition) were integrated together, which made it challenging to quantify water quality improvements attributable to GI at a watershed scale. Published performance values for GI could be used to estimate the benefits, but these are typically transferred from a different location, and there is uncertainty regarding the appropriateness of these values given potential differences in climate, soils, and other factors [65,66]. Stakeholders pointed out that few pilot scale studies have been conducted in southwestern Ohio and were uncomfortable applying general knowledge at site scale. Consequently, there is uncertainty regarding how individual GI projects impact the Mill Creek at a watershed scale. Understanding the watershed-scale effects of green infrastructure is an emerging science [67].

Evaluation of waterways in Mill Creek Watershed has been largely divided between many stakeholder organizations—a common trend in Midwestern cities [53]. The purposes of these sampling efforts vary, as do the methods, quality control, and data reporting procedures used. Organizations involved in watershed management and implementing GI (e.g., the Sewer District, Ohio EPA, and MCA) conducted longer-term monitoring of status and trends using standard methods over a broad spatial scale [19,68]. In contrast, research focused organizations (e.g., academic institutions and EPA Office of Research and Development) were more likely to use novel methods for ad hoc shorter-term, smaller-scale experiments examining the effectiveness of specific practices or elucidating mechanisms (e.g., [69]). The Sewer District has conducted long-term monitoring to demonstrate potential improvements from management activities throughout the sewershed. Ohio EPA has also routinely monitored surface waters throughout the watershed to understand if these resources are meeting CWA water quality criteria (https://www.epa.ohio.gov/dsw/bioassess/ohstrat). MCA's monitoring efforts have been supported by trained citizen scientist volunteers who conduct monthly water quality assessments for nine months per year on the following parameters: conductivity, pH, nitrates, total phosphorus, bacteria (*Escherichia coli* and total coliforms), Chlorophyll-a, and optical whiteners [68]. The goal of these assessments is to identify potential improvement in the watershed through GI and restoration efforts of the organization and provide some feedback on where to focus future efforts. However, these evaluations are not tied directly to a specific GI installation. EPA ORD and its contractors collect water samples in stream reaches of Mill Creek for method development and performance evaluation, but these efforts are not designed to inform the watershed planning process. The University of Cincinnati and other academic institutions also have student researchers seeking to evaluate GI in Mill Creek, but like ORD, these efforts are not intended to inform planning within the watershed. Though these organizations all have different motivations, improved collaboration between organizations may lead to a better understanding of how waterways in Mill Creek Watershed are changing as well as provide feedback mechanisms for how different land management practices, including GI, are affecting reaches within the watershed.

Several robust indicators comprised of chemical, toxicological, and ecological measures have been used to evaluate the Mill Creek watershed as various management actions such as sewer repairs and GI were applied (Table 2). Ohio EPA conducted surveys in 1992, 1997, and 2014, and the Midwest Biodiversity Institute (MBI) conducted surveys under contract with the Sewer District in 2011, 2013, and 2016 using these indicators to determine whether sites in the Mill Creek watershed were meeting their designated uses. Determining designated use attainment was done by examining whether environmental parameters and indicators met criteria specified by Ohio's Water Quality Standards under the Clean Water Act for aquatic life and recreation. Recreational use attainment status was determined by the fecal indicator bacteria, *E. coli*, which indicates the potential presence of pathogenic organisms associated with fecal contamination [19,70]. Multi-metric biological indicators were used to determine the health of the macroinvertebrate community {e.g., Invertebrate Community Index (ICI); [19,71,72]}, fish community {e.g., Index of Biotic Integrity for fish assemblages (IBI); [19,72–74] and the Modified Index of Well-Being (MIwb); [54,75,76]}, and habitat quality [Qualitative Habitat Evaluation Index (QHEI)] to assess whether the aquatic life criteria were being met. Additionally, MBI synthesized the results of multiple indicators and previous surveys into an area of degradation to attainment (ADV/AAV) ratio and restorability score [19,54]. The ADV/AAV ratio showed significant improvement in stream miles achieving use attainment in Mill Creek from 1992 to 2016 [19,54,77]. The purpose of the restorability score was to help prioritize sites for restoration; sites that are closer to use attainment have higher restorability scores [54]. These indicators are complex but have provided important insight to how watershed integrity is improving in the area. As a result, there is an opportunity to see if these metrics can capture the impact of GI.

Table 2. GI related indicators that have been used within Congress Run-Mill Creek Watershed [19,54].

Indicator	Description
Escherichia coli (E. coli) criterion	Recreational water quality criteria are based on the amount of *E. coli* bacteria present in a water sample. Levels of *E. coli* indicate fecal contamination and the potential presence of pathogenic organisms.
Invertebrate Community Index (ICI)	Aquatic life water quality criteria are based in part on the ICI, which uses the abundance and diversity of macroinvertebrates (such as mayflies, caddisflies, dipteran, and tolerant organisms) to determine the health of the macroinvertebrate assemblage.
Index of Biotic Integrity for fish assemblages (IBI)	Aquatic life water quality criteria are based in part on the IBI, which expresses the diversity and condition of a site's fish community.
Modified Index of Well-Being (MIwb)	Aquatic life water quality criteria are based in part on the MIwb, which is applicable in streams with drainage areas >20 miles^2 (52 km^2). This index is based upon fish assemblage measures including numbers, biomass, and two diversity indices (e.g., Shannon Index). The MIwb reflects the overall diversity and productivity of the fish population and frequently responds before the IBI to improvements in water quality and habitat.

Table 2. *Cont.*

Indicator	Description
Qualitative Habitat Evaluation Index (QHEI)	QHEI is a qualitative habitat evaluation assessment tool that identifies habitat variables that are important for attainment of the Ohio aquatic life criteria.
Area of Degradation to Attainment (ADV/AAV)	The ADV/AAV ratio can be used to demonstrate the magnitude and extent of changes in condition along segments between sampling years.
Restorability Score (1–100)	To ease the interpretation of complex environmental data, individual stressors and responses were ranked on a 1–100 scale linked to the tiered aquatic life uses codified in Ohio's water quality standards for prioritizing sites for restoration.

Though GI evaluation has not been prioritized in urban midwestern cities, substantial research has been conducted to show how effective various forms of GI can be in the Midwest and elsewhere [69,78–81]. In Shepherd Creek, a tributary of the Mill Creek watershed, Shuster and Rhea (2013) evaluated the effect of a rain barrel incentive program on local hydrology and found that parcel level GI could add small but significant decreases in runoff volume [78]. Efforts are ongoing in Congress Run to determine the impact of daylighting streams on surface water quality and hydrology. In northeastern Ohio, a paired sub-catchment study evaluating voluntary introduction of GI including rain barrels, bioretention cells, and rain gardens demonstrated that these efforts have a significant impact on peak flows in areas with small parcel sizes compared to larger lots [79]. In Missouri, a nine-month evaluation of green roofs planted with *Sedum* spp. demonstrated a 60% reduction in runoff volume relative to an impervious surface [69]. In the mid-Atlantic, watersheds with significantly more GI were found to have significantly lower (44%) peak runoff with less flashy hydrology [80]. Eckart, McPhee, and Bolisetti (2017) reviewed case studies of GI and found that many implementations can be successful depending on their location, local rainfall patterns, and relative area of contribution [81]. However, these authors also noted that there was still a significant amount of information unknown [81]. These efforts demonstrate that GI can have significant impact on watershed hydrology, and more research is needed to eliminate uncertainties. Whether this research will be driven by creating better GI plans, installing GI at locations where it can have the greatest impact, or increasing financial resources available for GI evaluation remains unclear.

3.4.2. Social Infrastructure and GI Evaluation

Because evaluation is largely focused on informing how GI has impacted local hydrology and water quality, less knowledge exists on whether this information is meaningful to communities that adopt or are considering adopting GI. Stakeholders noted that this information could be meaningful to decision makers and presented it as a future opportunity for GI research. Unfortunately, gathering these data by the federal government is limited by the Paperwork Reduction Act of 1980 and will remain a gap unless external stakeholders survey the communities in the planning process to understand what factors of GI effectiveness are important. Likewise, integrated socioecological metrics may be needed to integrate both community satisfaction with GI and use-attainment metrics. At the national scale, socioecological indices have been linked to watershed integrity assessments for ecosystem services and community well-being [82,83]. Future work evaluating how GI impacts social infrastructure could provide insight into what communities value in their green space and could better inform GI planning if they are sufficiently sensitive to urban watersheds.

3.4.3. Funding and Governance for GI Implementations

Stakeholders pointed to a lack of sufficient funding for evaluation of GI as a weakness, but because it has a lower priority when compared to expanding implementation, evaluation will likely continue to be underfunded in many Midwestern urban cities. This approach carries a general risk as GI is expanded as a watershed management strategy, but that may be acceptable if cities are meeting use-attainment goals. Because funding for GI has generally come from private sources, sewer districts, or non-profit groups, there is little impetus to evaluate how these structures impact the local environment and the watershed. Stakeholders generally agreed that improving waterways' use attainment status was more critical than understanding how green approaches affected the integrity of the entire watershed. In Mill Creek watershed, some funds have been available to academic institutions to conduct assessments of GI, but those projects are likely focused on individual installations and do not focus on improvement in the basin. Similarly, some research is completed by USGS and the USEPA Office of Research and Development, but this research is rarely intended to inform the planning process or provide insight on how GI may improve watershed integrity. Therefore, the impact of a mixed GI approach on watershed integrity in Midwestern urban cities is likely to continue being a knowledge gap. A potential solution to this problem would be to establish long-term study sites designed to understand the impacts of GI, as has been established for agricultural practices in the Chesapeake Bay (e.g., [84]). However, these efforts require dedicated funding and local support for long-term research.

4. Conclusions

GI continues to show promise as a watershed management strategy. In the Mill Creek watershed of southwestern Ohio, there are many stakeholders working to expand GI within the watershed through a strategic planning process. However, interdisciplinary collaboration between organizations to achieve a strategic implementation has had many hurdles. Effective planning and implementation of GI has required an established and positive relationship with multi-scale stakeholders such as individual landowners, community groups, and larger municipalities. Deploying GI opportunistically may be sufficient for improving use-attainment in local waterways, but it is difficult to distinguish grey infrastructure improvements in the watershed from the GI additions. These hurdles are symbolic of the disconnect among the planning process, the implementation, and the evaluation.

There are many tools available for planning, implementation, and evaluation of GI. However, it is unclear how these tools are being used by community planners and GI implementers to identify the ideal location for various GI installations. Strategic planning may require GI installations ubiquitously throughout the watershed, but not all communities may have the economic capacity to meet the costs of GI that could impact watershed integrity metrics. This contrast in strategic placement and community need demonstrates why GI has become more opportunistic. More information is needed to find an intersection where both use-attainment metrics and community needs are being met.

Similarly, a better confluence is needed between funding mechanisms and governance hierarchies for GI. Many midwestern urban cities implement GI as a watershed management strategy to achieve use attainment in urban waterways. However, fiscal austerity in these cities results in watershed managers seeking balance between what is feasible and what has been planned. Improved guidance on how GI is planned, managed, and evaluated has the potential to streamline some of the discordance between planning and implementing. However, this will require funding for GI evaluation to identify best practices.

Workforce development is a critical need for GI. More individuals with knowledge and technical skills specific to the planning, implementing, and evaluation of GI are needed to ensure GI can evolve with grey infrastructure improvements. Programs have begun to expand and provide opportunities in the planning of GI, but more research is needed to understand how emerging areas of water management and GI can be translated to institutionalized and accredited programs for improved workforce development. However, stakeholder organizations will need to be positioned to hire and expand their workforce to incorporate trained individuals.

Supplementary Materials: The following are available online at http://www.mdpi.com/2073-4441/11/4/738/s1, Figure S1: Summary of SWOT Analysis, Figure S2: Thirty-six different municipalities are fully or partially within the Mill Creek watershed, Figure S3: Within the largest municipality of Cincinnati, Hamilton County recognizes 47 different jurisdictions represented by 52 different community councils.

Author Contributions: S.D.S., S.J. contributed to paper conceptualization. S.D.S., T.N.-J., T.Y., and S.J. contributed to formal analysis, investigation, and writing-reviewing and editing.

Funding: This research was supported in part by the Oak Ridge Institute for Science Education through Interagency Agreement No. DW-8992433001 between the U.S. Department of Energy and the U.S. Environmental Protection Agency.

Acknowledgments: The authors thank Kate Mulvaney and Brian Dyson for technical review of this manuscript. The views expressed in this paper are those of the authors and do not necessarily represent the views or policies of the U.S. Environmental Protection Agency.

Conflicts of Interest: The authors declare no conflict of interest.

References

1. United Nations Department for Economic and Social Affairs. *Sustainable Development Goals Report*; United Nations: New York, NY, USA, 2017; ISBN 978-92-1-101368-9.
2. Kaushal, S.; McDowell, W.; Wollheim, W.; Johnson, T.; Mayer, P.; Belt, K.; Pennino, M. Urban Evolution: The Role of Water. *Water* **2015**, *7*, 4063–4087. [CrossRef]
3. Shuster, W.D.; Bonta, J.; Thurston, H.; Warnemuende, E.; Smith, D.R. Impacts of impervious surface on watershed hydrology: A review. *Urban Water J.* **2005**, *2*, 263–275. [CrossRef]
4. Kaushal, S.S.; Belt, K.T. The urban watershed continuum: Evolving spatial and temporal dimensions. *Urban Ecosyst.* **2012**, *15*, 409–435. [CrossRef]
5. Walsh, C.J.; Roy, A.H.; Feminella, J.W.; Cottingham, P.D.; Groffman, P.M.; Morgan, R.P. The urban stream syndrome: Current knowledge and the search for a cure. *J. N. Am. Benthol. Soc.* **2005**, *24*, 706–723. [CrossRef]
6. Schueler, T.R.; Fraley-McNeal, L.; Cappiella, K. Is impervious cover still important? Review of recent research. *J. Hydrol. Eng.* **2009**, *14*, 309–315. [CrossRef]
7. Emanuel, R.E.; Buckley, J.J.; Caldwell, P.V.; McNulty, S.G.; Sun, G. Influence of basin characteristics on the effectiveness and downstream reach of interbasin water transfers: Displacing a problem. *Environ. Res. Lett.* **2015**, *10*, 124005. [CrossRef]
8. Rexhausen, J. *Evaluation of Green Alternatives for Combined Sewer Overflow Mitigation: A Proposed Economic Impact Framework and Illustration of Its Application*; U.S. Environmental Protection Agency: Washington, DC, USA, 2013; p. 44.
9. Gibbs, B.H.R. *7279—115th Congress (2017–2018): Water Infrastructure Improvement Act*; 115th Congress: Washington, DC, USA, 2019.
10. Neale, M.W.; Moffett, E.R. Re-engineering buried urban streams: Daylighting results in rapid changes in stream invertebrate communities. *Ecol. Eng.* **2016**, *87*, 175–184. [CrossRef]
11. Wild, T.C.; Bernet, J.F.; Westling, E.L.; Lerner, D.N. Deculverting: Reviewing the evidence on the 'daylighting' and restoration of culverted rivers. *Water Environ. J.* **2011**, *25*, 412–421. [CrossRef]
12. Newcomer Johnson, T.; Kaushal, S.; Mayer, P.; Smith, R.; Sivirichi, G. Nutrient Retention in Restored Streams and Rivers: A Global Review and Synthesis. *Water* **2016**, *8*, 116. [CrossRef]
13. Thurston, H.W.; Taylor, M.A.; Shuster, W.D.; Roy, A.H.; Morrison, M.A. Using a reverse auction to promote household level stormwater control. *Environ. Sci. Policy* **2010**, *13*, 405–414. [CrossRef]
14. Hedeen, S. *The Mill Creek: An Unnatural History of an Urban Stream*; Blue Heron Press/The Rivers Unlimited Mill Creek Restoration Project: Cincinnati, OH, USA, 1994; ISBN 978-0-9643436-0-3.
15. National Weather Service. Cincinnati Climate Graphs. Available online: https://www.weather.gov/iln/climate_graphs_cvg# (accessed on 31 October 2018).
16. Homer, C.; Dewitz, J.; Yang, L.; Jin, S.; Danielson, P.; Xian, G.; Coulston, J.; Herold, N.; Wickham, J.; Megown, K. Completion of the 2011 National Land Cover Database for the conterminous United States–representing a decade of land cover change information. *Photogramm. Eng. Remote Sens.* **2015**, *81*, 345–354.
17. Soil Survey Staff Web Soil Survey. Available online: https://websoilsurvey.sc.egov.usda.gov (accessed on 29 October 2018).
18. Hamilton County. Cincinnati Area Geographic Information System (CAGIS) Internet Map Server. Available online: https://cagis.hamilton-co.org/ (accessed on 31 October 2018).
19. Midwest Biodiversity Institute (MBI). *Biological and Water Quality Assessment of Mill Creek 2016*; MBI: Hamilton County, OH, USA, 2016; p. 200.

20. Consent Decree on Sanitary Sewer Overflows, Wastewater Treatment Plants and Implementation of Capacity Assurance Program Plan for Sanitary Sewer Overflows (Civil Action No. C-1-02-107). Available online: http://www.msdgc.org/consent_decree/Global_consent_decree/index.html (accessed on 26 September 2017).

21. Metropolitan Sewer District of Greater Cincinnati. *Lick Run Watershed Master Plan*; Metropolitan Sewer District of Greater Cincinnati: Cincinnati, OH, USA, 2012; p. 115.

22. Clohessy, E.; Weidl, C.; Moyer, J.; Lehmann, A.; Koehler, B.; Wamsley, B.; Miller, M.; Merrill, K. *Nine-Element Nonpoint Source Implementation Strategic Plan (NPS-IS Plan)*; Congress Run—Mill Creek HUC-12 (05090203 01 04); Mill Creek Watershed Council of Communities: Cincinnati, OH, USA, 2017; p. 62.

23. Mill Creek Watershed Council of Communities. *Lower Mill Creek Watershed Action Plan*; Mill Creek Watershed Council of Communities: Cincinnati, OH, USA, 2014; p. 360.

24. Mills, M.; Álvarez-Romero, J.G.; Vance-Borland, K.; Cohen, P.; Pressey, R.L.; Guerrero, A.M.; Ernstson, H. Linking regional planning and local action: Towards using social network analysis in systematic conservation planning. *Biol. Conserv.* **2014**, *169*, 6–13. [CrossRef]

25. Mulvaney, K.K.; Lee, S.; Höök, T.O.; Prokopy, L.S. Casting a net to better understand fisheries management: An affiliation network analysis of the Great Lakes Fishery Commission. *Mar. Policy* **2015**, *57*, 120–131. [CrossRef]

26. Paletto, A.; Hamunen, K.; De Meo, I. Social network analysis to support stakeholder analysis in participatory forest planning. *Soc. Nat. Resour.* **2015**, *28*, 1108–1125. [CrossRef]

27. Faust, K. Centrality in affiliation networks. *Soc. Netw.* **1997**, *19*, 157–191. [CrossRef]

28. Field, S.; Frank, K.A.; Schiller, K.; Riegle-Crumb, C.; Muller, C. Identifying positions from affiliation networks: Preserving the duality of people and events. *Soc. Netw.* **2006**, *28*, 97–123. [CrossRef]

29. Scott, J.; Carrington, P.J. *The SAGE Handbook of Social Network Analysis*; SAGE Publications: London, UK, 2011.

30. Helms, M.M.; Nixon, J. Exploring SWOT analysis—Where are we now? A review of academic research from the last decade. *J. Strategy Manag.* **2010**, *3*, 215–251. [CrossRef]

31. Wasserman, S.; Faust, K. *Social Network Analysis: Methods and Applications*; Cambridge University Press: Cambridge, UK, 1994; Volume 8.

32. Prell, C.; Hubacek, K.; Reed, M. Stakeholder analysis and social network analysis in natural resource management. *Soc. Nat. Resour.* **2009**, *22*, 501–518. [CrossRef]

33. Janssen, M.A.; Bodin, Ö.; Anderies, J.M.; Elmqvist, T.; Ernstson, H.; McAllister, R.R.; Olsson, P.; Ryan, P. Toward a network perspective of the study of resilience in social-ecological systems. *Ecol. Soc.* **2006**, *11*. Available online: http://www.ecologyandsociety.org/vol11/iss1/art15/ (accessed on 8 April 2019). [CrossRef]

34. Borgatti, S.P.; Mehra, A.; Brass, D.J.; Labianca, G. Network analysis in the social sciences. *Science* **2009**, *323*, 892–895. [CrossRef] [PubMed]

35. Ernstson, H.; Barthel, S.; Andersson, E.; Borgström, S.T. Scale-crossing brokers and network governance of urban ecosystem services: The case of Stockholm. *Ecol. Soc.* **2010**, *15*, 28. [CrossRef]

36. Ohio Environmental Protection Agency Ohio Nonpoint Source Pollution Control Program; Nine-Element Nonpoint Source Implementation Strategies (9-Element NPS-IS) in Ohio. Available online: https://epa.ohio.gov/dsw/nps/index#120845160-9-element-nps-is (accessed on 31 December 2018).

37. *2018 Green Cincinnati Plan*; City of Cincinnati: Cincinnati, OH, USA, 2018; p. 273.

38. Wiggins, M.; Repasky, R.; Jeter, L.; Cogen, R. *A Framework for Ecosystem Restoration of the Ohio River and Its Watershed*; Ohio River Foundation: Blue Ash, OH, USA, 2004; p. 43.

39. Ohio-Kentucky-Indiana Regional Council of Governments (OKI). *Water Quality Management Plan for Butler, Clermont, Hamilton, and Warren Counties*; OKI Regional Council of Governments: Cincinnati, OH, USA, 2014; p. 722.

40. *Metropolitan Sewer District of Greater Cincinnati Final Wet Weather Improvement Plan*; Case 1:02-cv-00107-SAS-TSH.; Document 412-3; Metropolitan Sewer District of Greater Cincinnati: Cincinnati, OH, USA, 2010; p. 51.

41. Fuller, Mossbarger, Scott and May; Greenways Incorporated; Biohabitats, Inc.; Rhinoworks. *Mill Creek Watershed Greenway Master Plan*; Mill Creek Watershed Council: Cincinnati, OH, USA, 1999; p. 181.

42. *Plan Cincinnati; A Comprehensive Plan for the Future*; Cincinnati City Council: Cincinnati, OH, USA, 2012; p. 246.

43. Corburn, J. Confronting the challenges in reconnecting urban planning and public health. *Am. J. Public Health* **2004**, *94*, 541–546. [CrossRef]

44. Lane, M.B.; McDonald, G. Community-based environmental planning: Operational dilemmas, planning principles and possible remedies. *J. Environ. Plan. Manag.* **2005**, *48*, 709–731. [CrossRef]

45. Hum, T. Planning in neighborhoods with multiple publics: Opportunities and challenges for community-based nonprofit organizations. *J. Plan. Educ. Res.* **2010**, *29*, 461–477. [CrossRef]

46. Heckert, M.; Rosan, C.D. Creating GIS-based Planning Tools to Promote Equity through Green Infrastructure. *Front. Built Environ.* **2018**, *4*, 27. [CrossRef]

47. Schilling, J.; Logan, J. Greening the rust belt: A green infrastructure model for right sizing America's shrinking cities. *J. Am. Plann. Assoc.* **2008**, *74*, 451–466. [CrossRef]

48. Doucet, B.; Smit, E. Building an urban 'renaissance': Fragmented services and the production of inequality in Greater Downtown Detroit. *J. Hous. Built Environ.* **2016**, *31*, 635–657. [CrossRef]

49. Meerow, S.; Newell, J.P. Spatial planning for multifunctional green infrastructure: Growing resilience in Detroit. *Landsc. Urban Plan.* **2017**, *159*, 62–75. [CrossRef]

50. Safransky, S. Greening the urban frontier: Race, property, and resettlement in Detroit. *Geoforum* **2014**, *56*, 237–248. [CrossRef]

51. Wolch, J.R.; Byrne, J.; Newell, J.P. Urban green space, public health, and environmental justice: The challenge of making cities 'just green enough'. *Landsc. Urban Plan.* **2014**, *125*, 234–244. [CrossRef]

52. Hamilton County Rate Affordability Task Force 2016. *Hamilton County Affordability Task Force*; Hamilton County: Hamilton County, OH, USA, 2016; p. 22.

53. Keeley, M.; Koburger, A.; Dolowitz, D.P.; Medearis, D.; Nickel, D.; Shuster, W. Perspectives on the use of green infrastructure for stormwater management in Cleveland and Milwaukee. *Environ. Manag.* **2013**, *51*, 1093–1108. [CrossRef]

54. Midwest Biodiversity Institute (MBI). *User Manual for the MSDGC Integrated Prioritization System (IPS) and Data Exploration Tool, Version 2.0*; Technical Report MBI/2015-10-10; MSD Project Number 10180900; Midwest Biodiversity Institute: Columbus, OH, USA, 2015; Available online: www.midwestbiodiversityinst.org/publications/ (accessed on 8 April 2019).

55. Kiker, G.A.; Bridges, T.S.; Varghese, A.; Seager, T.P.; Linkov, I. Application of multicriteria decision analysis in environmental decision making. *Integr. Environ. Assess. Manag.* **2005**, *1*, 95–108. [CrossRef]

56. Roe, M.; Mell, I. Negotiating value and priorities: Evaluating the demands of green infrastructure development. *J. Environ. Plan. Manag.* **2013**, *56*, 650–673. [CrossRef]

57. Hansen, R.; Pauleit, S. From multifunctionality to multiple ecosystem services? A conceptual framework for multifunctionality in green infrastructure planning for urban areas. *Ambio* **2014**, *43*, 516–529. [CrossRef]

58. Wright, H. Understanding green infrastructure: The development of a contested concept in England. *Local Environ.* **2011**, *16*, 1003–1019. [CrossRef]

59. Young, R.F.; McPherson, E.G. Governing metropolitan green infrastructure in the United States. *Landsc. Urban Plan.* **2013**, *109*, 67–75. [CrossRef]

60. Kane, J.; Tomer, A. *Renewing the Water Workforce: Improving Water Infrastructure and Creating a Pipeline to Opportunity*; Metropolitan Policy Program at Brooking Institute: Washington, DC, USA, 2018; p. 72.

61. Nichols, E.G.; DeLuca, W.V.; Ebersohl, R.D. Multi-Institutional Collaboration for a Shared Renewable Energy Assessment Curriculum. *Nat. Sci. Educ.* **2015**, *44*, 34–42.

62. Water Environment Federation National Green Infrastructure Certification Program. *Body of Knowledge, Version 2.0*; National Green Infrastructure Certification Program: Alexandria, VA, USA, 2016; p. 32.

63. Rubin, Z.; Kondolf, G.; Rios-Touma, B. Evaluating stream restoration projects: What do we learn from monitoring? *Water* **2017**, *9*, 174. [CrossRef]

64. Bernhardt, E.S.; Palmer, M.A.; Allan, J.D.; Alexander, G.; Barnas, K.; Brooks, S.; Carr, J.; Clayton, S.; Dahm, C.; Follstad-Shah, J.; et al. Synthesizing US river restoration efforts. *Science* **2005**, *308*, 636–637. [CrossRef]

65. Peters, D.P.; Loescher, H.W.; SanClements, M.D.; Havstad, K.M. Taking the pulse of a continent: Expanding site-based research infrastructure for regional-to continental-scale ecology. *Ecosphere* **2014**, *5*, 1–23. [CrossRef]

66. U.S. EPA. *Stormwater Best Management Practices (BMP) Performance Analysis*; United States Environmental Protection Agency—Region 1: Boston, MA, USA, 2010; p. 232.

67. Golden, H.E.; Hoghooghi, N. Green infrastructure and its catchment-scale effects: An emerging science: Green infrastructure and its catchment-scale effects. *Wiley Interdiscip. Rev. Water* **2018**, *5*, e1254. [CrossRef]

68. Green Umbrella StreamBank Regional Water Quality Database. Available online: https://greenumbrella.org/StreamBank-Database (accessed on 15 February 2019).

69. Harper, G.E.; Limmer, M.A.; Showalter, W.E.; Burken, J.G. Nine-month evaluation of runoff quality and quantity from an experiential green roof in Missouri, USA. *Ecol. Eng.* **2015**, *78*, 127–133. [CrossRef]

70. Dufour, A.P. Escherichia coli: The fecal coliform. In *Bacterial Indicators/Health Hazards Associated with Water*; ASTM International: West Conshohocken, PA, USA, 1977.

71. DeShon, J.E. Development and Application of the Invertebrate Community Index (ICI). In *Biological Assessment and Criteria-Tools for Water Resource Planning and Decision Making*; Davis, W.S., Simon, T.P., Eds.; Lewis Publ.: Boca Raton, FL, USA, 1995; Volume 2, pp. 7–244.

72. Ohio Environmental Protection Agency. *Biological Criteria for the Protection of Aquatic life: Volume II: Users Manual for Biological Field Assessment of Ohio Surface Waters*; Ohio Environ Protection Agency—Division of Water Quality Planning and Assessment: Columbus, OH, USA, 1987; p. 288.

73. Karr, J.R. Assessment of biotic integrity using fish communities. *Fisheries* **1981**, *6*, 21–27. [CrossRef]

74. Fausch, K.D.; Karr, J.R.; Yant, P.R. Regional application of an index of biotic integrity based on stream fish communities. *Trans. Am. Fish. Soc.* **1984**, *113*, 39–55. [CrossRef]

75. Gammon, J.R. *The effects of thermal inputs on the populations of fish and macroinvertebrates in the Wabash River*; Indiana Water Resources Research Center, Perdue University: West Lafayette, IN, USA, 1973; p. 123.

76. Gammon, J.R.; Spacie, A.; Hamelink, J.L.; Kaesler, R.L. Role of electrofishing in assessing environmental quality of the Wabash River. In *Ecological Assessments of Effluent Impacts on Communities of Indigenous Aquatic Organisms*; ASTM International: West Conshohocken, PA, USA, 1981.

77. Yoder, C.O.; Rankin, E.T.; Smith, M.A.; Alsdorf, B.C.; Altfater, D.J.; Boucher, C.E.; Miltner, R.J.; Mishne, D.E.; Sanders, R.E.; Thoma, R.F. Changes in fish assemblage status in Ohio's nonwadeable rivers and streams over two decades. In Proceedings of the American Fisheries Society Symposium; American Fisheries Society: Bethesda, MD, USA, 2005; Volume 45, p. 399.

78. Shuster, W.; Rhea, L. Catchment-scale hydrologic implications of parcel-level stormwater management (Ohio USA). *J. Hydrol.* **2013**, *485*, 177–187. [CrossRef]

79. Jarden, K.M.; Jefferson, A.J.; Grieser, J.M. Assessing the effects of catchment-scale urban green infrastructure retrofits on hydrograph characteristics. *Hydrol. Process.* **2016**, *30*, 1536–1550. [CrossRef]

80. Pennino, M.J.; McDonald, R.I.; Jaffe, P.R. Watershed-scale impacts of stormwater green infrastructure on hydrology, nutrient fluxes, and combined sewer overflows in the mid-Atlantic region. *Sci. Total Environ.* **2016**, *565*, 1044–1053. [CrossRef] [PubMed]

81. Eckart, K.; McPhee, Z.; Bolisetti, T. Performance and implementation of low impact development—A review. *Sci. Total Environ.* **2017**, *607*, 413–432. [CrossRef] [PubMed]

82. Scown, M.W.; Flotemersch, J.E.; Spanbauer, T.L.; Eason, T.; Garmestani, A.; Chaffin, B.C. People and water: Exploring the social-ecological condition of watersheds of the United States. *Elementa* **2017**, *5*, 64. [CrossRef] [PubMed]

83. Thornbrugh, D.J.; Leibowitz, S.G.; Hill, R.A.; Weber, M.H.; Johnson, Z.C.; Olsen, A.R.; Flotemersch, J.E.; Stoddard, J.L.; Peck, D.V. Mapping watershed integrity for the conterminous United States. *Ecol. Indic.* **2018**, *85*, 1133–1148. [CrossRef] [PubMed]

84. Lizotte, R.E.; Shields, F.D.; Knight, S.S.; Cooper, C.M.; Testa, S.; Bryant, C.T. Effects of Artificial Flooding on Water Quality of a Floodplain Backwater. *River Res. Appl.* **2012**, *28*, 1644–1657. [CrossRef]

water

MDPI

Article

Actors in Water Governance: Barriers and Bridges for Coordination

Eva Lieberherr [1,*] and Karin Ingold [2,3]

[1]   Institute for Environmental Decisions, Swiss Federal Institute of Technology, 8092 Zürich, Switzerland
[2]   Institute of Political Science and Oeschger Center for Climate Change Research, University of Bern,
      3012 Bern, Switzerland; karin.ingold@ipw.unibe.ch
[3]   Department of Environmental Social Sciences, Swiss Federal Institute of Aquatic Science and Technology,
      8600 Dübendorf, Switzerland
*   Correspondence: eva.lieberherr@usys.ethz.ch; Tel.: +41-44-632-93-36

Received: 15 January 2019; Accepted: 11 February 2019; Published: 14 February 2019

Abstract: Multiple actors across different institutional levels play a role in water governance. The coordination of these actors is important for effective water governance. However, the joining together of multiple actors can have several implications, such as a redistribution of power across actors, a change in democratic control and citizen influence as well as shifting accountability structures. These implications can involve different barriers and bridges that might impede or foster coordination. Through qualitative and quantitative methods, we assess the following barriers and bridges for coordination: (1) reputational power in terms of who is perceived as important for coordination in the water sector; (2) democratic legitimacy in terms of actors' value of local control of water services; and (3) accountability in terms of the regional actors' capacity to steer in the water sector. This article focuses on three cases in a Swiss region that has experienced water provision challenges due to its highly fragmented water supply structures. We find that reputational power serves as a bridge in our three cases: when the actors responsible for water supply regard potential coordination partners as important, then we observe coordination. In contrast, we do not find conclusive evidence to support the assumption that a fear of losing local control is a barrier for coordination. Instead, our results indicate that accountability, in the form of vertical steering by the regional actors, serves as a bridge for coordination, and that this could help mitigate some of the potentially negative effects of democratic legitimacy perceptions: through convening local actors or providing positive incentives to municipalities to work together, regional actors can foster coordination.

Keywords: coordination; water supply; social network analysis

---

1. Introduction

Water systems are experiencing highly dynamic and conflicting societal demands and environmental pressures [1]. Many of the prevailing and looming water crises are not necessarily due to water scarcity or a lack of technological expertise, but rather because of poor management and governance of available water resources and infrastructure [2,3]. The inability of current systems to address these issues has led to calls for increased coordination among key actors [4]. Multiple actors, both public and private, across different institutional levels, from local to international, play a role in water governance. On the one hand, typically municipalities have been delegated the responsibility to provide water services in cities [1]. On the other hand, regional, national and international as well as private actors may play an important role in terms of oversight, technical knowledge and funding.

In this paper, we consider coordination as the joining of two or more actors in providing water services. Coordination often initially involves actors at the same jurisdictional level [5]. Going beyond this, coordination typically entails a whole network of organizations and entities that decide on and provide public tasks, rather than single public actors. This network of actors belongs to the public and private sphere and includes actors from different jurisdictional levels [6–10]. In short, coordination aims at establishing structures to enhance cooperation among a variety of actors to co-manage service provision [11–13]. We focus on "observable" coordination in the form of existing organizational structures, such as inter-municipal associations, through which actors jointly provide water services.

Different barriers and bridges might impede or foster coordination across multiple actors in the water sector. In this contribution, we focus on three aspects: reputational power, democratic legitimacy and accountability.

Firstly, coordination involves a redistribution of power amongst actors. Either municipal actors have to delegate a degree of power to an inter-municipal form of coordination or they might delegate this to regional (e.g., constituent state), private or even national actors. Therefore, a potentially important precondition for coordination among actors is the degree to which actors perceive each other as important and, hence, powerful [14–16]. Ideally, coordination could occur across levels (municipal, regional or national as well as public and private). However, as municipal actors historically play an important role in the provision of water provision [1], a first step would be the inter-municipal coordination. Hence, regarding fellow municipalities as important would arguably play a critical role for coordination. Following this, our first expectation is that if actors responsible for water provision regard other municipalities as important, then this serves as a bridge for coordination.

Second, coordination can be opposed at the local level, often because of citizens' and local politicians' participation concerns [17,18]. Indeed, urban water provision has traditionally enjoyed a high degree of citizen influence. With coordination reforms, the concern emerges that decision-making is no longer democratically legitimated if citizens can neither directly make decisions nor elect the people who manage their utilities [19,20]. This relates to democratic legitimacy, where direct and representative democratic elements such as citizens' influence on public decision-making are considered key for acceptable policies [21]. Hence, strong democratic legitimacy, that is local actors' value of democratic control, may hinder coordination [17,18]. A perceived democratic deficit at the local level has been found to be a barrier for coordination [18]. This leads us to our second expectation: Strong democratic legitimacy, and thus, actors' value of local influence on service provision is a barrier for coordination.

Third, accountability mechanisms, such as the legal mandates traditionally fulfilled by public authorities, play an important role for coordination reforms [22,23]. The broader water governance literature typically defines accountability as a determinant of water sector performance [24]. Generally, the literature links accountability to hierarchical control, that is, the accountability of water officials to democratic institutions, as specified in water laws [23]. Hence, accountability is often tied to the degree of regulation for water quality, infrastructure maintenance, etc. [25]. The literature on accountability identifies many other (less hierarchical and less formal) accountability mechanisms, based on reputation and internal discretion such as accountability through the professional norms, media, and consumer choice [22,26,27]. For the purposes of this article, we focus on formal, hierarchical means of accountability. In federalist systems, the locus of accountability is typically at the regional level, as the municipalities must account to the regional water laws and officials [28]. We thus argue that the regional level has a capacity to steer the local or municipal level. Through this steering role, regional actors can support coordination through subsidies or mandates. Hence, a bridge for coordination may be strong accountability, where the government's capacity to steer is desired and it has the necessary tools to do so [6,7,18]. Our third expectation is thus: Accountability, and thus a strong (perceived) capacity of the region to steer serves as a bridge for coordination.

Empirically, we assess these expectations in the water supply sector in Switzerland, which provides an interesting case to study coordination between municipalities and the regions (cantons) in a federal system. Switzerland's key peculiarities of direct democracy make it an extreme case to study democratic legitimacy and accountability issues. The water sector is salient for such an analysis of coordination, as it is characterized by high fragmentation [29].

The aim of this article is firstly to identify the degrees of coordination in the Swiss water supply sector and then shed light onto the bridges and barriers for such coordination. To do so, we proceed in four parts. We first specify our case selection and methods. In the empirical part, we then identify observed coordination by local and regional actors. Next, we outline the reputational power constellations using Social Network Analysis (SNA), and then assess legitimacy and accountability in the case studies. After discussing these results comparatively, we finally conclude with some suggestions for future research trajectories.

2. Materials and Methods

We have purposively selected the jurisdiction of the canton (region) Basel-Landschaft, in Switzerland, to conduct an analysis of the water supply sector because it has fragmented structures, but pursues coordination processes. Increasing water quality and quantity demands as well as climatic variations have led to growing challenges for these small water supply organizations to meet policy goals. In an effort to improve their performance, the region has driven processes of coordination, but the municipalities have largely resisted these reforms.

We have selected three case study areas within Basel-Landschaft that face different challenges: Case 1 includes 18 municipalities and is peri-urban with water quality concerns in the context of flooding and droughts. Case 1 spans a hydrogeological "water region" that cuts across canton Basel-Landschaft and the neighboring canton Solothurn. This is why we have included the canton of Solothurn in the interviews (see Appendix B). Case 2 encompasses eight municipalities and is rural with water quantity challenges, as hillside municipalities experience water shortages during droughts and are dependent on the water-rich valley municipalities. Case 3 spans 19 municipalities, is urban, and faces water quality concerns due to the river Rhine and nearby industrial deposits. Case 3 spans a hydrogeological "water region" that cuts across canton Basel-Landschaft and neighboring cantons, Basel City and Solothurn. This is why we have included the latter two cantons in the interviews (see Appendix B).

Our data comprise archival and documentary sources (laws and regulations) on the one hand, and 23 semi-structured, in-person interviews on the other hand. Appendix B contains the list of the interviewees. The interviews were conducted between June 2013 and February 2014, lasted between 60 and 90 min, and were recorded and transcribed.

To operationalize our dependent variable, we assessed the extent to which coordination is occurring in the cases. We did this based on an analysis of the existing organizational structures in terms of legal forms. That is, we first looked at the number of inter-municipal associations or joint-stock corporations through which the local actors jointly provide water services. Second, we also asked interview partners about ongoing processes to increase coordination. Third, we conducted a qualitative assessment about who retains decision-making competence. A high degree of coordination is defined as a situation with only a single inter-municipal organization where the local, municipal level retains little control. A low level of coordination is when the municipalities each operate their own water service system. A moderate degree of coordination involves a mix of inter-municipal associations, joint-stock-corporations and individual municipal waterworks.

The subsequent research steps involved document analysis and semi-structured interviews. To assess reputational power, we identified the key actors in all three cases. We did this by pinpointing the leading agency responsible for water services at the regional level, as well as one project leader at the local level as the two starting points for our snowball sampling. Through scoping interviews with these actors per region, we identified additional actors, whom we also interviewed. This resulted in 8 interview partners for Case 1 and 9 interview partners in both Cases 2 and 3. We systematically asked these actors (called "the core") whom they perceived as important actors for water supply. In this way, we were able to compile a list of 52, 46 and 42 collective actors and organizations, respectively, coding them by their jurisdictional level (local, regional, national) and sector (private or public).

To assess the reputational power of single actors in the network, we based our analysis on two measures borrowed from Social Network Analysis (SNA). SNA assumes that formal and informal arrangements and the involvement of a variety of actors are crucial for coordination [30]. In this context, we define nodes in the network as public and private, including local, regional and national organizations that are involved in the water supply sector (be it through management, regulation, or supply and consumption). A tie in the network is drawn as soon as an organization mentions another organization as important. It is thus a so-called directed network. We first calculate network density per case and, thus, the share of observed relations compared to all possible relations in one network [31]. Yet density is very sensitive to response rates in a network and to node-number per network [32]. This is why we predominantly concentrated on the densities within our core networks, as we have full data on these (all interview partners). Densities of the core indicate whether what we initially identified as the most important actors in the cases are also perceived as such by the concerned actors. In-degree centrality, that is the number of times the same actor is mentioned several times by various interview partners, is the second indicator for reputational power. The more the same set of actors are mentioned as being important (having high in-degree centrality), the more they are, thus, tightly linked to the core of the network. Network densities and in-degree centralities were calculated using UCINET 6.0 software.

We first measured democratic legitimacy in terms of citizens' rights to elect those responsible for service provision into office and to vote on projects and expenditure [20,33]. This provides us with an indication of the formal democratic legitimacy as specified in the legislation. We then further assessed legitimacy by asking interview partners how they perceive citizen influence and local competence in relation to water service provision. That is, whether they value having a high degree of democratic influence, which includes local competences and control over water services, and how this serves as a bridge or barrier for coordination. This second component of the legitimacy analysis enables us to identify the perceived democratic legitimacy in relation to bridges and barriers for coordination. We realize that the formal and informal (values) may be interlinked, but what is interesting is that the formal democratic legitimacy is constant across the three cases and the values are different.

Finally, we assessed accountability in terms of the degree of the regional government's capacity to steer [21]. Again, to identify the formal mechanisms, we first analyzed the relevant laws to pinpoint the type of steering tools that the regional (cantonal) actors have vis-à-vis municipalities. We defined steering tools as including planning processes, making recommendations, issuing mandates and fines, denying concessions, or providing subsidies. We then asked interview partners about which instruments are actually employed and how they perceive the regional government's capacity to steer, whether they prefer stronger or weaker steering and how the regional government's capacity to steer serves as bridges or barriers for coordination. We define stronger tools as those with more force (e.g., denying concessions, giving fines or subsidies) and weaker tools as those with less force (e.g., making recommendations, planning processes). Again, we realize that the formally possible and the actually employed and perceived accountability mechanisms may be interlinked. However, and again, the former is constant across the three cases and the latter are different.

3. Results

3.1. Coordination

The organizational analysis shows that Case 1 has a moderate degree of coordination, as there are five municipal water suppliers, one municipal waterworks that supplies additional municipalities, three inter-municipal associations and one inter-municipal joint-stock corporation (see Table 1). According to our interviews, we found that the inter-municipal organizations have been constructed in such a way that the municipalities have retained a high degree of control, as the individual municipalities can strongly affect the decisions. The municipalities running their own waterworks have been involved in a process to join in a new organizational form, but this has not materialized thus far.

Table 1. Overview of water suppliers and types in Case 1.

Name	Type
1. Wahlen Municipality	Municipal water works
2. Dittingen Municipality	Municipal water works
3. Kleinlützel Municipality	Municipal water works
4. Röschenz Municipality	Municipal water works
5. Liesberg Municipality	Municipal water works
6. City of Laufen	Municipal water works, but supplies three municipalities
7. Water Federation Birstal	Inter-municipal association (task-specific association for water supply)
8. Lüsseltaler Water Association	Inter-municipal association (task-specific association for water supply)
9. Water Supply Gilgenberg	Inter-municipal association (task-specific association for water supply)
10. Regional Water Supply Birstal-Thierstein	Inter-municipal joint-stock corporation (stockholders are all public)

Case 2 involves high coordination: there is one inter-municipal joint-stock corporation that serves as a water distributor between the municipalities across the entire case area (see Table 2). The municipalities retain only a minimal degree of influence as delegates sit on the board of directors. The corporation owns the reservoirs and regional distribution pipes, while the municipalities own their own water supply structure and retain water rights. There has been a process to integrate the municipalities further into the corporation (through joint-ownership and joint water rights) but this has not been successful thus far.

Table 2. Overview of water suppliers and types in Case 2.

Name	Type
Water Supply Waldenburgertal	Inter-municipal joint-stock corporation (all stockholders are public) Redistribution of water between municipalities as follows: Water supplying municipalities: Hölstein, Oberdorf, Niederdof Water receiving municipalities: Arboldswil, Bennwil, Lampenberg, Ramlinsburg, Waldenburg

Like Case 1, Case 3 has a moderate degree of coordination, as there are four municipal waterworks, two inter-municipal associations, and one inter-municipal joint-stock corporation (see Table 3). A process to increase coordination by joining all waterworks together has been underway, but has failed to date. Comparatively, the degree of coordination is slightly higher in Case 3 than Case 1, as there are fewer organizational forms, despite having slightly more municipalities (19 vs. 18 municipalities).

Table 3. Overview of water suppliers and types in Case 3.

Name	Type
1. Münchenstein Municipality	Municipal water works
2. Arlesheim Municipality	Municipal water works
3. Birsfelden Municipality	Municipal water works
4. Muttenz Municipality	Municipal water works
5. Task-specific association Waterworks Reinach Region	Inter-municipal association (task-specific association for water supply)
6. Task-specific association Aesch, Dornach and Pfeffingen (Basel-Landschaft and Solothurn)	Inter-municipal association (task-specific association for water supply)
7. Joint-stock company Hardwasser AG	Inter-municipal joint-stock corporation (all stockholders are public)

3.2. Reputational Power

By asking core actors at the local and regional level about whom they perceive as important in the water supply sector of Basel-Landschaft, we found that national actors are not perceived as playing a prominent role in water service provision. The regional actors, the municipalities and inter-municipal organizations enjoy the highest reputational power. Private actors, such as consultancies, do not play a central role. However, both the regional actors and the municipalities contract them to accomplish tasks and enjoy a degree of reputational power.

Table 4 summarizes data about the core actors in each case who we interviewed as well as all the actors named as important in the case. We specify the share of local, regional, national and public–private actors as well as two network measures, in-degree centrality and densities. Density is an appropriate measure for the core of the network: All interview partners comprising this core had the opportunity to mention all other interview partners as important actors. In this local and very sector-specific context, one can expect that potentially important actors know each other. This is why a density in the core below 50% can be considered as low, and everything above 50% can be considered as high. The maximum density would be 1. The in-degree centrality then shows how many times local actors, such as municipalities or waterworks, were mentioned to be important and are, thus, connected to the core of the network. One word of caution concerning the density measure of the whole network: these are not complete networks, meaning that only one-fifth of the actors in the network actually gave their judgments about whom they perceive as important (but we controlled for this when calculating densities). This picture could look very different if at least 50% of all actors in these networks had answered the question about whom they perceive as important. However, these network constellations provide us with insight regarding who the relevant actors are for coordination processes.

Table 4. Reputational power in three cases.

	Total Number of Actors	Local	Regional	National	Private	Avg. In-Degree Local	Density
Case 1							
Core	8	6 (75%)	2 (25%)	0	0	5.5	80%
All	52	26 (50%)	10 (23%)	6 (14%)	7 (16%)	2.4	30%
Case 2							
Core	9	8 (88%)	1 (11%)	0	0	5.6	65%
All	46	21 (46%)	12 (26%)	6 (13%)	7 (15%)	3.71	32%
Case 3							
Core	9	7 (78%)	2 (22%)	0	0	4.86	50%
All	42	21 (50%)	12 (29%)	5 (12%)	4 (9%)	3.14	26%

Case 1 is the biggest network compared to the others. Fifty percent of the actors perceived as important in the water sector are local actors (red nodes) and 23% are regional (cantonal) actors (turquoise nodes). The local actors are only loosely connected to the core (average in-degree centrality of 2.4, Table 4, Case 1; see also Table A1 in the Appendix A). Interestingly, the interview partners prominently perceive regional actors (Table A1 in Appendix A; and turquoise nodes in Figure 1) as important players in the water sector in Case 1. The national actors (blue nodes) and the private actors (orange nodes) are far fewer and generally (with two exceptions) at the periphery of the network and, hence, have a low degree of reputational power.

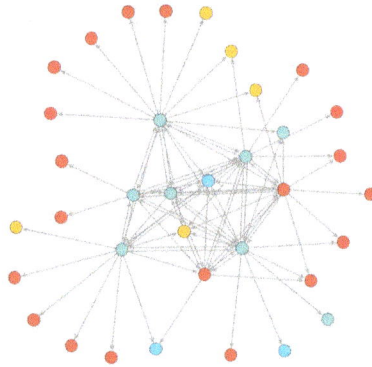

Figure 1. Power in Case 1. Legend: Having 8 interview partners and 52 actors in this network, 408 connections are possible, and 124 were observed, which results in a density of 0.30. The overall density (considering all potential relations if all the nodes had been active) is 0.047. Red = municipal/local; Turquoise = cantonal/regional; Blue = national; Orange = private.

Similar to Case 1, in Case 2 almost half of the actors that were mentioned as being important for water supply were local actors and 26% (red nodes) were regional actors (turquoise nodes; Figure 2). In contrast to Case 1, a key aspect in Case 2 is that the local actors are well connected to the core (average in-degree centrality of 3.71, Table 4, Case 2; see also Table A2 in Appendix A and red nodes at the center of Figure 2) whereas regional actors (turquoise nodes; Figure 2) are not considered very important. In contrast to Case 1, more national actors (blue nodes) are identified as important, but generally at the periphery. The private actors (orange nodes) are fewer and at the periphery with lower importance.

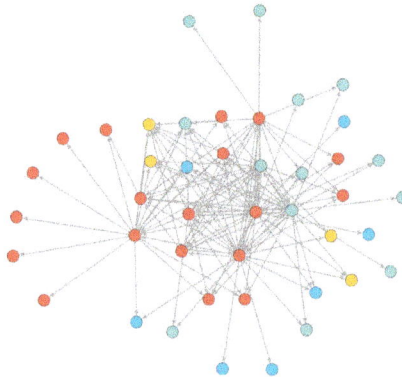

Figure 2. Power in Case 2. Legend: Having 9 interview partners and 46 actors in this network, 414 connections are possible, and 132 were observed, which results in a density of 0.32. The overall density (considering all potential relations if all the nodes had been active) is 0.064. Red = municipal/local; Turquoise = cantonal/regional; Blue = national; Orange = private.

Like the other two cases, the interview partners mainly named local actors (red nodes) as important for water provision (50%; Table 4, Case 3) in Case 3. However, there is a separation into a divided core–periphery structure (see Figure 3): one consisting of municipal actors (red nodes) and the other of mainly regional actors (turquoise nodes). Municipal actors in the periphery only link to municipal actors in the core, and regional actors in the periphery only link to regional actors in the core. Furthermore, regional actors also build a cluster internally linked to each other. Here we find that regional and local actors do not seem to perceive each other as important for water provision.

The in-degree centrality is between that of Case 1 and 3 (average in-degree centrality of 3.14, Table 4, Case 3; see also Table A3 in Appendix A). In addition, and in contrast to the other two cases, the national actors (blue nodes) are most clearly at the periphery. Private actors (orange node) have the least importance in this case.

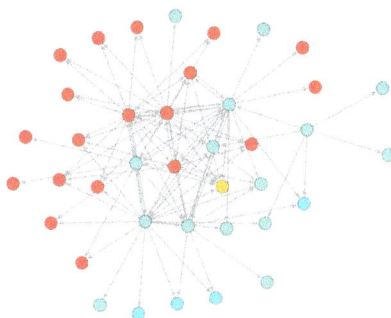

Figure 3. Power in Case 3. Legend: Having 9 interview partners and 42 actors in this network, 378 connections are possible, and 104 were observed, which results in a density of 0.26. The overall density (considering all potential relations if all the nodes had been active) is 0.060. Red = municipal/local; Turquoise = cantonal/regional; Blue = national; Orange = private.

In sum, we find that municipalities and inter-municipal organizations are both perceived as the most important across all cases (approximately 50%). As such, municipalities and inter-municipal organizations have the highest reputational power, regional actors have the second highest, and national and private actors have the least.

3.3. Democratic Legitimacy

All three cases have historically had strong democratic legitimacy. According to the cantonal law, the regional government has the formal decision-making rights and discretion to intervene at the local level to safeguard water supply in its territory. However, the regional government has delegated operational competence to the municipalities. Hence, municipal councils control the operational decisions about water supply when the municipality provides water (through municipal waterworks). Citizens have historically had a high degree of influence on the current structures, both indirectly by electing their municipal councilors and directly by voting on large financial projects in the context of municipal waterworks. Even in the case of inter-municipal organizations, the municipal councilors sit on the board and make decisions. However, in such arrangements, the citizens can no longer vote on large financial projects. In Switzerland, the legal form of joint-stock corporations means that the municipalities have less control than inter-municipal associations. Hence, a joint-stock corporation means less democratic legitimacy than inter-municipal associations.

In Case 1, interview data indicate that the actors' values of democratic legitimacy are the lowest, compared to the other two cases. That is, the interview partners do not place a high level of importance on having an influence on their water provision. Instead, interviewees state that the difficulty in deciding which competences the individual municipalities should give up and the need to clarify the financial redistribution are the main factors that have created barriers for reforms (Interviews 3, 4, 5, 8). While the former concern links to democratic legitimacy, as there is fear of losing local competences and responsibilities, the latter has more to do with financial fairness. This latter aspect can potentially be better solved at the regional rather than the local level, as the higher institutional level is responsible for deciding redistribution across subsidiary actors. Hence, this also serves as an indicator that these actors are rather open to the involvement of actors at the regional level and place a low value on democratic legitimacy.

In Case 2, we find democratic legitimacy values to fall between Cases 1 and 3. This is primarily because we find a more differentiated perception of democratic legitimacy here than in the other two cases. The values of citizen influence and local competences remain high in the water-rich valley municipalities, as interviewees from the valleys express concern of losing their direct control over decision-making (Interviews 9, 11, 13, 15). Conversely, the hillside municipalities prioritize engaging in inter-municipal coordination with the valley municipalities over direct democratic influence, to safeguard their water supply (Interviews 10, 12, 14; see above). Overall, the hillside and valley municipalities tend to block in each other in decision-making processes.

In Case 3, values of citizen influence and local competences are the highest in comparison to the other two cases. Here, several interviewees state that the municipalities want to retain sovereignty over local water supply and that citizens' ability to influence their local water supplier is considered a high priority (Interviews 19, 21, 22, 23). One interviewee explains that "citizens' ability to directly influence their local water supply is fundamental, as this water is a key public service" (Interview 21). What is more, this interviewee continues, is that people in this area want their water to come from their municipality and not from a neighbor or distant water supplier. Several interviewee partners were concerned about the quality of water provided by the joint-stock company, which draws its water from infiltrated Rhine groundwater (Interviews 19, 21, 22). Accordingly, the high priority of democratic control alongside water quality concerns has been found to be a barrier for coordination.

3.4. Accountability

Basel-Landschaft has a range of steering tools: the ability to grant or deny concessions for water use, issue mandates and fines, and create Regional Water Supply Plans. However, a regional interviewee says, "The canton lacks strong policy instruments to put pressure on the municipalities (...)" (Interview 3) and, hence, cannot strongly promote coordination. Moreover, instead of using mandates and denying concessions, the region has employed weaker steering tools, such as funding planning processes for coordination.

Overall accountability is found to be moderate in Case 1; the regional actors have neither engaged in a planning process nor in providing subsidies or issuing mandates to coordinate. The local actors in this case wish that the regional actors would take on a stronger steering role, as they perceive this as a bridge for coordination. Comparatively, the capacity to steer has been the strongest in Case 2, where the region has been more active in convening the local actors in planning processes than in the other two cases. This has served as a bridge for coordination (Interview 1.2). Similar to Case 1, the canton has not engaged actively in steering Case 3. However, unlike Case 1, in Case 3 the, actors are most averse to stronger regional steering and, hence, accountability is the weakest here, which has blocked coordination reforms.

Interviewees in all three cases have indicate that the region particularly lacks incentive-based steering tools. They argue that strong financial incentives could serve as an important means for the region to steer. All local actors would rather be encouraged by incentives and want the regional government to serve as facilitators for coordination processes; they do not want the regional government to give them mandates (Interviews 3–23).

4. Discussion

We now discuss the observed coordination in relation to the bridges and barriers of reputational power, democratic legitimacy and accountability for our three cases.

In Case 1, we observe a moderate degree of coordination in comparison with the other two cases, as there are both municipal waterworks and inter-municipal organizations. Our analysis of reputational power helps to explain a lack of further coordination within Case 1: the regional actors have the highest reputational power in this network, as the interview partners regard these as the most important. In line with our first expectation, this form of reputational power seems to be a barrier for coordination as the key partners for inter-municipal coordination are the municipalities, but the municipalities do not all regard each other as important. However, coordination by a regional actor (that is the canton) could be an option in this region, as the regional actors are considered the most important. Similarly, the results on democratic legitimacy show that the actors do not place a high value on having local influence on their water provider. These actors also regard financial redistribution as a barrier toward coordination, which can potentially be better solved at the regional rather than the local level. Hence, it is not surprising that the interview partners find regional actors more important for water provision than local actors. The low degree of importance placed on democratic legitimacy has not led to a high degree of coordination, but rather a moderate degree of coordination. We thus find inconclusive evidence for our second expectation. Finally, we turn to accountability, where we find that the regional actors have had only a moderate steering capacity, despite the wishes of the local actors that the regional actors take a stronger steering stance, as they regard stronger accountability as a bridge for coordination. This positive perception of accountability for coordination supports our third expectation.

Case 2 is the most coordinated of all the three cases, as there is one inter-municipal organization distributing water across all the municipalities. Our analysis of reputational power helps to explain this. While all the cases have a considerable overall density, only in Case 2, we find that a variety of municipalities are regarded as important and connect to the core of the network. This supports our first expectation that if fellow municipalities are considered as important for water provision, then this is a bridge for coordination. In contrast, the coordination in Case 2 cannot be conclusively explained by our analysis of democratic legitimacy, which we find to be moderate, due to the diverging interests of the hillside and valley municipalities who block each other. Hence, we can neither support nor counter our expectation. Finally, in line with our third expectation, accountability seems to be an important bridge for coordination in this case. Accountability is the strongest in Case 2 and we observe the highest coordination.

In Case 3, coordination is also moderate albeit slightly higher than in Case 1, as there are fewer organizational forms. Nevertheless, several individual municipal waterworks and inter-municipal organizations exist. Reputational power helps to explain this form of coordination, as we find a clear core–periphery structure where municipal and regional actors only link to their peers (municipal versus regional actors, respectively). In contrast to Case 2, there is no strong integrated core of municipal actors in this network. This provides a negative affirmation for our first expectation, as this lack of regarding fellow municipalities as important seems to impede coordination. Our results on democratic legitimacy for Case 3 also support our second expectation, as we find a high concern (the highest of the three cases) about a loss of democratic legitimacy in this case, which is a barrier for coordination. Conversely, and in line with our third expectation, we find that accountability in terms of the capacity to steer is the weakest in this case, which impedes coordination: the local actors in this case place the least amount of emphasis on the value of the regional government's steering capacity. This is also supported by the findings on reputational power, which shows that the local actors do not regard the regional actors as important.

5. Conclusions

The current and often highly fragmented water supply structures face different socio-economic, technical and biophysical challenges, even in highly industrialized countries [29]. To address these difficulties and in order to deliver water of good quality and sufficient quantities, reforms have called for municipalities to join forces and coordinate. In this article, we focused on coordination as involving the joining of two or more actors to provide water services. To effectively (or not) provide public services through coordination, such reforms first need to be realized, which often depends on power relations as well as procedural aspects such as democratic legitimacy and accountability. We have thus addressed three central bridges and barriers in the context of coordination: reputational power, democratic legitimacy and accountability. We then studied coordination in the Swiss water supply sector in Basel-Landschaft via social network analysis and interviews.

As coordination involves a redistribution of power across actors, we first argued that reputational power matters for coordination. Put differently, if actors regard potential coordinating partners—that is, the municipalities—as important, then this is a bridge for coordination. We indeed found that municipal as well as regional actors enjoy the highest reputational power across our three cases and that it is the municipal actors that have engaged in coordination. In line with our first expectation, we found that if municipal actors responsible for water service provision regard other municipalities as important, then they are more likely to engage in actual coordination. The negative affirmation of this was confirmed in Cases 1 and 3. In the former, the interview partners mainly perceive the regional actors, rather than the municipal actors, as important and we found moderate coordination. In the latter, we found a core-periphery structure without integration of the municipal actors. Finally, our expectation on reputational power was affirmed positively in Case 2, where a variety of actors is regarded as important and where we found the strongest coordination.

Second, we posited that a high preference for democratic legitimacy, with strong values of local control and influence, is a barrier for coordination. Case 3 provides the strongest evidence to support this, as our expected negative link between local actors' prioritization of democratic control and effects on coordination reforms was largely confirmed. However, despite a high concern about democratic legitimacy, Case 3 has slightly higher coordination than Case 1, where we observe low concern about democratic legitimacy. Indeed, our other two cases provide inconclusive evidence on for this link between democratic legitimacy and coordination: In Case 1, we found that the actors do not prioritize democratic legitimacy, but also consider arguments of financial redistribution in the context of coordination, which might be more aptly solved at the regional rather than the local level. Despite low values of democratic legitimacy, we only found moderate coordination in Case 1, rather than the expected high level. In Case 2, we found that the hillside and valley municipalities largely blocked each other with their contrasting viewpoints on democratic legitimacy, leading to moderate legitimacy, but high coordination. In contrast to this lack of explanatory power of democratic legitimacy as a bridge or barrier, we found that accountability is valuable. Indeed, our third expectation was that accountability, in terms of the region's capacity to steer, serves as a bridge for coordination. In all three cases, our results support the expected positive link between accountability and coordination: the stronger the observed or desired region's capacity to steer, the stronger the coordination. Concretely, the regional actor serving as a convener of the local actors in planning processes is a bridge for coordination (Case 2). Moreover, local actors in all cases say that they regard positive incentives (payments) by the regional actors as bridges for coordination. In contrast, an aversion to steering from the regional level seems to be a barrier for coordination (Case 3). This concurs with previous studies that have found that steering from a higher institutional level is often required to foster coordination [28].

To summarize, we have found the following bridges and barriers for coordination in this study. Reputational power serves as a bridge, as we found that if actors responsible for water provision regard potential coordination partners as important, then we observe a higher degree of coordination. We have not found conclusive evidence that high democratic legitimacy (strong local values of control and influence) is a barrier for coordination. In contrast, we have found that accountability, in the form of vertical steering by the regional actors, serves as a bridge for coordination.

These findings from the Swiss case of water supply in Basel-Landschaft are relevant to other federal countries where local autonomy is juxtaposed with the clout of regions (in this case cantons), albeit perceptions of democratic legitimacy might vary between localities. Moreover, democratic legitimacy, and particularly the value of municipal control, is not only salient in Switzerland but also across Western Europe, where we have seen recent citizen-initiated waves of re-municipalization of outsourced public services [34].

Author Contributions: E.L. and K.I. conceived and designed the research approach; K.I. conceptualized the reputational power part. E.L. conceptualized the coordination, democratic legitimacy and accountability parts. E.L. conducted the empirical analysis. Both co-authors wrote the paper.

Funding: This research was funded by the government of Basel-Landschaft, Switzerland and the Competence Centre for Drinking Water at the Swiss Federal Institute of Aquatic Research.

Conflicts of Interest: The authors declare no conflict of interest.

Appendix A

Table A1. In-degree centralities in Case 1.

Actor Type	Actor (Total n = 52)	In-Degree Centrality (Max. 8)
Regional actors N = 14	Bureau of Environmental Protection and Energy Basel-Landschaft	6
	Drinking Water Inspector of the region Basel-Landschaft	6
	Bureau of Environment Solothurn	6
	Drinking Water Inspector of the region Solothurn	3
	Building Insurance Solothurn	3
	University of Basel	3
	Bureau for Military and Population Protection Basel-Landschaft	2
	Industrial Works Basel (City)	2
	Government of the Region Basel-Landschaft	1
	Fire Inspector and Building Insurance Basel-Landschaft	1
	Region Basel-City	1
	Government of the region Solothurn	1
	Building, Planning and Environmental Director Conference	1
	Association of region Chemists	1
National actors N = 5	Swiss National Association for Gas and Water	6
	Swiss Association of Drinking Water Technicians	2
	Federal Office of Food Security and Veterinary Issues	1
	Federal Office of the Environment	1
	Swiss Federal Institute of Aquatic Science and Technology	1
Municipal actors N = 26	Water network "Birstal"	7
	Drinking water technicians for the joint-stock corporation regional water supply "Birstal-Thierstein" (RWV) and Water network "Birstal"	6
	Joint-stock corporation regional water supply "Birstal-Thierstein"	6
	Water supply "Lüsseltal" (Solothurn and Basel-Landschaft)	5
	City of "Laufen"	5
	Drinking water technicians City of "Laufen"	4
	Water network "Gilgenberg"	4
	Drinking water technician Water supply "Lüsseltal"	3
	Municipality of Röschenz	3
	Drinking water technician Water Supply "Gilgenberg"	2
	Municipality of Dittingen	2
	Municipality of Zwingen	2
	Municipality of Wahlen	1
	Municipality of Liesberg	1
	Municipality of Kleinlützel	1
	Municipality of Blauen	1
	Municipality of Nenzlingen	1
	Municipality of Brislach	1
	Municipality of Breitenbach	1
	Municipality of Büsserach	1
	Municipality of Erschwil	1
	Municipality of Meltingen	1
	Municipality of Fehren	1
	Municipality of Zullwil	1
	Municipality of Nunningen	1
	Municipality of Himmelried	1

Table A1. *Cont.*

Actor Type	Actor (Total n = 52)	In-Degree Centrality (Max. 8)
Private actors N= 7	Consultancy Sutter	5
	Accountancy firm BDO Visura (joint-stock corporation)	2
	Consultancy Holinger	2
	Water technician joint-stock corporation Heinis	1
	Consultancy Schmidlin und Partner	1
	Consultancy Lienhard	1
	Industry and Chamber of Commerce	1

Table A2. In-degree centralities in Case 2.

Actor Type	Actor (Total n = 46)	In-Degree Centrality (Max. 9)
Regional actors N = 12	Bureau of Environmental Protection and Energy Basel-Landschaft	8
	Region Drinking Water Inspector Basel-Landschaft	7
	Bureau of Industrial Services Basel-Landschaft	2
	Directorate of Finances and Church	2
	Government of the region Basel-Landschaft	2
	University of Basel	2
	Bureau for Spatial Planning Basel-Landschaft	1
	Region Basel-City	1
	Regional Leadership and Civil Protection	1
	Bureau of Statistics Basel-Landschaft	1
	Bureau for Military and Population Protection Basel-Landschaft	0
	Industrial Works Basel (City)	0
National actors N = 6	Swiss National Association for Gas and Water	4
	Swiss Federal Institute of Aquatic Science and Technology	3
	Swiss Association of Drinking Water Technicians	2
	"Aqua Viva" Swiss Action Group for the Protection of Rivers and Lakes	1
	Federal Office of the Environment	1
	Federal Office of Food Security and Veterinary Issues	1
Municipal actors N =21	Municipality of Niederdorf	7
	Municipality of Arboldswil	6
	Municipality of Hölstein	6
	Municipality of Oberdorf	6
	Joint-stock corporation water supply Waldenburgertal	6
	Drinking water technician Joint-stock corporation water supply Waldenburgertal	5
	Municipality of Ramlinsburg	5
	Municipality of Waldenburg	5
	Municipality of Bennwil	4
	Drinking water technician of Niederdorf	4
	Drinking water technician of Oberdorf	4
	Municipality of Lampenberg	4
	Drinking water technician of Hölstein	3
	Municipality of Langenbruck	3
	Association of Basel-Landschaft Municipalities	3
	Drinking water technician of Arboldswil	2
	Drinking water technician of Bennwil	1
	Drinking water technician of Lampenberg	1
	Drinking water technician of Langenbruck	1
	Drinking water technician of Ramlinsburg	1
	Drinking water technician of Waldenburg	1

Table A2. *Cont.*

Actor Type	Actor (Total n = 46)	In-Degree Centrality (Max. 9)
Private actors N = 7	Consultancy Sutter	9
	Consultancy GRG	5
	Consultancy Holinger	3
	Accountancy firm Schneeberger	2
	Aqua Plus (private partnership firm)	1
	Elektra Basel-Landschaft (cooperative)	1
	Industry and Chamber of Commerce	0

Table A3. In-degree centralities in Case 3.

Actor Type	Actor (Total n = 42)	In-Degree Centrality (Max. 9)
Regional actors N = 12	Bureau of Environmental Protection and Energy	9
	Industrial Works Basel (City)	5
	Region Drinking Water Inspector Basel-Landschaft	5
	Building Insurance Solothurn	2
	University of Basel	2
	Bureau of Environment Solothurn	1
	Bureau of Industrial Services Basel-Landschaft	1
	Bureau for Military and Population Protection	1
	Association of drinking water technicians Basel-Landschaft	1
	Region Drinking Water Inspector Solothurn	1
	Government of the region Basel-Landschaft	1
	Region Basel-City	0
National actors N = 5	Swiss National Association for Gas and Water	3
	Swiss Federal Institute of Aquatic Science and Technology	2
	Swiss Association of Drinking Water Technicians	2
	Federal Office of the Environment	0
	Federal Office of Food Security and Veterinary Issues	0
Municipal actors N = 21	Joint-stock company Hardwasser AG	6
	Municipality of Münchenstein	6
	Task-specific association Waterworks Reinach Region	6
	Municipality of Aesch	4
	Municipality of Arlesheim	4
	Municipality of Birsfelden	4
	Municipality of Dornach Solothurn	4
	Municipality of Muttenz	4
	Municipality of Pfeffingen	4
	Task-specific association Aesch, Dornach and Pfeffingen (Basel-Landschaft and Solothurn)	4
	Municipality of Bottmingen	3
	Municipality of Biel-Benken	2
	Drinking water technician of Waterworks Reinach Region	2
	Drinking water technician of task-specific association Aesch, Dornach and Pfeffingen (ZV ADP) (Basel-Landschaft and Solothurn)	2
	Municipality of Ettingen	2
	Municipality of Oberwil	2
	Municipality of Reinach	2
	Municipality of Therwil	2
	Municipality of Allschwil	1
	Drinking water technician of Hardwasser AG	1
	Drinking water technician of Industrial Works Basel	1
Private actors N = 4	Consultancy Holinger	7
	Consultancy Heinis AG	1
	Pharmaceutical Industry	1
	Industry and Chamber of Commerce	1

Appendix B

Interviewee list

No.	Actor	No. of People	Date
Region			
1.1 1.2	Bureau of Environmental Protection and Energy Basel-Landschaft Representatives	2	2013
2.1	Bureau of Environment Solothurn Representative	1	2013
2.2	Industrial Works Basel (City)	2	2014
Case 1			
3	City councilor of the city of Laufen	1	2013
4	President of water network Gilgenberg	1	2013
5	President of joint-stock corporation regional water supply Birstal-Thierstein	1	2013
6	President of waterworks Birstal	1	2013
7	Drinking water technician for the joint-stock corporation regional water supply Birstal-Thierstein and Waterworks Birstal	1	2013
8	President of water supply Lüsseltal (Solothurn and Basel-Landschaft)	1	2013
Case 2			
9	President of joint-stock corporation water supply Waldenburgertal	2	2013
10	City councilor of Arboldswil municipality	1	2013
11	City councilor of Niederdorf municipality	1	2013
12	City councilor of Waldenburg municipality	1	2013
13	City councilor of Hölstein municipality	1	2013
14	City councilor of Lampenberg municipality	1	2014
15	City councilor of Oberdorf municipality	1	2013
16	City councilors of Ramlinsburg municipality	3	2013
Case 3			
17	President of joint-stock company Hardwasser AG	1	2014
18	President of task-specific association Waterworks Reinach Region	1	2014
19	City councilor of municipality of Münchenstein	2	2014
20	President of task-specific association Aesch, Dornach and Pfeffingen (Basel-Landschaft and Solothurn)	1	2014
21	City councilors of Arlesheim municipality	2	2014
22	City councilors of Muttenz municipality	3	2014
23	City councilors of Birsfelden municipality	2	2014

References

1. Larsen, T.A.; Hoffmann, S.; Lüthi, C.; Truffer, B.; Maurer, M. Emerging Solutions to the Water Challenges of an Urbanizing World. *Science* **2016**, *352*, 928–933. [CrossRef]
2. Cosgrove, W.J.; Rijsberman, F.R. *World Water Vision: Making Water Everybody's Business*; Routledge: Abingdon, UK, 2014.
3. Ménard, C.; Saleth, R.M. The Effectiveness of Alternative Water Governance Arrangements. In *Investing in Water for a Green Economy*; United Nations Environment Program/Routledge: New York, NY, USA, 2012; pp. 152–174.
4. Leirens, S.; Zamora, C.; Negenborn, R.R.; De Schutter, B. Coordination in Urban Water Supply Networks Using Distributed Model Predictive Control. In Proceedings of the 2010 American Control Conference, Baltimore, MD, USA, 30 June–2 July 2010.
5. Bezes, P.; Fimreite, A.L.; Le Lidec, P.; LÆGreid, P.E.R. Understanding Organizational Reforms in the Modern State: Specialization and Integration in Norway and France. *Governance* **2013**, *26*, 147–175. [CrossRef]
6. Vangen, S.; Hayes, J.P.; Cornforth, C. Governing Cross-Sector, Inter-Organizational Collaborations. *Public Manag. Rev.* **2015**, *17*, 1237–1260. [CrossRef]
7. Kraak, A. Horizontal Coordination, Government Performance and National Planning: The Possibilities and Limits of the South African State. *Politikon* **2011**, *38*, 343–365. [CrossRef]

8. Rethemeyer, R.K.; Hatmaker, D.M. Network Management Reconsidered: An Inquiry into Management of Network Structures in Public Sector Service Provision. *J. Public Admin. Res. Theory* **2008**, *18*, 617–646. [CrossRef]

9. Raab, J.; Kenis, P. Heading toward a Society of Networks. *J. Manag. Inq.* **2009**, *18*, 198–210. [CrossRef]

10. Shrestha, M.K. Self-Organizing Network Capital and the Success of Collaborative Public Programs. *J. Public Admin. Res. Theory* **2013**, *23*, 307–329. [CrossRef]

11. Benz, A.; Fürst, D.; Kilper, H.; Rehfeld, D. *Regionalisierung: Theorie—Praxis—Perspectiven*; VS Verlag: Wiesbaden, Germany, 1999.

12. Fürst, D. Regional Governance. In *Handbuch Governance: Theoretische Grundlagen Und Empirische Anwendungsfelder*; Benz, A., Lütz, S., Schimank, U., Simonis, G., Eds.; VS Verlag: Wiesbaden, Germany, 2007; pp. 353–365.

13. Frisken, F.; Norris, D.F. Regionalism Reconsidered. *J. Urban Aff.* **2001**, *23*, 467–478. [CrossRef]

14. Ingold, K.; Fischer, M. Drivers of Collaboration to Mitigate Climate Change: An Illustration of Swiss Climate Policy over 15 Years. *Glob. Environ. Chang.* **2014**, *24*, 88–98. [CrossRef]

15. Fischer, M.; Sciarini, P. Unpacking Reputational Power: Intended and Unintended Determinants of the Assessment of Actors' Power. *Soc. Netw.* **2015**, *42*, 60–71. [CrossRef]

16. Ingold, K.; Leifeld, P. Structural and Institutional Determinants of Influence Reputation: A Comparison of Collaborative and Adversarial Policy Networks in Decision Making and Implementation. *J. Public Admin. Res. Theory* **2014**, *26*, 1–18. [CrossRef]

17. Cheng, K.-T. Governance Mechanisms and Regulation in the Utilities: An Investigation in a Taiwan Sample. *Util. Policy* **2013**, *26*, 17–22. [CrossRef]

18. Kübler, D.; Schenkel, W.; Leresche, J.-P. Bright Lights, Big Cities? Metropolisation, Intergovernmental Relations, and the New Federal Urban Policy in Switzerland. *Swiss Polit. Sci. Rev.* **2003**, *9*, 261–282. [CrossRef]

19. Papadopoulos, Y. Cooperative Forms of Governance: Problems of Democratic Accountability in Complex Environments. *Eur. J. Polit. Res.* **2003**, *42*, 473–501. [CrossRef]

20. Lieberherr, E. Trade-Offs and Synergies: Horizontalization and Legitimacy in the Swiss Wastewater Sector. *Public Manag. Rev.* **2016**, *18*, 456–478. [CrossRef]

21. Schmidt, V.A. Democracy and Legitimacy in the European Union Revisited: Input, Output and 'Throughput'. *Political Stud.* **2013**, *61*, 2–22. [CrossRef]

22. Michels, A.; Meijer, A. Safeguarding Public Accountability in Horizontal Government. *Public Manag. Rev.* **2008**, *10*, 165–173. [CrossRef]

23. Pahl-Wostl, C. *Water Governance in the Face of Global Change: From Understanding to Transformation*; Springer: Berlin, Germany, 2015.

24. Araral, E.; Yu, D.J. Comparative Water Law, Policies, and Administration in Asia: Evidence from 17 Countries. *Water Resour. Res.* **2013**, *49*, 5307–5316. [CrossRef]

25. Furlong, K. Good Water Governance without Good Urban Governance? Regulation, Service Delivery Models, and Local Government. *Environ. Plan. A* **2012**, *44*, 2721–2741. [CrossRef]

26. Van de Meene, S.J.; Brown, R.R.; Farrelly, M.A. Towards Understanding Governance for Sustainable Urban Water Management. *Glob. Environ. Chang.* **2011**, *21*, 1117–1127. [CrossRef]

27. Klenk, T.; Lieberherr, E. *Autonomy in Public Service Provision and the Challenge of Accountability: Insights from German Policy Fields*; The Hebrew University: Jerusalem, Israel, 2014.

28. Jordan, A.; Lenschow, A. Environmental Policy Integration: A State of the Art Review. *Environ. Policy Gov.* **2010**, *20*, 147–158. [CrossRef]

29. Teisman, G.R.; Edelenbos, J. Towards a Perspective of System Synchronization in Water Governance: A Synthesis of Empirical Lessons and Complexity Theories. *Int. Rev. Adm. Sci.* **2011**, *77*, 101–118. [CrossRef]

30. Adam, S.; Kriesi, H. The Network Approach. In *Theories of the Policy Process*; Sabatier, P., Ed.; Westview Press: Boulder, CO, USA, 2007; pp. 129–154.

31. Wasserman, S.; Faust, K. *Social Network Analysis: Methods and Applications*; Cambridge University Press: Cambridge, UK, 1994; Volume 8.

32. Scott, J. Social Network Analysis: Developments, Advances, and Prospects. *Soc. Netw. Anal. Min.* **2011**, *1*, 21–26. [CrossRef]

33. Scharpf, F. Governing in Europe: Effective and Democratic? Oxford University Press: New York, NY, USA, 1999.

34. Wollmann, H.; Balboni, E.; Gaudin, J.-P.; Marcou, G. The Multi-Level Institutional Setting in Germany, Italy, France and the Uk: A Comparative Overview. In *The Provision of Public Services in Europe*; Wollmann, H., Marcou, G., Eds.; Edward Elgar Publishing: Cheltenham, UK, 2010.

![water logo] *water*

MDPI

Article

Aligning Climate Governance with Urban Water Management: Insights from Transnational City Networks

Jale Tosun [1,2,*] and Lucas Leopold [1]

[1] Institute of Political Science, Heidelberg University, Bergheimer Straße 58, 69115 Heidelberg, Germany; l.leopold@stud.uni-heidelberg.de

[2] Heidelberg Center for the Environment, Heidelberg University, Im Neuenheimer Feld 229, 69120 Heidelberg, Germany

* Correspondence: jale.tosun@ipw.uni-heidelberg.de; Tel.: +49-6221-54-3726

Received: 27 February 2019; Accepted: 2 April 2019; Published: 4 April 2019

Abstract: A growing number of cities in different world regions are forming transnational networks in order to mitigate and adapt to climate change. In this study, we are interested in the nexus between climate change and urban water management. How do transnational city networks for climate action perceive urban water management? What kind of activities do they adopt for improving urban water management? How effective are these in practice? This study maps 17 transnational city networks that primarily work on climate governance, assesses whether they formally embrace urban water management as a field of activity, and analyzes the extent to which they influence local climate action regarding water-related issues. Our descriptive analysis reveals that the great majority of transnational city networks has embraced goals related to urban water management, mostly framed from the perspective of adaptation to climate change. However, our in-depth analysis of two frontrunner cities in Germany shows that membership in ICLEI (Local Governments for Sustainability) has only limited influence on the initiation and implementation of water-related policy measures.

Keywords: city networks; climate change; ICLEI; Sustainable Development Goals (SDGs); urban water management

1. Introduction

Climate change represents one of the fundamental global challenges of modern society and requires swift and coordinated action [1]. The Paris Agreement adopted in 2015, which builds on the United Nations Framework Convention on Climate Change (UNFCCC), represents the most recent commitment of all nations to undertake ambitious efforts to combat climate change and to adapt to its effects. Another feature of the Paris Agreement is its recognition that climate protection requires not only state actors (most importantly national governments) but also subnational and non-state actors to take action, which corresponds to the analytical concept of polycentrism [2–4]. Polycentrism perceives governance processes to involve multiple governing authorities at different scales, which are mostly or completely independent from each other [5–8]. By acting simultaneously at different levels, global challenges such as climate change can be addressed both more effectively and efficiently [7].

The growing body of research on polycentric climate governance ascribes cities and transnational city networks an important role in climate governance [9,10]; however, this does not mean that international and national policy-making bodies can be regarded as superfluous or unimportant [11]. On the one hand, cities are the main sources of greenhouse gas emissions since local policymakers decide on industry-related questions as well as energy, housing, infrastructure, land use, and traffic policy [8]. On the other hand, cities are also affected considerably by climate change and must spend vast amounts on implementing the necessary adaptation and/or prevention measures [12]. The policy competences of cities also put them in a position, whereby they can engage in policy experiments and design climate policy inventions [13–17], which are likely to produce even better policy outcomes if they are coordinated with the national governments [18].

One set of challenges related to climate change is urban water management, which concerns providing water security as well as mitigating flood risk and damage. Water insecurity and urban floods are two of the most pressing water-related challenges presented by climate change [19]. The exceptionally hot and dry summer of 2018 has shown policymakers across northern and central Europe how vulnerable they are to water insecurity. For example, the water levels of the Rhine River in Germany were alarmingly low for around six months, which forced freight vessels to sail only partly loaded, and consequently increased the cost of delivering critical commodities such as mineral oil [20]. In response to the unusual weather conditions, many German local authorities restricted or even banned the withdrawal of surface water for irrigation and other purposes.

In this study, we examine the nexus between climate change and urban water management, which is also acknowledged by the Sustainable Development Goals (SDGs) adopted in 2015 [21]. More precisely, we are interested in how transnational city networks for climate action perceive urban water management, what kind of activities they adopt for improving urban water management, and how effective these are in practice. These three questions underlie this study. The latter of these three is particularly relevant as earlier studies (mostly focusing on cities in developing countries) have noted the difficulties of assessing the true impact of city networks on the initiation and implementation of climate-related policies, as well as the various local obstacles that have hindered positive outcomes of networking approaches [22,23]. We will assess this aspect for two frontrunner cities on climate action in Germany as a representative case of a developed country. For answering the first two questions, we rely on a content analysis of official documents published by the city networks. Turning to the third question, we provide two in-depth studies that complement the first analysis by revealing insights that could not be yielded from the descriptive content analysis.

Our analyses provide four intriguing insights. First, transnational city networks on climate action are indeed committed to goals related to urban water management. Second, measures related to urban water management are mostly framed from the perspective of adapting to climate change–a finding that concurs with the literature (see, e.g., [24]). Third, transnational city networks do not only make mere reference to urban water management but also commit themselves to specific targets. Fourth, we can show that participation in a network is not the primary motivation for adopting and implementing water-related policy measures.

While we believe this study provides some important preliminary and indicative insights, we are aware that it suffers from a number of limitations, of which four appear particularly pertinent. First, the number of networks scrutinized in this study is small and the analysis would have benefitted from increasing the sample size. Second, because of the small number of cases, we had to limit our analysis to very basic techniques for presenting and exploring the data. We invite scholars to build on our study in future research and to advance our knowledge by rectifying these flaws. Third, we could not offer a detailed analysis of policy integration, but we examined which city networks recognize water management as an area of climate action. We consider the mentioning of water management a necessary condition for determining whether climate policy and water policy are integrated. Yet, strictly speaking, we did not provide insights into how policy integration is achieved.

Fourth, our initial descriptive analysis relies exclusively on a multitude of materials that were published by the city networks themselves, and the information provided therein was not always satisfactory, which is also reflected in the empirical basis of this present analysis. The websites of the city networks often do not contain detailed information on how they function, how they are funded, and what their governance mechanisms are. The difficulties in obtaining this information make it difficult to provide a scientifically sound assessment of the roles of city networks for governing the challenges of urban water management and climate change in a comparative fashion. Of course, good examples of disclosing organizational information also exist. On average, however, the information provided is superficial.

The remainder of this study unfolds as follows: first, we base this research in the context of the literature on policy integration. Second, we offer some theoretical considerations before moving on to the section on our materials and methods. We then turn to the presentation and discussion of our findings, which we summarize before coming to our conclusions.

2. Aligning Climate Action with Urban Water Management

We have learned from numerous studies that solving policy problems requires the integration of policies originating from different sectors [25], which is also recognized by the SDGs [21]. The literature on the nexus approach, for example, elaborates on this need to coordinate or integrate policies from different sectors, mostly concentrating on the energy–water nexus (e.g., [26]) or on the water–energy–food nexus (e.g., [27–29]).

A useful way of thinking about integration, especially in the context of the SDGs, was put forward by Nilsson and Persson [30], who differentiated between harmonization and coordination. Of these two, harmonization is the more demanding concept as it concerns bringing different policy objectives onto equal terms across sectoral policies. Coordination is the weaker form and is about "avoiding contradictory sectoral policies or mitigating adverse spill-over effects from sectoral policies" [30] (p. 37). Concentrating on the SDGs as well, Nilsson et al. [31] argued that the interactions between the different sectoral goals can be both negative and positive. Negative interactions refer to trade-offs between sectoral goals that require a different policy approach than positive interactions between individual SDGs, which produce co-benefits. While both Nilsson and Persson [30] and Nilsson et al. [31] concentrated on the SDGs, their overall judgment holds true for the feasibility of policy integration in general: in some instances, integration is easy to achieve since sectoral goals align naturally with each other; whereas, in other instances, it would be more challenging to attain integration. The nexus between water management and climate change in the urban context is one that is comparatively easier to address than other constellations [32] since many facets of water management are related to climate governance.

So far, the extensive literature on policy integration has consistently shown that policy integration tends to fall short, especially in terms of delivering the intended policy outcomes (e.g., [33,34]). The potential role of transnational city networks as facilitators of policy integration has not been a focus of the literature. Therefore, this study presents one of the first attempts to combine the study of policy integration with the study of transnational city networks. In doing so, it concentrates on the joint realization of water management with climate action in the urban context. Since some of the water-related issues are directly linked with climate change, whereas others are less so, the climate–water nexus provides an interesting source of variation for this analysis.

3. Theoretical Considerations

In this section, we outline our theoretical considerations for explaining under which conditions transnational city networks on climate governance embrace organizational goals related to urban water management. It should be noted that our analytical interest in transnational city networks is more specific than that of other studies in the pertinent body of literature (e.g., [35])

There is good reason to expect city governments to be more willing than central governments to engage in policy integration if the policy targets concern water and climate goals. For one, cities are affected by climate change and are also the setting in which climate-relevant activities take place (e.g., [8]). As both the source and potential victim of climate change, the incentive is strong for cities to become active in climate change. Cities can become active by themselves as well as form networks able to lobby policymakers at the national and international levels, for example, by participating at the annual Conference of the Parties (COP) of the UNFCCC [8] (p. 81).

Why would transnational city networks commit themselves to integrating climate and water management concerns? Before turning to this question, we first need to clarify how this study conceives policy integration. We do not examine whether transnational city networks commit themselves to policy integration as defined above (i.e., in the sense of coordination or harmonization). Instead, we develop a simpler understanding of policy integration and analyze whether city networks recognize the need for action in multiple sectors, which we consider a necessary condition for policy integration. Therefore, we examine whether transnational city networks on climate action mention urban water management as a field of activity. In doing so, we concentrate on three dimensions:

- First, whether transnational city networks mention urban water management as one of their organizational goals;
- Second, the framing of urban water management as an activity related to climate change mitigation, adaptation, or both;
- Third, the number and type of specific fields of activity related to urban water management.

The first dimension refers to the literature on the role of cities in climate governance [8,12–17,36]. The dual role of cities as the actors both causing and being affected by climate change puts them in a position, whereby they will push for designing and implementing effective means of dealing with climate change when they are participating in transnational networks. From this perspective, it is reasonable to expect that transnational city networks will support measures that adopt a holistic understanding of the causes and consequences of climate change and therefore commit themselves to policies that cut across policy domains. For networks that focus on climate change in all its facets, this expectation is even more plausible as climate change policies span numerous policy domains and include urban water management as well. Therefore, we expect that transnational city networks committed to mitigating and/or adapting to climate change are likely to embrace measures related to urban water management as a rational decision for lowering the anticipated or actual costs that incur from climate change (**Expectation 1**).

Considering the differences in climate change mitigation and adaptation policies, it is reasonable to expect that the latter will dominate the way in which measures related to urban water management are framed. Adaptation policies consist of concrete measures to inhibit the adverse effects of climate change [37], for which cities tend to have the necessary formal competences. In this context, Massey and Huitema [38] (p. 347), for example, showed that local governments in England have become increasingly aware of the need for committing to adaptation. Of course, cities are also effective in mitigating climate change, but, given their (limited) resources as well as their knowledge and experience with planning, we expect that transnational city networks are more likely to frame measures related to urban water management from the perspective of adaptation (**Expectation 2**).

Lastly, we expect that if transnational city networks commit themselves to urban water management, they are likely to support actions in multiple fields rather than in only one [26,32], since climate change has several implications for the management of water resources (**Expectation 3**).

4. Materials and Methods

In order to analyze the accuracy of our three expectations, first, we carried out a descriptive analysis of the transnational city networks' acknowledgment of urban water management. The descriptive nature of the analysis resulted from a very small number of cases and the limited availability of comparative data. More precisely, we analyzed 17 of 24 networks identified by Lee and Jung, which defined transnational city networks for climate change as institution-led or decentralized and multilateral platforms that enable cities and local entities to promote local climate action by performing at least one of the following functions: information sharing, networking, research, target setting, funding, lobbying, planning, and monitoring [17]. In doing so, we applied a broad definition of transnational city networks. However, the analyzed networks varied to a considerable degree;

some of the cases included members other than local authorities (e.g., the CDP included private sector businesses; see, e.g., [39]).

As the focus of our analysis was on functions rather than the institutional designs of the networks, the case selection put forward by [17] seemed justified for the sake of our argument. We limited ourselves to those networks that demonstrated activity in 2018; this was determined by the presence or absence of an up-to-date website (see Table 1). We consulted every resource accessible on the networks' websites to derive information on their goals and activities. As Acuto and Rayner [40] showed, reports were the most frequent outputs of transnational city networks and therefore provided a good—albeit not ideal—empirical basis for this analysis. The material consulted was appropriate since our analytical interest was in the official acknowledgment of goals related to urban water management. The diverse set of documents included annual or periodical reports, project descriptions, online presentations (prepared for conferences and workshops), statements, and press releases. We used graphs and tables to illustrate the plausibility of our theoretical considerations as set out in the previous section.

Table 1. Transnational city networks included in the analysis.

City Networks	
1. Asian Cities Climate Change Resilience Network (ACCRN): 2328 individuals in Bangladesh, India, Indonesia, Thailand, and Vietnam **2. C40 Cities:** 94 cities in various countries in Africa, Asia, Australia/Oceania, Europe, North America, and South America **3. CDP (formerly: Carbon Disclosure Project):** more than 620 cities (and more than 7000 companies) in various countries in Africa, Asia, Australia/Oceania, Europe, North America, and South America **4. Clean Air Asia:** more than 1000 cities in Asia **5. Climate Alliance of European Cities with Indigenous Peoples:** 1739 cities in 26 countries in Europe **6. Connecting Delta Cities:** Eight cities in China, Indonesia, Japan, The Netherlands, United Kingdom, United States of America, and Vietnam **7. Covenant of Mayors:** 7755 cities in 53 countries in Europe and Asia **8. Energy Cities:** More than 1000 cities and local authorities in 30 countries in Europe and Asia **9. EUROCITIES Declaration on Climate Change:** More than 140 largest cities and more than 45 partner cities in 39 countries in Europe and Asia	**10. ICLEI—Local Governments for Sustainability:** More than 1500 cities in 124 countries in Africa, Asia, Australia/Oceania, Europe, North America, and South America **11. Renewable Energy and Energy Efficiency Partnership (REEEP):** 359 governments, international and multilateral organizations, non-governmental institutions, foundations and private sector actors in various countries in Africa, Asia, Australia/Oceania, Europe, North America, and South America **12. Sustainable Cities International:** Projects in cities in Bolivia, Canada, Cuba, Indonesia, Mexico, Philippines, Senegal, South Africa, Sri Lanka, Tanzania, and Thailand **13. The Climate Group:** Projects in regions in various countries in Africa, Asia, Australia/Oceania, Europe, North America, and South America **14. The Climate Registry:** 300 organizations in Canada and the United States of America **15. United Cities and Local Governments (UCLG):** More than 240000 cities in 140 countries in Africa, Asia, Australia/Oceania, Europe, North America, and South America **16. The US Mayors Climate Protection Agreement:** 1060 mayors in the United States of America **17. Western Climate Initiative:** Various cities in **Québec** and Nova Scotia (Canada), and the state of California (United States of America)

Notes: Case selection based on [17] (pp. 102–103).

To systematize the materials collected, we concentrated on three aspects. To gain a basic understanding of the networks' general approaches, we first used the materials to determine whether a given network indicated commitment to urban water management. To this end, we searched the materials for specifications of the networks' main objectives. We constructed a measurement that assigned a value of 1 in either of the following cases: first, if a network stated that it had in place one organizational goal related to water management; second, if it presented on its website concrete projects related to urban water management. In all other cases, the variable was coded 0.

Second, we examined whether the networks' objectives related to urban water management were framed as measures to mitigate climate change, to adapt to it, or both. We differentiated between mitigation and adaptation activities for two reasons: first, this distinction was frequently made in the literature on climate policy (e.g., [41]); second, it was instructive to see whether different perceptions across the networks existed regarding these two approaches to climate action. In a nutshell, adaptation covers all measures that are taken to reduce the negative consequences of climate change on the environment, such as the hydration of drained areas (e.g., [37]). Mitigation, in contrast, means taking steps against the causes of climate change and, therefore, attempting to inhibit further deterioration (e.g., [42]). In the context of urban water management, mitigation policies can consist, for example, of measures for increasing the efficiency of water use or energy generation from water resources. For this purpose, we constructed a variable that assigned a value of 1 if the identified objectives or projects on the networks' homepages referred to adaptation strategies only. In cases where all the projects and objectives could be subsumed under mitigation strategies exclusively, the variable was coded 2, and when they referred to both adaptation and mitigation, the variable was coded 3.

In a third step, we employed an even more nuanced coding scheme and differentiated between the five most important fields of activity in urban water management as indicated by the transnational city networks. These included the following categories:

- Water supply and access to drinking water
- Water quality and hygiene
- Efficient water use
- Flood resilience
- Electricity generation from water

The coding categories generated inductively from the analysis of the materials provided by the networks corresponded to categories identified in the relevant literature. For a network to be assigned to any of these categories, we looked for an explicit acknowledgment or a pertinent project. We were aware that the types of targets mentioned would vary according to the vulnerability of the cities that participated in the respective network as well as the cities' economic and organizational capacities. A source of bias we were confident to be able to control was that the commitment to (different activity fields of) water management resulted merely from local and/or temporary circumstances in the member cities [43]. We examined the networks' collective commitments only, which meant that defining water management as an area of action was the outcome of a collective deliberation process, and it was unlikely to be changed based on local or temporary conditions.

After assessing the aims of the city networks, as derived from the analyzed materials, we complemented our descriptive analysis of 17 transnational city networks with two in-depth studies on how effective these networking approaches were in reality. Therefore, we concentrated on how ICLEI membership had helped launch and implement water-related adaptation measures in the German cities of Hamburg and Heidelberg. We decided to focus on the effects of membership in ICLEI since that network was committed to the comparatively broad incorporation of water management in climate action (see Table 2). We concentrated on Germany as a developed state in order to afford a hard test of the potential on-the-ground effects of membership in transnational city networks.

Since both cities were in Germany, they acted in the same political context. This helped us to rule out contextual factors that may have affected the results to be obtained from the analysis. Although both Heidelberg and Hamburg are situated in the same polity and are strongly committed to fighting climate change, these cities differed in several significant ways, which made them insightful for a comparative, in-depth analysis. While Hamburg is a city located in a coastal region, Heidelberg is an inland city located along the Neckar River. Furthermore, Hamburg, unlike Heidelberg, is a port city, which makes water management vital for the city's economy and represents an important task for the future. Hamburg also has a long history of storm surges and flooding along the Elbe River, with one extreme example in 1962 still remembered today, whereas Heidelberg has been spared from major flood catastrophes. Hamburg has a population of around 1.8 million and is more than 11-fold the size of Heidelberg; the population size makes the preservation of water supply in times of drought an even more pressing issue. Lastly, Hamburg represents a state of its own and benefits from the legal competence to adopt and implement policy measures, whereas Heidelberg is part of the state of Baden-Württemberg and is constrained in policy-making by state rules.

Table 2. Fields of activity of the city networks.

Fields of activity	Supply	Quality	Efficiency	Flood Resilience	Electricity	Number
Energy Cities	-	-	-	-	-	0
The US Mayor Climate Protection Agreement	-	-	X	-	-	1
The Climate Registry	-	-	X	-	-	1
C40 Cities	X	-	X	-	-	2
CDP	-	X	X	-	-	2
Connecting Delta Cities	X	-	-	X	-	2
Renewable Energy and Energy Efficiency Partnership	X	-	-	-	X	2
The Climate Group	X	-	X	-	-	2
United Cities and Local Governments	X	-	-	X	-	2
Asian Cities Climate Change Resilience Network	X	X	-	X	-	3
ICLEI-Local Governments for Sustainability	X	-	X	X	-	3
Sustainable Cities International	X	X	X	-	-	3
Climate Alliance of European Cities with Indigenous Peoples	X	X	X	X	-	4
EUROCITIES Declaration on Climate Change	X	X	X	X	-	4

For the purpose of data generation, we conducted interviews with experts of both city administrations, who either acted as intermediaries for the transnational networks or worked in the planning department where the formulation of climate-related targets took place. The interview-based approach provided us with interesting insights, which cannot be drawn from studying written documents only.

5. Transnational City Networks' Commitment to Water Management

We begin the presentation of our results by differentiating between transnational city networks on climate action that acknowledged water management as one of their organizational goals and those that abstained from doing so. Of the 17 networks analyzed, 14 mentioned water management, which supported Expectation 1. The networks that did not mention urban water management as one of their organizational goals were Clean Air Asia, the Covenant of Mayors for Climate and Energy, and the Western Climate Initiative. In the case of Clean Air Asia, the lack of recognition was unsurprising given the regional network's narrow focus on improving air quality. The same held true for the Western Climate Initiative, a network comprised of representatives from the Canadian provinces of Québec and Nova Scotia as well as the state of California, which provided administrative and technical services to support the implementation of subnational greenhouse gas emission trading programs. The Covenant of Mayors for Climate and Energy was founded in 2008 when the European Commission introduced the Climate and Energy Package 2020 [44]. Membership is open to all local authorities in Europe and Asia, which are democratically constituted. Considering that the participating entities were committed to adopting and implementing an integrated approach to climate change mitigation and adaptation, it was surprising that water management did not constitute one of this network's organizational goals. We considered this a first important finding.

Turning to the second dimension of our descriptive analysis, Table 3 shows which networks framed water management from the perspective of mitigation, adaptation, or both. Of the 14 networks that mentioned water management as an organizational goal, eight networks framed it in the context of adaptation to climate change, four as an issue related to mitigation, and two—C40 Cities and The Climate Group—used both frames. Under the maxim of "Cities get the job done", the C40 network promotes the crucial role of municipal climate action, and for this purpose it unites cities from all around the globe, representing 25% of the global economy and more than 650 million people. The relatively high capacity of the participating cities might be a driving force in the adoption of broader climate change approaches and increased policy integration. The same goes for the cities, which work under the guidance of The Climate Group.

Table 3. Assignment of city networks to different types of climate action.

Types of climate action	Adaptation	Mitigation	Both
Asian Cities Climate Change Resilience Network			
Climate Alliance of European Cities with Indigenous Peoples			
Connecting Delta Cities			
Energy Cities			
EUROCITIES Declaration on Climate Change			
ICLEI - Local Governments for Sustainability			
Sustainable Cities International			
United Cities and Local Governments			
CDP			
Renewable Energy and Energy Efficiency Partnership			
The Climate Registry			
The US Mayor Climate Protection Agreement			
C40 Cities			
The Climate Group			

Notes: Illustration is based on original data from the networks' websites and our own coding.

When inspecting Table 3, we can see that the majority of transnational city networks framed the embracement of water management from the perspective of adaptation to climate change, as postulated by Expectation 2.

Table 2 presents the five main fields of activity related to urban water management. The greatest number of networks adopted goals related to water supply and access, which was plausible given the importance of this issue for adapting to climate change. The second greatest number of networks mentioned the efficient use of water, followed by measures to protect cities against flooding. Less attention was paid to issues related to water quality and the generation of electricity from water resources.

The fields of activity of the individual networks varied as shown in Table 2, which also revealed that one network—Energy Cities—did not mention any specific fields related to urban water management. On the one hand, it was surprising that Energy Cities did not mention water management since the network demonstrated a great level of commitment to climate action in the past and is consisted of many high-capacity cities, which are mostly based in Europe. On the other hand, the main focus of Energy Cities was to mitigate climate change by realizing the energy transition at the local level. From that perspective, it was plausible that this network was committed to water management as an adaptation strategy, but without specifying fields of action. After all, water management is not closely related to energy transition.

Two networks were active in one field, but most addressed two fields. While this empirical pattern supported Expectation 3, we would have expected more networks to flag three or more fields of activity. EUROCITIES and Climate Alliance of European Cities with Indigenous Peoples were the two networks that showed the broadest range of fields of activity. The EUROCITIES Declaration on Climate Change was one of several joint letters of intent agreed on by the members of the EUROCITIES network. After being founded in 1986 by six ambitious European cities, it has now grown to include the mayors of more than 140 local entities. In addition to the network's general climate action approach, EUROCITIES had (jointly with ICLEI) been selected by the EU Commission for assuming co-leadership in the development of the Urban Water Agenda 2030. This process gave birth to the Porto Declaration on the Urban Water Agenda 2030, which addresses a wide variety of challenges, such as water scarcity, the combatting of water pollution, and flood resilience. In contrast to EUROCITIES, the Climate Alliance of European Cities follows a slightly different path by not only setting targets that concern the climate performance at home but also abroad. In addition to limiting their own carbon emissions, the European members agreed to abstain from the use of tropical timber in order to preserve the rainforests and to support the indigenous communities living inside them.

6. Effects of ICLEI (Local Governments for Sustainability) Membership on Water Management

What are the effects of membership in city networks on the initiation and implementation of water-related adaptation measures? To address this question, we examined the local policies adopted by Hamburg and Heidelberg. In addition to adaptation, both cities were strongly committed to climate change mitigation, with Hamburg aiming at a long-term reduction of carbon emissions by 80% and Heidelberg by 95% [45,46]. Of the different networks in which these two cities were active, we focused on the effects of their membership in ICLEI—a network that was particularly committed to water-related policy goals in climate actions (more generally on ICLEI, see [47]).

6.1. Case Study 1: Hamburg

Coastal flood protection, a measure made necessary by the accelerated rising of sea levels, is a central challenge Hamburg faces in the context of climate change [45]. Since the city's economy relies heavily on direct and indirect port-related activities (of which the production value was as high as 7.6 billion EUR in 2010 [48] (p. 26)), improving resilience against storm surges is a top priority. A comprehensive resilience strategy also includes the protection of inland areas from floods. The city has already experienced severe floods. In the future, climate change is likely to bring heavy rainfalls, which are expected to increase the risk of floods. Yet the competent authorities do not only concentrate on the potential damages caused by large amounts of water but also prepare for more intense and more frequent drought periods. Consequently, another key policy goal concerns the securing of water supply.

As a response to these threats, the Senate of Hamburg adopted the Hamburg Climate Plan in 2015, which contains the long-term vision for both adaptation and mitigation to be reached by 2050 with interim milestones set for 2020 and 2030. The Hamburg Climate Plan replaced the Climate Action Master Plan and the Adaptation to Climate Change Action Plan, which had been in place since 2013. Central mitigation targets of the plan include the 80% reduction of carbon emissions by 2050 as well as making the city administration carbon-neutral by 2030 through increasing energy efficiency and reducing energy consumption. Water-related adaptation strategies are comprised of:

- the strengthening of public flood protection measures
- designating building plots in light of flood risks
- improved risk communication
- rainwater retention, and
- improved drainage.

Furthermore, the adaptation strategy relies on a climate impact monitoring scheme, whose indicators measure changes in the city's climate resilience over time. The so-called IMPACT indicators capture, among other variables, the run-off characteristics of the Elbe River and its tidal range at a checkpoint in the borough of Hamburg-St. Pauli [49] as well as data on the frequency and intensity of storm surges [50]. The funding of the climate impact monitoring was increased immediately following the adoption of the Climate Plan, totaling 320,000 EUR in 2016. The mean financial expenditure was only as high as ca. 67,000 EUR per annum between 2013 and 2015 [45].

The impact of ICLEI membership becomes manifest in the CLEVER Cities project, which is currently being implemented in the borough of Hamburg-Harburg and is funded by Horizon 2020, an initiative of the European Union which aims to make nearly € 80 billion available for research and innovation projects between 2014 and 2020 [51]. Alongside Hamburg, the project is being implemented in several ICLEI member cities, such as London and Milan, to pursue the goal of urban regeneration with nature-based solutions, that is, implementing measures inspired by processes that can be found in natural ecosystems [52]. Among others, the activity in Hamburg-Harburg focuses on the management of rainwater, which shall be utilized in irrigation systems for planting areas, thereby increasing the efficiency of rainwater re-use.

According to one interviewee from the Coordination Centre for Climate Issues of the City of Hamburg, however, the actual impact of membership with ICLEI in particular and city networks in general on policy initiation and implementation is rather limited. The main reason for this is that the city administration is already active in climate action; another is that climate action has become a necessity, given the city's vulnerability to climate change. The main impact of ICLEI membership concerns networking events as their ideas are exchanged and discussed among network partners, including the implementation of specific collaborative projects such as CLEVER Cities. ICLEI membership does not bring new policies to the political agenda, and the formulation of policy targets is also not affected directly. The city network does not demand specific policy action and, therefore, competent authorities in the city are swifter in realizing the need for action and designing corresponding policies and measures. Yet, exchange with other cities within the ICLEI context is helpful for validating the measures and objectives adopted by local authorities. Thus, a central input provided by ICLEI is its effectiveness in expanding the capacity for learning from others and for optimizing the policies and measures that are in place in the individual cities.

The network's impact on the actual advances towards becoming a climate-resilient city is even more limited. Network-driven projects like CLEVER Cities are helpful for the implementation of the city's adaptation targets, but participation in such collaborative projects is often not feasible due to a lack of organizational capacity as well as the limited number of funding opportunities. The overall number of realized projects, thus, remains relatively low.

Nevertheless, the membership in city networks does help foster Hamburg's image as a frontrunner and reliable partner in transnational climate action. According to our interviewee, this offers a strategic advantage. In the interviewee's opinion, having a strong reputation as a frontrunner in climate governance provides a wide range of indirect benefits. Most notably, it reduces the costs of getting in touch with other relevant actors, such as other cities, private businesses, or academic institutions. This became remarkably apparent when Energy Cities, a network in which Hamburg has never been a member, reached out to our interviewee's authority and requested an opportunity to benefit from Hamburg's experience with specific policies.

6.2. Case Study 2: Heidelberg

The Office of Environmental Protection, Trade Supervision and Energy of the City of Heidelberg explained that its main water-related adaptation challenge is the protection against floods caused by heavy precipitation. The city expects such events to occur more frequently in the future due to rising temperatures. In May 2016, this threat became apparent when the Neckar River was flooded after approximately 90 liters of rain per square meter hit the city area within just a couple of hours, causing approximately 1 million EUR worth of damage to property [53]. This event helped the authorities realize that the need for appropriate management approaches was urgent, which resulted in action being taken by the city government and administration. As a direct consequence, the city council of Heidelberg adopted a resolution that initiated an encompassing heavy rain management analysis. The resulting technical report contained risk maps for different flood scenarios and identified vulnerable areas, which are given priority in management efforts. To prepare for the management of heavy rain incidents, regular workshops are held for responsible agencies, such as the fire departments and the Office for Landscape and Forestry. The involvement of different agencies also sought to develop an integrated management approach. At the time of writing, the city authorities were in the process of developing a risk communication concept. Despite the city's own initiatives and efforts, the specific design of the risk management is shaped by the provisions of the state of Baden-Württemberg and is dependent upon eligible funding. Compared to the influence of the state level, the impact of city networks is limited.

The city's membership in ICLEI has inspired activities related to water management. ICLEI has made an impact by facilitating occasional contact between the network and city administration to discuss Heidelberg's participation in specific projects. In late 2018, Heidelberg was asked to join the Climate Resilient Cities and Infrastructures (RESIN) initiative funded by Horizon 2020. One of the components of RESIN was to connect cities and the Fraunhofer Institute for Intelligent Analysis and Information Systems. Its aim was to carry out vulnerability assessments of water-related climate issues according to a novel methodology and to share knowledge on adaptation strategies. Apart from city governments, the RESIN partners are comprised of private sector businesses and several academic institutions. The role of ICLEI was to identify cities that were still in the early stages of water management preparation and to induce them to participate in RESIN. Heidelberg's participation in the project was finally called off because of the city's high level of existing preparations, which had been established after the city council's resolution. Therefore, Heidelberg no longer corresponded to the category of cities 'still in [the] early stages' of implementing resilience measures and so did not qualify for project participation. Besides, according to our interviewee, Heidelberg did not want to 'over-academize' its adaptation strategies, for this could have impeded the broader participation of the general public. The exchange of ideas and communication among the network partners was appreciated nonetheless and provided valuable indirect benefits to the city of Heidelberg. As in the case of Hamburg, these indirect benefits are, above all, rooted in the validation of the city's climate-related objectives through communication with other network members, as well as in the establishing of its reputation as a forerunner in adaptation and mitigation—a status which makes it easier for the city to find partners in climate governance.

Overall, the direct impact of ICLEI membership can be considered limited: it neither contributed new themes to the political agenda nor shaped adaptation strategies at the local level. Similar to Hamburg, Heidelberg by itself is already active in mitigating and adapting to climate change.

6.3. Summary

Despite the differences between Hamburg and Heidelberg, the two case studies revealed that the impact of city networks in general and ICLEI in particular was limited, at least in terms of initiating new policies and policy implementation. The city governments and agencies are the ones who identify water-related issues and place them on the political agenda. They are also the ones who design and implement the corresponding measures. In this context, ICLEI matters for validating the approaches selected and exploring possibilities for collaborative implementation projects. From that perspective, ICLEI offers the benefit of exchanging knowledge and expertise as well as increasing the reputation of the participating cities.

ICLEI might exert a more direct impact on cities that are less active and possess lower financial and/or organizational capacity. However, given our empirical focus on two cities that are both committed to fighting climate change and which have the financial and organizational means for doing so, we cannot make any strong claims. Therefore, we invite future research to concentrate on cities that do not consider themselves frontrunners and that are limited in their capacities.

7. Discussion of the Findings

In Section 3, we formulated a number of expectations, which we can now revisit in light of the findings presented in the previous sections. Our first expectation postulated that transnational city networks on climate action regard urban water management as one component of their organizational goals. Considering that only three networks did not mention urban water management, we interpreted our findings as indicative support of this expectation. In our second expectation, we postulated that transnational city networks framed water management from the perspective of adaptation. Our findings were certainly in line with our initial reasoning, as a clear imbalance in favor of adaptation efforts was identified. Even though other studies found that the distinction between these two frames was becoming increasingly elusive (e.g., [41]), the specific knowledge and capabilities developed in local communities indeed seemed to favor the adoption of adaptation strategies in polycentric systems, as forecasted by the literature on adaptive (co-)management [54]. The third expectation postulated that transnational city networks supported actions in multiple fields rather than in only one. As put forth by this expectation, we were able to show that the majority of networks was active in two or more fields related to urban water management, which we found plausible since water problems related to climate change were unlikely to be limited to one aspect only (e.g., [55]).

What conclusions can we draw from this analysis with regard to the potential of transnational city networks to promote policy integration? As explained previously, we did not analyze the willingness of transnational city networks to engage in the integration of policies that originated from different sectors in the sense of coordination or harmonization (see [30]). Rather, we examined the recognition of the urban water sector in the networks' approaches to climate action, which is a first indication of the willingness to promote policy integration, though it requires a more refined analysis. At any rate, with the exception of three networks, there seems to exist a common understanding that climate governance requires multi-sectoral action—an observation that reveals the potential for attaining policy integration. Of course, the formal commitment to multi-sectoral action and the actual behavior of the network members are two different things and again need to be investigated in detail.

How effective are city networks in initiation and implementation of policies that align climate action with water management? By analyzing two cities in Germany (Hamburg and Heidelberg) and concentrating on one particular network (ICLEI), we could show that the city network concerned was not decisive in putting new policies on the agenda or for the implementation of policies that combined goals related to climate change with water management. However, this does not mean that network membership does not have any effects. Indeed, the communication processes that take place within such networks help to assess the design of the policies adopted. Further, network membership facilitates the implementation of integrative policies through the possibility of forming collaborative implementation regimes. In addition, membership in city networks increases the profile and reputation of the cities and makes them more attractive for collaboration. All these points can be conducive to developing strategies for better aligning climate governance with water management. Of course, we expect the effects of membership in transnational city networks to vary across developed and developing countries. We expect to encounter more impeding factors in the latter than the former [22,23,27,29,56].

8. Conclusions

Many academic observers (e.g., [4]) argue that climate governance has moved from being state-centric to a form labeled as polycentric [5–7]. In this study, we took polycentricism as a starting point to argue that, with the growing relevance of such arrangements, we need to broaden our analytical perspective on the actors that can help to promote policy integration. Among the set of 'new' transnational actors in climate governance, we decided to concentrate on transnational city networks since they provide particularly instructive insights regarding the implementation of SDGs 6, 11 and 13, which lie at the heart of this special issue. Our descriptive findings revealed that, in an increasingly polycentric world, transnational city networks indeed have the potential to formally promote the integration of water management with climate change in the urban context. In practice, however, we found the actual impacts of city networks on the climate agendas and progresses in water-related adaptation to be rather limited, at least in cities and communities with high administrational capacity and strong, long-standing commitments to climate change mitigation and adaptation. The main impact of city networks is to provide a forum for validating and optimizing the design of policies and measures and to exchange experiences regarding their implementation. Concerning the crucial role of the validation of policy initiatives on the local level, it becomes apparent that transnational city networks provide a form of "normative incentives" to the participating cities, which are not connected with economic appeals (see, e.g., [57]). Investigating the constituents of these kinds of incentives and their implications for policy success in participating cities could provide an interesting foundation for further studies on polycentric governance.

It remains unclear whether the impact of such networks is greater on cities that are less equipped with administrational capabilities. Our in-depth analysis only focused on two cities in Germany and, therefore, only applies to countries with similar economic advancement. We invite further research to focus more on cities in developing regions. It could prove fruitful to examine whether the lack of administrational capacity is limited by hampered networking activities, or whether it is higher because of lower levels of preparation in the field of water-related adaptation and its potential for improvement.

Author Contributions: Conceptualization: J.T.; Methodology: J.T.; Resources: L.L.; Data Curation: J.T. and L.L.; Formal analysis: J.T. and L.L.; Validation: J.T. and L.L.; Visualization: J.T. and L.L.; Writing – original draft: J.T. and L.L.; Project administration: J.T.

Funding: This research was funded by the Water Research Network Baden-Württemberg of the Baden-Württemberg Ministry of Science, Research and the Arts within the context of the collaborative research project Effect-Net and the Heidelberg Center for the Environment. The authors acknowledge financial support from the Deutsche Forschungsgemeinschaft within the Open Access Publishing Program of the Baden-Württemberg Ministry of Science, Research and the Arts and from Ruprecht-Karls-Universität Heidelberg.

Acknowledgments: We would like to thank Kristina Wetzel and Kai Schaupp from the Office of Environmental Protection, Trade Supervision and Energy of the City of Heidelberg and one anonymous interviewee from the Coordination Centre for Climate Issues of the City of Hamburg for their kind support and interesting insights from their work, which contributed fundamentally to our research. We are grateful for helpful comments and suggestions by four anonymous reviewers and Kees van Leeuwen. We thank Laurence Crumbie for language editing.

Conflicts of Interest: The authors declare no conflict of interest.

References

1. Levin, K.; Cashore, B.; Bernstein, S.; Auld, G. Overcoming the tragedy of super wicked problems: Constraining our future selves to ameliorate global climate change. *Policy Sci.* **2012**, *45*, 123–152. [CrossRef]
2. Bulkeley, H. Can cities realise their climate potential? Reflections on COP21 Paris and beyond. *Local Environ.* **2015**, *20*, 1405–1409. [CrossRef]
3. Hale, T. "All hands on deck": The Paris agreement and nonstate climate action. *Glob. Environ. Polit.* **2016**, *16*, 12–22. [CrossRef]
4. Jordan, A.J.; Huitema, D.; Hildén, M.; Van Asselt, H.; Rayner, T.J.; Schoenefeld, J.J.; Tosun, J.; Forster, J.; Boasson, E.L. Emergence of polycentric climate governance and its future prospects. *Nat. Clim. Chang.* **2015**, *5*, 977. [CrossRef]
5. Ostrom, V.; Tiebout, C.M.; Warren, R. The organization of government in metropolitan areas: A theoretical inquiry. *Am. Politics Sci. Rev.* **1961**, *55*, 831–842. [CrossRef]
6. Ostrom, V.; Ostrom, E. Public goods and public choices. In *Alternatives for Delivering Public Services: Toward Improved Performance*; Savas, E., Ed.; Westview Press: Boulder, CO, USA, 1977; pp. 7–49.
7. Ostrom, E. Polycentric systems for coping with collective action and global environmental change. *Glob. Environ. Chang.* **2010**, *20*, 550–557. [CrossRef]
8. Van der Heijden, J. City and subnational governance: High ambitions, innovative instruments and polycentric collaborations. In *Governing Climate Change: Polycentricity in Action*; Jordan, A.J., Huitema, D., Van Asselt, H., Forster, J., Eds.; Cambridge University Press: Cambridge, UK, 2018; pp. 81–96.

9. Dolšak, N.; Prakash, A. Join the club: How the domestic NGO sector induces participation in the Covenant of Mayors program. *Int. Interact.* **2017**, *43*, 26–47. [CrossRef]

10. Fuhr, H.; Hickmann, T.; Kern, K. The role of cities in multi-level climate governance: Local climate policies and the 1.5 C target. *Curr. Opin. Environ. Sustain.* **2018**, *30*, 1–6. [CrossRef]

11. Jordan, A.; Huitema, D. Innovations in climate policy: The politics of invention, diffusion, and evaluation. *Environ. Politics* **2014**, *23*, 715–734. [CrossRef]

12. Bulkeley, H.; Betsill, M.M. Revisiting the urban politics of climate change. *Environ. Politics* **2013**, *22*, 136–154. [CrossRef]

13. Kern, K.; Bulkeley, H. Cities, Europeanization and multi-level governance: Governing climate change through transnational municipal networks. *J. Common. Mark. Stud.* **2009**, *47*, 309–332. [CrossRef]

14. Hoffmann, M.J. *Climate Governance at the Crossroads: Experimenting with a Global Response after Kyoto*; Oxford University Press: Oxford, UK, 2011.

15. Broto, V.C.; Bulkeley, H. A survey of urban climate change experiments in 100 cities. *Glob. Environ. Chang.* **2013**, *23*, 92–102. [CrossRef] [PubMed]

16. Croci, E.; Lucchitta, B.; Janssens-Maenhout, G.; Martelli, S.; Molteni, T. Urban CO_2 mitigation strategies under the Covenant of Mayors: An assessment of 124 European cities. *J. Clean. Prod.* **2017**, *169*, 161–177. [CrossRef]

17. Lee, T.; Jung, H.Y. Mapping city-to-city networks for climate change action: Geographic bases, link modalities, functions, and activity. *J. Clean. Prod.* **2018**, *182*, 96–104. [CrossRef]

18. Chan, S.; Boran, I.; van Asselt, H.; Iacobuta, G.; Niles, N.; Rietig, K.; Scobie, M.; Bansard, J.S.; Delgado Pugley, D.; Delina, L.L. Promises and risks of nonstate action in climate and sustainability governance. *Wiley Interdiscip Rev. Clim Chang.* **2019**, e572. Available online: https://doi.org/10.1002/wcc.572.. [CrossRef]

19. Hunt, A.; Watkiss, P. Climate change impacts and adaptation in cities: A review of the literature. *Clim. Chang.* **2011**, *104*, 13–49. [CrossRef]

20. Rhine River Shipping Returns to Normal as Water Levels Rise. Availabe online: https://www.reuters.com/article/us-shipping-rhine-germany/rhine-river-shipping-returns-to-normal-as-water-levels-rise-idUSKBN1O91K4 (accessed on 14 December 2018).

21. Tosun, J.; Leininger, J. Governing the interlinkages between the sustainable development goals: Approaches to attain policy integration. *Glob. Chall.* **2017**, *1*, 1700036. [CrossRef]

22. Hickmann, T.; Fuhr, H.; Hoehne, C.; Lederer, M.; Stehle, F. Carbon governance arrangements and the nation-state: The reconfiguration of public authority in developing countries. *Public Adm. Dev.* **2017**, *37*, 331–343. [CrossRef]

23. Elsässer, J.P.; Hickmann, T.; Stehle, F. The role of cities in South Africa's energy gridlock. *Case Stud. Environ.* **2018**. [CrossRef]

24. Chu, E.; Anguelovski, I.; Roberts, D. Climate adaptation as strategic urbanism: Assessing opportunities and uncertainties for equity and inclusive development in cities. *Cities* **2017**, *60*, 378–387. [CrossRef]

25. Peters, B.G. *Pursuing Horizontal Management: The Politics of Public Sector Coordination*; University Press of Kansas: Lawrence, KS, USA, 2015.

26. Hussey, K.; Pittock, J. The energy-water nexus: Managing the links between energy and water for a sustainable future. *Ecol. Soc.* **2012**, *17*, 31. [CrossRef]

27. Endo, A.; Burnett, K.; Orencio, P.; Kumazawa, T.; Wada, C.; Ishii, A.; Tsurita, I.; Taniguchi, M. Methods of the water-energy-food nexus. *Water* **2015**, *7*, 5806–5830. [CrossRef]

28. Biggs, E.M.; Bruce, E.; Boruff, B.; Duncan, J.M.A.; Horsley, J.; Pauli, N.; McNeill, K.; Neef, A.; Van Ogtrop, F.; Curnow, J. Sustainable development and the water–energy–food nexus: A perspective on livelihoods. *Environ. Sci. Policy* **2015**, *54*, 389–397. [CrossRef]

29. Tosun, J.; Lang, A. Policy integration: Mapping the different concepts. *Policy Stud.* **2017**, *38*, 553–570. [CrossRef]

30. Nilsson, M.; Persson, Å. Policy note: Lessons from environmental policy integration for the implementation of the 2030 Agenda. *Environ. Sci. Policy* **2017**, *78*, 36–39. [CrossRef]

31. Nilsson, M.; Chisholm, E.; Griggs, D.; Howden-Chapman, P.; McCollum, D.; Messerli, P.; Neumann, B.; Stevance, A.-S.; Visbeck, M.; Stafford-Smith, M. Mapping interactions between the sustainable development goals: Lessons learned and ways forward. *Sustain. Sci.* **2018**, *13*, 1489–1503. [CrossRef] [PubMed]

32. Beck, M.B.; Walker, R.V. On water security, sustainability, and the water-food-energy-climate nexus. *Front. Environ. Sci. Eng.* **2013**, *7*, 626–639. [CrossRef]

33. Brouwer, S.; Rayner, T.; Huitema, D. Mainstreaming climate policy: The case of climate adaptation and the implementation of EU water policy. *Environ. Plann. C Gov. Policy* **2013**, *31*, 134–153. [CrossRef]

34. Runhaar, H.; Wilk, B.; Persson, Å.; Uittenbroek, C.; Wamsler, C. Mainstreaming climate adaptation: Taking stock about "what works" from empirical research worldwide. *Reg. Environ. Chang.* **2018**, *18*, 1201–1210. [CrossRef]

35. Lee, T. Global cities and transnational climate change networks. *Glob. Environ. Politics* **2013**, *13*, 108–127. [CrossRef]

36. Bulkeley, H. *Cities and Climate Change*; Routledge: Abingdon, UK, 2013.

37. Biesbroek, R.; Lesnikowski, A.; Ford, J.D.; Berrang-Ford, L.; Vink, M. Do Administrative Traditions Matter for Climate Change Adaptation Policy? A Comparative Analysis of 32 High-Income Countries. *Rev. Policy Res.* **2018**, *35*, 881–906. [CrossRef]

38. Massey, E.; Huitema, D. The emergence of climate change adaptation as a policy field: The case of England. *Reg. Environ. Chang.* **2013**, *13*, 341–352. [CrossRef]

39. Bansard, J.S.; Pattberg, P.H.; Widerberg, O. Cities to the rescue? Assessing the performance of transnational municipal networks in global climate governance. *Int. Environ. Agreem. Politics Law Econ.* **2017**, *17*, 229–246. [CrossRef]

40. Acuto, M.; Rayner, S. City networks: Breaking gridlocks or forging (new) lock-ins? *Int. Aff.* **2016**, *92*, 1147–1166. [CrossRef]

41. Fleig, A.; Schmidt, N.M.; Tosun, J. Legislative dynamics of mitigation and adaptation framework policies in the EU. *Eur. Policy Anal.* **2017**, *3*, 101–124.

42. Hatfield-Dodds, S.; Schandl, H.; Newth, D.; Obersteiner, M.; Cai, Y.; Baynes, T.; West, J.; Havlik, P. Assessing global resource use and greenhouse emissions to 2050, with ambitious resource efficiency and climate mitigation policies. *J. Clean. Prod.* **2017**, *144*, 403–414. [CrossRef]

43. Wagner, M.; Mager, C.; Schmidt, N.M.; Kiese, N.; Growe, A. Conflicts about Urban Green Spaces in Metropolitan Areas under Conditions of Climate Change: A Multidisciplinary Analysis of Stakeholders' Perceptions of Planning Processes. *Urban. Sci.* **2019**, *3*, 15. [CrossRef]

44. Pablo-Romero, M.d.P.; Sánchez-Braza, A.; Manuel González-Limón, J. Covenant of Mayors: Reasons for being an environmentally and energy friendly municipality. *Rev. Policy Res.* **2015**, *32*, 576–599. [CrossRef]

45. Report by the Senate to the Hamburg Parliament. Hamburg Climate Plan. Availabe online: https://www.hamburg.de/contentblob/9051304/754a498fcf4e4bbf9516e1f9a99e2bfe/data/d-21-2521-hamburg-climate-plan.pdf (accessed on 6 February 2019).

46. Heidelberg auf dem Weg zur klimaneutralen Kommune. Availabe online: https://www.heidelberg.de/hd,Lde/HD/Leben/Masterplan+Klimaschutz.html (accessed on 6 February 2019).

47. Zeppel, H. The ICLEI Cities for Climate Protection programme: Local government networks in urban climate governance. In *Climate Change and Global Policy Regimes: Towards Institutional Legitimacy*; Cadman, T., Ed.; Palgrave Macmillan: New York, NY, USA, 2013; pp. 217–231.

48. The Competitiveness of Global Port-Cities: Synthesis Report. Availabe online: https://www.oecd.org/cfe/regional-policy/Competitiveness-of-Global-Port-Cities-Synthesis-Report.pdf (accessed on 6 February 2019).

49. Klimafolgen-Monitoring Hamburg. Kennblatt zum IMPACT-Indikator Tideverhältnisse (KH-I-1). Availabe online: https://www.hamburg.de/contentblob/8138008/d6eedde8e37041541ee197a6565a7324/data/d-kfm-hf-k-tideverhaeltnisse.pdf (accessed on 6 February 2019).

50. Gesamtübersicht aller mit Daten unterlegten Klimafolgen-Indikatoren Stand 2015. Availabe online: https://www.hamburg.de/contentblob/6642006/c07478f9361e4fbb33465ef496536bf2/data/d-kfm-gesamtuebersicht-impact-indikatoren.pdf (accessed on 4 February 2019).

51. Horizon 2020 in brief. The EU Framework Programme for Research & Innovation. Availabe online: https://ec.europa.eu/programmes/horizon2020/sites/horizon2020/files/H2020_inBrief_EN_FinalBAT.pdf (accessed on 18 March 2019).

52. Nature-Based Solutions. Availabe online: https://ec.europa.eu/research/environment/index.cfm?pg=nbs (accessed on 4 February 2019).

53. Klimawandel-Anpassungskonzept für Heidelberg. Beschlussvorlage. Availabe online: https://ww1.heidelberg.de/buergerinfo/getfile.asp?id=275987&type=do& (accessed on 6 February 2019).

54. Huitema, D.; Mostert, E.; Egas, W.; Moellenkamp, S.; Pahl-Wostl, C.; Yalcin, R. Adaptive water governance: Assessing the institutional prescriptions of adaptive (co-) management from a governance perspective and defining a research agenda. *Ecol. Soc.* **2009**, *14*, 26. [CrossRef]

55. Chindarkar, N.; Howlett, M.; Ramesh, M. Introduction to the Special Issue:"Conceptualizing Effective Social Policy Design: Design Spaces and Capacity Challenges". *Public Adm. Dev.* **2017**, *37*, 3–14. [CrossRef]

56. Rahman, S.; Tosun, J. State bureaucracy and the management of climate change adaptation in Bangladesh. *Rev. Policy Res.* **2018**, *35*, 835–858. [CrossRef]

57. Wang, Y.; Ching, L. Institutional legitimacy: An exegesis of normative incentives. *Int. J. Water Resour. Dev.* **2013**, *29*, 514–525. [CrossRef]

MDPI

St. Alban-Anlage 66

4052 Basel

Switzerland

Tel. +41 61 683 77 34

Fax +41 61 302 89 18

www.mdpi.com

Water Editorial Office

E-mail: water@mdpi.com

www.mdpi.com/journal/water

www.ingramcontent.com/pod-product-compliance
Lightning Source LLC
Chambersburg PA
CBHW051716210326
41597CB00032B/5496